THE CARDAMOM CONUNDRUM

Sponsored by:

Conservation International

Ministry of Agriculture, Forestry and Fisheries, Kingdom of Cambodia

The Agence Française de Développement (AFD)

The Gordon and Betty Moore Foundation via its commitment to Conservation International

The Oxford Centre for Tropical Forest, The Oxford Martin School, Environmental Change Institute, Oxford University

THE CARDAMOM CONUNDRUM

Reconciling Development and Conservation in the Kingdom of Cambodia

Timothy J. Killeen

NUS PRESS
SINGAPORE

CONSERVATION INTERNATIONAL

FFEM
Fonds Français pour l'Environnement Mondial

afd
AGENCE FRANÇAISE DE DÉVELOPPEMENT

LIST OF FIGURES

PREFACE

Cambodia is a remarkable, fascinating and admirable nation. It has a unique and rich cultural heritage and a modern history that provides important lessons for the planet. The Khmer have recovered from an extraordinarily brutal period of despotic governance and civil war and, in the process, shown the world how a post-conflict society can pursue a development agenda that simultaneously addresses human rights, poverty alleviation, the consolidation of democratic institutions, sustainable development and environmental conservation. Cambodia has made massive strides in creating a modern economy, starting without any significant infrastructure, capital assets, financial resources, or even a tradition of entrepreneurial innovation. The global community, recognizing its debt to the Khmer people, invested tens of billions of dollars in direct development aid and by providing preferential market access for industrial goods. Cambodia's ability to sustain economic growth and reduce poverty can be attributed partly to this generous assistance, but also to its own open door policy towards foreign investment and the persistent hard work of its citizens. Its example provides important lessons for other resource-rich nations of Southeast Asia, Africa and Latin America where development often lags behind large potential.

Cambodia now has the opportunity to show another kind of global leadership by choosing an innovative development trajectory based on the precepts of a low carbon or "green" economy. The country faces what many would consider a mutually exclusive choice: pursuing economic development and poverty reduction versus the conservation of its natural resources, particularly the biodiversity and native habitats that characterize the landscapes of the Cardamom Mountains. The term Cardamom Conundrum captures the essence of that choice and this book seeks to highlight the strategic nature of the decisions facing Cambodia at this crucial stage in its development trajectory. The purpose of this book is to show that the perceived tradeoffs between development and conservation represent a false dichotomy: the construction of a modern economy — or more precisely a post-modern economy — is entirely dependent on the sustainable exploitation of a country's renewable natural resources and the wise investment of the revenues from the exploitation of those that are not.

The narrative focuses on development options in the Greater Cardamom Region, an interconnected series of landscapes that includes the Cardamom and Elephant Mountains, the Cambodian Coast and the littoral plain of the Tonle Sap, the great lake situated in the center of Cambodia. The lessons to be learned from a careful consideration of the risks and opportunities that accompany different development choices in this region are particularly germane, because like the country itself, the Greater Cardamom Region is experiencing massive change. Preserving the status quo is not an option, but unplanned development could seriously degrade the natural resources that underpin the current economy and future productivity of the region. The citizens of Cambodia expect an end to entrenched poverty, and demographic growth dictates the adoption of policies that will create jobs and foster economic opportunity. Meanwhile, development and change is being accelerated by the ongoing transformation and integration of the Southeast Asian regional economy. If change is not managed with strategic forethought, the long-term economic growth that Cambodia's leaders hope to consolidate over the next decade will be compromised or, at the very least, seriously constrained. The long-term prosperity of the nation is contingent on the adoption of production paradigms that are genuinely sustainable when considered in their economic, social and environmental dimensions. In this book, examples and models show how and why Cambodia can pursue a green development pathway based on five main investment strategies: renewable energy, forest conservation, sustainable fisheries, agricultural intensification and a diversified tourist industry.

The country already has a fairly low carbon economy due, in part, to its delayed development trajectory. Abundant forest resources have allowed the Khmer to rely on wood biomass as the principal source of renewable energy for both rural and urban households. A relatively small transportation sector, an impediment in many ways, also means Cambodia is not excessively dependent on petroleum or the suite of technologies that come with that addiction. The country's abundant water resources are responsible both for its phenomenally productive natural fisheries and the inherently stable and resilient rice farming system that has sustained the nation for centuries. Looking forward, the ongoing and planned development of hydropower facilities in the Cardamom Mountains can be the foundation for the next phase of Cambodia's renewable energy strategy. Hydropower will provide affordable energy to the nation's rapidly expanding industrial sector and urban communities, and allow it to avoid the carbon-intensive energy systems that characterize more developed countries.

The strategic importance of hydropower dictates that farsighted watershed management philosophies must be integrated into the regulatory framework of the energy sector. To be effective, these authorities must closely coordinate with the other important agencies that oversee the use of water for agriculture and fisheries. Coordination and dialogue are of extraordinary importance when determining how water is used at the national, regional and local level. Water is the most precious of all natural resources and it can only rarely be substituted, but it can be conserved, recycled and managed to maximize the benefits it brings to society. Estimating the full costs of water use and identifying alternatives should be an integrated part of the national development strategy. For example, fast-tracking hydropower development in the Cardamom Mountains can provide the nation with the gift of time, which will allow authorities to postpone decisions to construct dams on the main stem of the Mekong River that might negatively impact the freshwater fisheries of the Tonle Sap. In the intervening period, research will improve the knowledge base regarding the resilience of the functional attributes of that complex ecosystem and allow Cambodian society to make a more informed decision. In the intervening period, technologies such as drip irrigation and solar energy can be explored and implemented, which may reduce, or even eliminate, the need for large-scale hydropower and irrigation projects.

Cambodia's dependence on forest biomass can also be converted into a new opportunity in the transition to a green economy. Forest resources are also carbon assets and these will have increasing value over the medium-term as sources of finance linked to climate change mitigation and adaptation become available, which can be used to supplement traditional development assistance and, even more importantly, to leverage investments in the private sector. It has yet to be determined how the world's community of nations will organize the global response to climate change, but there is increasing confidence that global society will rise to the challenge and provide a realistic path for avoiding the worst impacts of global warming. This global agreement, whatever its final shape, will provide significant opportunities to countries that have identified a clear vision of a sustainable future based on a low emissions development strategy. In the case of developing countries, this will include financial compensation for countries that "reduce emissions from deforestation and forest degradation." Known as REDD+, this policy proposal has been embraced by the United Nation's Framework Convention on Climate Change (UNFCCC) and tens of billions of dollars have been committed to implement what is viewed as a cost effective mechanism to mitigate climate change over the medium-term.

Cambodia is well placed to exploit the opportunities presented by the REDD+ mechanism and can move aggressively to offer itself as a testing ground for strategies in a yet-to-be-defined system that will include a combination of regulatory and market incentives. Cambodia can influence how the mechanism evolves, because test cases and pilot projects will inform international negotiations to create a REDD+ framework. As an early actor, Cambodia can ensure that the new rules contemplate its specific needs and conditions. The Greater Cardamom Region contains the nation's largest and most prominent forest assets; these carbon stocks can be used to access REDD+ resources and leveraged with other investments to foster sustainable production models that benefit the lives of its inhabitants. By moving quickly and decisively, the Cambodia government can be a global agent of change and enjoy the economic benefits of global leadership.

Revenues from REDD+ will make forest management synonymous with forest conservation and provide significant revenues to stabilize land-use on the agricultural frontier. Over the next few years, other forms of climate finance will emerge that have equal or even greater potential to promote sustainable land use. Traditional forms of agriculture are responsible for approximately 15 percent of global greenhouse gas emissions and climate change negotiators have begun to focus their attention on this massively important sector. Improving the sustainability of agriculture is vital, because it will bring benefits to the planet in the form of food security and, in most cases, as a form of climate change adaptation. The projected demand for food can be a green economic motor that lifts hundreds of millions out of poverty. Poised at the brink of significant agricultural expansion, Cambodia has the opportunity to develop innovative incentive frameworks to support agricultural production systems that sequester carbon and resolve some of the long-standing constraints to tropical agriculture.

The Greater Cardamom Region provides an excellent venue to roll out innovative paradigms for sustainable agriculture, most of which are the result of decades of research and testing, including plantation forestry, mixed agroforestry systems, drip irrigation technology, integrated pest management, and the cultivation of specialty crops such as of bamboo and silk. All of these options are appropriate for the region, both ecologically and socially, and can complement and diversify the rice farming, which is the most resilient agricultural system on the planet. New technologies and markets will bring challenges; for example, biofuels will compete with food production and may exacerbate land tenure conflicts. These and other issues must be dealt with equitably to ensure social stability, but also to reduce risk for investors and promote a robust agribusiness sector that adds value to farm production. Cambodia can provide leadership and policy innovation for sustainable low-carbon agriculture at precisely the moment the world looks beyond forest conservation for other ways to sequester carbon and mitigate climate change.

The opportunities associated with by climate finance provide Cambodia with many attractive development options, but climate change itself has the potential to upend the best laid plans. Cambodia needs to better understand the uncertainty and risk associated with global warming. The monsoon is inherently stable, because it is a climatic phenomena linked to the rotation of the planet and the progression of the seasons. Nonetheless, one of the most commonly predicted impacts of global warming is an increase in extreme events, which in the monsoonal climates of Southeast Asia translates into more severe drought and floods. These "natural disasters" might threaten Cambodia's food security, either by disrupting crop production or the ecosystem functions that underpin natural fisheries. Perhaps less well appreciated, drought can wreak havoc on the efficient operation of a hydropower asset, leading to power blackouts and industrial slowdowns. Efforts to make hydropower systems more resilient to climate change will be largely dependent on "ecosystem based adaptation," which is based on the conservation of habitat on both natural and productive landscapes. Once again, the Greater Cardamom Region presents opportunities for the Cambodian state to invest in land-use planning and watershed management policies. Foresight and good governance are essential for organizing how land is used, and the best possible outcome probably is one where forest conservation, plantation forestry, and sustainable agriculture share common landscapes. If done well, the integration of land-use planning and watershed management will contribute to the growth of other economic sectors, particularly the country's vibrant tourist industry.

This book has been written from a philosophical orientation that avoids the mentality of "trade-offs" where pro-development advocates view conservation efforts as impediments, and conservationists are convinced that development inevitably leads to a loss for nature. This mentality all too often leads to a discussion plagued by conflict or, more commonly, to a process where the protagonists pursue parallel agendas that are neither coordinated nor coherent and, consequently, miss opportunities where development and conservation are not only compatible but entirely synergistic. This study hopes to demonstrate that the conservation of natural ecosystems is beneficial to the national economy and that the wise management of the natural resources of the region will promote better economic growth in the short, medium and long-term. Resolving the Cardamom Conundrum will open a pathway towards sustainable development and the eradication of poverty, but it will require that all parties avoid value judgments of what is right and wrong, and instead focus on a shared vision of a prosperous future where communities reside on intact and healthy landscapes.

ACKNOWLEDGEMENTS

I would like to express my sincere appreciation to the Cambodian people for their openness and hospitality during my sojourn in their beautiful country. I am indebted to the Ministry of Agriculture, Forestry and Fisheries and the Ministry of the Environment of the Kingdom of Cambodia for their support and access to vital information sources. My research and writing was supported by several institutions, most notably the Fonds Français pour l'Environnement Mondial (FFEM), the Agence Française de Développement (AFD) with whom the idea of the book was born, the Gordon and Betty Moore Foundation via its commitment to Conservation International (CI), and the Oxford Martin School and the Oxford Centre for Tropical Forests at Oxford University, where most of the text was actually written. I am beholden to Conservation International for providing me with the freedom to pursue an in-depth study of this nature. I am personally indebted to Claude Gascon, who sent me to Cambodia and commissioned the study; David Emmett and Bunra Seng, who provided essential insight into Cambodian culture and biodiversity; Ty Sokhun and Keo Omaliss, who welcomed our efforts on behalf of the Cambodian government; Hervé Conan, Audrey Rousson and Sideth Muong, who supported us from the AFD in Phnom Penh, and Ghislain Rieb and Constance Corbier Barthaux respectively from the FFEM and AFD in Paris. I received terrific support from many individuals in Cambodia, including Ouk Kimsan, Pheakrith Vong, Sokunthea Ros, Clarisse Reiter, Annette Olsson, Emmeline Johansen and Joshua Weil, as well as the rest of the staff in the CI office in Phnom Penh and Mathilde Iweins from afar. Most of the photographs are mine, but the ones that seem professionally taken were provided by Alan Michaud; the maps were prepared by Kellee Koenig at CI's geography lab. All authors rely on colleagues for critical comment and I benefitted from the insight provided by Taber Hand, Thomas Cochrane, Mauricio Arias, Keith Lawrence, Michael Totten, Tracy Farrell and Larry Gorenflo for their knowledge on wildlife, water and people, as well as editorial expertise from Suzanne Zweizig and an anonymous reviewer who provided valuable insight on Khmer culture. Finally, none of this would have transpired without the support and patience of my beloved wife, Laura Limpias.

CHAPTER 1

Introduction: The Cardamom Conundrum

The word "conundrum" connotes a puzzle whose solution also represents a paradox. Resolving a conundrum requires thinking outside of the box, in untraditional ways, or in ways that may seem counter-intuitive. Cambodia's "Cardamom Conundrum" represents the challenges inherent in identifying and implementing development strategies based on the exploitation of a country's renewable and non renewable resources. The Cardamom Mountains and surrounding landscapes have abundant natural resources, including arable land to clear and cultivate for agriculture, fresh running water to harness for electricity and industry, tremendously productive natural fisheries, large expanses of native forest with important timber reserves and a beautiful coastline to develop for large-scale tourism, as well as mineral resources that include precious stones, industrial metals, and hydrocarbons. These natural assets are already key components of Cambodia's national economy, but as their exploitation and utilization intensify, they will pass through a transition as passive assets to dynamic components of a modern economy. That transition is inherently risky: As passive assets, these resources provide strategically important goods and services to the national economy and their over-exploitation via an unsustainable development strategy that focuses on short-term benefits would compromise the long-term economic security of the country. Balancing the Khmer people's legitimate aspirations for development and the equally urgent need to conserve the nation's vital natural heritage is at the heart of that paradox.

The Cardamom Conundrum encapsulates the challenges and opportunities facing developing nations across the planet. Change is driven by local, regional and global phenomena, but this is manifested by the actions and investments of individuals and corporations who act locally, migrating into formerly remote regions in search of economic opportunity. On one hand, the Cambodian state is encouraging this development through a wide range of policies and considers the exploitation of the region's natural resources strategically important for the economic future of the nation. Simultaneously, important sectors from civil society are urging the government to conserve the biodiversity and natural habitats of the region, and the government has responded by creating protected areas and implementing policies to manage renewable natural resources. Many view these two policy agendas — conservation and development — to be contradictory. Few would deny that both goals — conservation and economic growth — are overriding national priorities; however, they are traditionally seen as mutually exclusive and there is broad disagreement on specific priorities when development and conservation agendas meet on the same landscape. But the conservation of natural ecosystems is — or should be — beneficial to the national economy, and the wise management of the natural resources will, in actuality, promote economic growth in the short, medium and long term. The solution to this "Cardamom Conundrum" is akin to solving a complex puzzle — arranging the "pieces" (policies and actions) that create wealth and generate jobs so that they are complimentary with the "pieces" (policies and actions) that protect the goods and services provided by natural ecosystems. At stake in this conundrum is the wellbeing of the Cambodian people, who will either prosper or suffer, depending on how these resources are used and allocated.

The landscape in question is the "Greater Cardamom Region," which includes not only the Cardamom Mountains (Chuor Phnom Krâvanh), but also the associated massifs of Phnom Aural, Phnom Samkos, and Phnom Kran, as well as the Soi Dao mountains along the border with Thailand, the Mountain of Joy (Phnom Kirirom) that overlooks the Mekong River, and the Elephant Mountains (Chuor Phnom Damrei). The highlands overlook and dominate the Cambodian Coast, while numerous streams and small rivers connect the Cardamom Mountains to the floodplains of Tonle Sap, a unique freshwater ecosystem linked to the Mekong River (Figure 1.2). The Greater

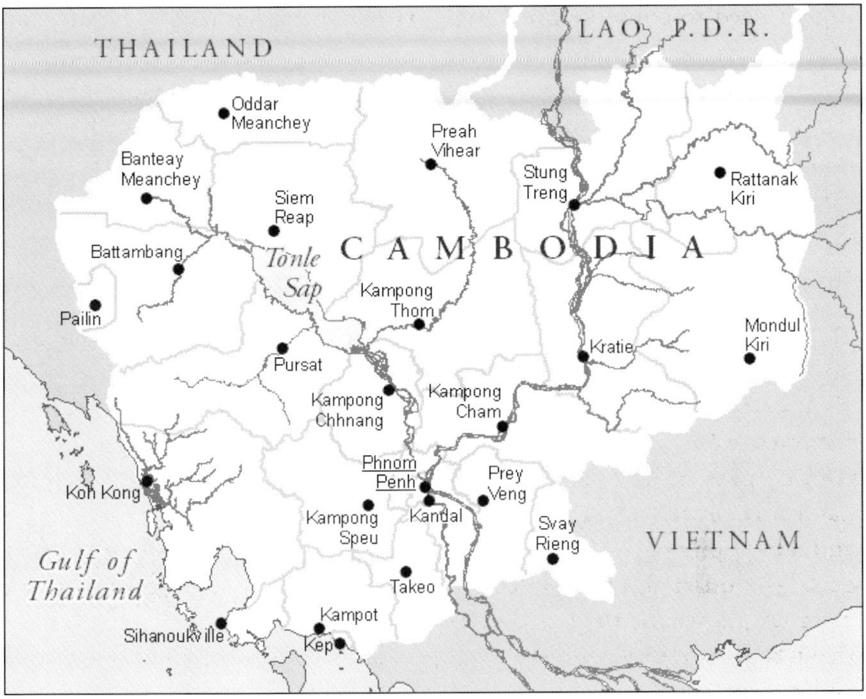

Figure 1.1. The Kingdom of Cambodia.

Cardamom Region accounts for 30 percent of Cambodia's physical area and will play an important role in the future of Cambodia due to its abundant natural resources — timber, minerals, biodiversity, scenic beauty, and the most important natural resource of all — water. These resources have been preserved over centuries due to the region's relative isolation, low human population density, and the historical phenomena (including the recent civil war) that have impeded access to them and development in the region.

This state of affairs is now changing, however. Unparalleled economic growth from exports and tourism over the last decade has transformed the Cambodian nation and created a foundation for sustained economic growth. Although the global economic recession has brought a pause in this growth, it has also demonstrated the benefits of a diversified economy and has motivated the Cambodian government to encourage agricultural expansion and production — a sector of the economy less susceptible to the fluctuations of global markets — on the upland landscapes adjacent to the floodplains the Tonle Sap and

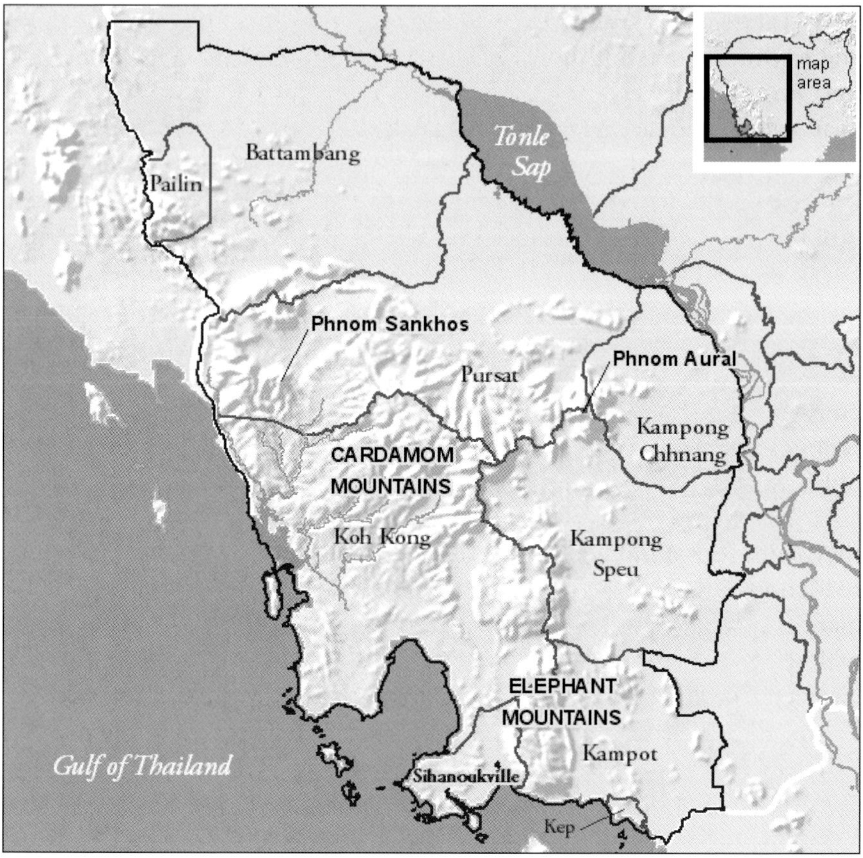

Figure 1.2. The Greater Cardamom Region is situated in the southeast sector of Cambodia between the Gulf of Thailand and the Tonle Sap (Great Lake), stretching from the Thai border on the west to the Vietnam border on the southeast. The region accounts for approximately 30 percent of Cambodia's physical area and will play a vital role in Cambodia's future due to the abundant natural resources found in its mountains, on the coast and in the alluvial plains of the Tonle Sap.

Mekong River. It has also motivated authorities to seek strategic investments in transportation and hydropower infrastructure that will integrate Cambodia into the regional economy of Southeast Asia and improve the competitiveness of its production in global markets. At the same time, private sector investments in mass tourism and industrial development are occurring in the coastal provinces and on their offshore islands.

The future of the Greater Cardamom Region will depend on how the environmental and social impacts of these investments are managed and on whether this management embraces an underlying philosophy that values and accounts for the ecosystem services provided by the region's forest, freshwater and coastal ecosystems. Within this context, the management and conservation of the region's natural resources are absolutely essential for the long-term sustainable development of industries such as hydropower, forest products, agriculture and tourism — all of which are key components in Cambodia's national development strategy. They also establish the foundation for organizing a green economy or "low carbon" economy that will ensure Cambodia's global competitiveness in the twenty-first century. Such an economic vision requires an understanding that conservation and development are not only compatible, but that they are also strategically linked.

This lesson has repercussions far beyond Cambodia. Creating the necessary conditions for sustainable economic growth and poverty reduction is a challenge that confronts developing nations across the planet. Likewise, the consequences of achieving, or failing at this vision, will also be felt not only by Cambodia, but by the entire planet. The Cardamom Conundrum is a global paradox that confronts all nations: Poor undeveloped countries face the biggest challenges and most difficult choices, but resolving the paradox of true sustainability also confronts emerging economies and, in their own fashion, rich nations.

Historical Background

The Cardamom Mountains have played an important role in the history of the Cambodian nation for millennia. The oldest known record of human occupation in Cambodia is a megalithic sandstone structure recently discovered in the Central Cardamoms. It is composed of two massive upright pillars that support a large horizontal slab across their top — an arrangement that archeologists refer to as a *dolmen*. Similar structures in Europe date from the Neolithic period (2000–5000 BC), but in Asia the estimated age of megalithic structures tends to be around 1000 BC or later. Regardless of their age, archaeologists hypothesize they had a religious or ceremonial function linked to celestial observation. Who these people were, where they came from, and what relationship they might have with the Khmer people will likely never be known, but their presence is a testament to the bond between nature and human civilization that spans centuries.

By the thirteenth century, we know that the Cardamom Mountains were populated by several ethnic groups referred collectively to as the *Khmer Daeum* or ancestral Khmer.[1] Historical accounts of the cardamom spice trade* by a Chinese diplomat in the thirteenth century mention these people, and recent archaeological discoveries have shed light on their burial practices, which involved preserving the mortal remains of their dead in ceramic jars, in sharp contrast to the cremation practices of the Hindu and Buddha traditions that dominated the lowland cultures during much of this historical period. This practice apparently survived well into the next millennium, as evidenced by the numerous burial sites located in caves near the summits of remote mountains on the Cardamom Plateau.[2] The *Khmer Daeum* have been greatly reduced in number and are now restricted to two small villages deep in the heart of the Cardamom Mountains (see Chapter 3).

The Cardamom Mountains have acted as a barrier to communications for centuries. Although the Khmer Empire was a flourishing urban civilization by the year 1000 AD, its communication with the outside world was conducted largely via the Tonle Sap and Mekong River. The name Tonle Sap is translated as "Great Lake" and refers both to the lake and the river that originates on the eastern margin of the lake and joins the Mekong at Phnom Penh. The mountain barriers to the south and west and the complex river morphology of the lower Tonle Sap functioned as effective protection for the Khmer civilization against the attacks of Malay pirates and neighboring states.[3] The combination of river access and the benefits of protection obviated the need to build an expensive road across the Cardamom highlands in order to establish trade with other nations. The Khmer Empire ruled large parts of mainland Southeast Asia from its capital near the well-known temple complex of Angkor Wat (Figure 1.3), whose inhabitants developed a sophisticated water management system that included large reservoirs to store water during the long dry season and a network of canals integrated within the extensive urban and temple complex on the north side of the Tonle Sap. The economy of this obviously wealthy medieval state depended on the extraordinarily productive natural fisheries of the

---

* The cardamom of the Cardamom Mountains is *Amomum kravanh* Pierre ex Gagnep., also known as black cardamom; it is used as a medicinal herb in traditional Chinese medicine, while the cardamom used as a culinary spices, *Elettaria cardamomum* L., is cultivated in India, Sri Lanka and Guatemala. <http://en.wikipedia.org/wiki/Cardamom> [accessed 15 June 2011].

Figure 1.3. The Temple of Angkor Wat, a cultural landmark recognized as a UNESCO World Heritage site, is a cornerstone of the Cambodian national economy receiving more than two million tourists visits each year (top). The Cardamom Mountains were home to an ethnic group that buried their dead in jars, which is distinct from the Buddhist and Hindu traditions of present and ancient Khmer culture (bottom). (Photographs provided by D. Emmett)

Tonle Sap and the equally productive rice paddy agriculture system that spread across the fertile alluvial plains surrounding the lake. The uninterrupted cultivation of this landscape for more than a millennium is a testament to the inherent resiliency and sustainability of traditional rice farming technology.

To the south of the Cardamom Mountains, the coastline stretches for approximately 350 km along the Gulf of Thailand. In spite of its isolation from the main Khmer population centers, the presence of Khmer culture on the coast dates from the height of the medieval empire and was a point of contact with the Malay cultures that dominated insular Southeast Asia. Under the French, the Cambodian Coast became an early manifestation of what is now known as the "tourist industry," as colonial administrators and entrepreneurs built summer villas in the coastal villages of Kep and Kampot. Kampot also became famous for its high quality black and white pepper, as French chefs incorporated it and the name of its village of origin into their culinary inventions.[4] A luxury casino was built on the summit of Bokor Hill at the southern tip of the Elephant Mountains, providing spectacular views of the Gulf of Thailand to visitors from around the world. Not surprisingly, all of the villas were abandoned or destroyed during the long period of civil conflict dating from the 1970s, and the strategic lookout at Bokor Hill was the site of numerous battles. The casino is being rebuilt and the abandoned villas are attractions for a new group of tourists as the region is experiencing an economic revival. Following Independence, the Cambodian government constructed a rail line from Phnom Penh to the newly constructed port of Kampong Som, now known as Sihanoukville (Figure 1.4) as part of its national development plan and to establish an ocean port that was not dependent upon the Mekong River.

The economy of the Cardamom Coast was traditionally dominated by subsistence fishing and agriculture, although shrimp farming has been adopted over the past few decades, and salt extraction has always been an important activity on the low lying coastal swamps near Kampot. Many of the current residents migrated into the coastal region after the civil war when large-scale shifts in populations characterized the country as a whole. People moved to areas because the prospect of combining rice farming with subsistence coastal fishing seemed like a good and familiar strategy to people who had previously found their subsistence on freshwater fisheries. This model still predominates among the many poor villages of the Cambodian Coast, although what will seem most predominant to the outsider are the large stretches of

Figure 1.4. The Greater Cardamom Region is flanked by two recently upgraded highways: Route 5 connects Phnom Penh with the provincial capitals of Pursat and Battambang, which function as the gateway to the north slope of the Cardamom Mountains. Route 48 is an entirely new highway that transects large tracts of natural forest landscapes and integrates the port city of Koh Kong with the rest of the country, as well as providing a direct connection to Bangkok.

unspoiled beaches and mangrove swamps. The lure of these beaches is now fueling a real estate boom based on tourism that will transform this region and the lives of the people who live there.

As for the mountain region and its rainforests, the isolation of the Cardamoms provided an ideal haven for the Khmer Rouge before its brutal three-year rule during the late 1970s. Following Vietnam's invasion of Cambodia in 1982, the Khmer Rouge retreated into the Cardamom Mountains and waged a civil war that lasted another decade until the Paris Peace Accords of 1991. The Samlaut District on the border with Thailand was among the first regions to come under Khmer Rouge domination in the early 1970s and also the last to cease hostilities in the 1990s; the last holdouts in the Pailin enclave were not fully integrated into national life until 1998. The impact on the country and the Greater Cardamom Region of this tragic and unfortunate period acts as a constant backdrop to all discussions that describe the region and has been the subject of numerous books,[5,6,7,8] documentary films,[9,10] and internet resources.[11]

During this long period of strife, the villages of the Cardamoms were subject to domination by the Khmer Rouge and, eventually, evacuated by the national government to villages on the periphery of the mountains in order to deny the insurgents support and sustenance. This traumatic period had a long-term impact on the region and its inhabitants; the indiscriminant use of landmines by both sides continues to wreak havoc and impede development.[12] Moreover, *Khmer Daeum* communities suffered a significant decline in their population and a near calamitous disruption of their traditional customs, including the loss of traditional languages and the decimation of their domesticated elephant herds. As the war came to an end, the survivors returned to their villages to reclaim abandoned farmlands, many of which are still covered with second-growth forest or *Imperata* grasslands.* Returning refugees were also accompanied by lowland Khmer seeking economic opportunities, thus accelerating the changes that are threatening the traditional ethnic communities that had dominated the highlands prior to the 1970s.

After peace returned to Cambodia, the government and its international partners made transportation infrastructure a priority investment in order to integrate the country and stimulate economic growth. This has included renovating the national highway network, including two highways between Phnom Penh and the coast, as well as three separate trunk highways that link the country with Thailand and Vietnam. One of these transportation corridors follows the coast via the previously isolated district capital, Koh Kong, passing through large tracts of intact tropical forest. The other two major highways flank the Tonle Sap, while the road situated on the south side of the lake also

* The term second-growth forest refers to pioneer forest that has colonized abandoned farmland. *Imperata* is a genus of grasses characterized by coarse foliage that is unpalatable to livestock; extremely resistant to fire and very difficult to eradicate.

parallels the northern slope of the Cardamom Mountains (Figure 1.4). Basic improvements have also been made in the secondary road network that penetrates the Cardamoms to ensure peace in this conflict-ridden region and to facilitate its integration into the national economy. Not surprisingly, this improved access has led to increased migration and settlement.

The end of hostilities in the 1990s also led to a boom in scientific research and biological inventory, which very soon identified the Cardamom region as a conservation priority. The Cardamom Mountains remain the largest wilderness area in mainland Southeast Asia and are home to some of Cambodia's most spectacular and endangered biodiversity, including elephants, tigers, and banteng, wild relatives of domesticated cattle. These charismatic wildlife species, together with the region's majestic forests and unspoiled beaches, are key components of a nascent tourist industry.[13] To conserve the biodiversity of Cambodia, the government created a national system of protected areas in the early 1990s, twelve of which were located in the Cardamom Mountains and surrounding areas.[14] The area allocated to conservation was subsequently expanded in 2002 with the creation of eight protected forests. Altogether these areas cover 4.76 million hectares, which corresponds to 26 percent of the total area of the country (Figure 1.5).

Future Scenarios

The resilience and persistence of the Khmer people, who have benefitted from the generous assistance of the international community, are transforming Cambodia's once ravaged economy. Throughout most of the past fifteen years, the country has enjoyed annual increases between 5 and 13 percent in the rate of growth of gross domestic production (GDP). Surprisingly, the government and the country's private sector have also managed to contain the damage from the global recession that started in 2008, and robust economic growth seems likely in the near term. Regional integration, direct foreign investment, the ongoing improvement in infrastructure, and the development of a strengthened institutions are all part of a positive national panorama reflected in the bustling streets of Phnom Penh.

Nonetheless, rapid economic growth comes with risks, as evidenced by the recent global recession: Economic booms lead to financial bubbles that eventually burst once a nation overshoots its capacity to productively invest what is essentially speculative capital. This phenomenon

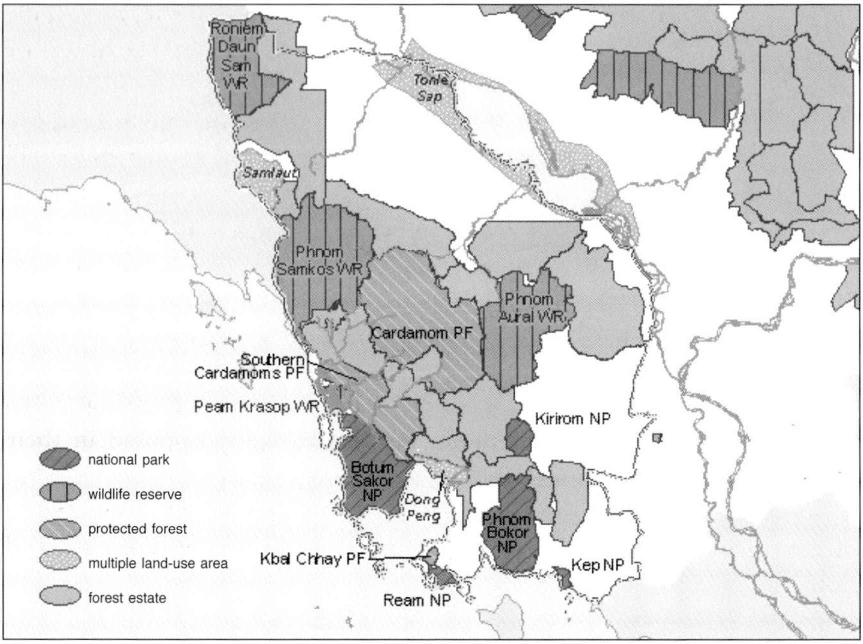

Figure 1.5. The Cambodian government created a system of protected areas in the early 1990s to conserve the nation's biodiversity. Ten of these areas are located within the Greater Cardamom Region, including five national parks, three wildlife sanctuaries, three multiple-use areas. These protected areas and their connecting landscapes are collectively referred to as the Cardamom and Elephant Mountain Conservation Corridor.

is by no means novel and is a largely cyclical and self-correcting process, if somewhat painful for the individuals and nations involved. Lessons are learned (or not), the cycle begins again, and life goes on. Rapid economic growth and many of the investment decisions that occur in this type of hot-house environment bring other risks, which are neither cyclical nor self-correcting — they have long-term economic impacts that are analogous to a one-way street of no return. These include poor planning and a reliance on feasibility studies that do not consider the full social and economic costs and benefits of development. Environmental degradation is an insidious side effect of many investment decisions, and even though investors may profit, local communities may suffer, especially if they lose access to resources they were accustomed to using. The exploitation of both renewable and non-renewable natural resources are necessary steps for economic growth, but how that exploitation is managed will determine whether these natural assets are capitalized so as to invest in a sustainable future or consumed in a quest for short-term profits.

The Greater Cardamom Region will play an important role in the future development of Cambodia, but exactly how that transpires will depend on the Cambodian people and how they choose to organize their country, especially the role of the government and the decisions of its elected officials. Although it is impossible to predict this future, it is possible to envision various scenarios that might result from different policies and market phenomena. The following sections describe three scenarios that illustrate how the Greater Cardamom Mountain Landscape might look after a century of human-induced change. Two are decidedly optimistic in outlook, but diametrically opposed in their philosophical underpinnings, whereas the third scenario is a somewhat more pessimistic view of what is likely to occur without a change in current public policies and more effective regulation by domestic and global markets.

The Utilitarian Scenario

In a utilitarian scenario, the natural resources of the Cardamom Mountains would be used to create a diversified economy based on maximizing the return on investments over decades. It assumes that a long-term national investment strategy would be implemented that extracted the greatest value from the region's natural assets by converting them into traditional commodities. This vision is sustainable if implemented properly, but requires a technologically sophisticated production model that is efficiently managed by an educated populace. In this scenario, urbanization and industrialization transform Cambodia and the country leverages its strategic location, abundant natural resources and rich cultural heritage to develop a diversified economy based on the export of commodities and industrial goods, as well as on being a world-class tourist destination (Figure 1.6).

The keystone investment in the Cardamom Mountains under this scenario is the full implementation of 22 hydroelectric power facilities on nine watersheds. These provide a total of 1700 MW of generating capacity, meeting about 50 percent of the national electricity demand in 2050.[15] This investment requires the establishment of a paved road network during the construction phase, which is then used to maintain and operate the hydropower facilities. This road network is also used

Figure 1.6. In a Utilitarian Scenario, the Cambodian state chooses to exploit the region's natural resource to create a diversified economy. The keystone investment are the construction of hydroelectric facilities (top), a modern transportation system (bottom), while mines and intensive agriculture are mainstays of the economy (middle). (Top photograph provided by Bruce Levick)

to launch an intensive forest management model that monetizes the hardwood timber assets on state lands. Over a few years, lands are clear-cut, and concessionaires plant fast-growing tropical timber species in blocks. These species have harvest maturities that match the requirements of processing facilities located in the expanded port town of Koh Kong. Fast-growing eucalyptus species are preferred because they can be harvested as pulp for paper mills or as feedstock for a biofuel processing facility that produces ethanol for domestic and international markets. Industrial minerals such as titanium, tungsten, chromium, nickel and copper are mined and concentrated into enriched ore at local mills prior to export as bulk commodities.

Hydropower facilities on the main stem of the Mekong River are completed and provide an even greater share of country's power needs; consequently, the electrical generation of the Cardamom watersheds is used to supply industrial processing facilities at Koh Kong, thus adding value to Cambodian production and increasing revenues for the state. The oil and gas reserves located offshore are exploited, as are similar hydrocarbon discoveries in the geological formations in the Tonle Sap depression.[16] These too are exported, creating a significant new revenue stream for the national government, although a decentralization law diverts 25 percent of royalty revenues to investments in the province of origin, with another 25 percent to be shared by the other 21 provinces. Regional governments have authority over the distribution of these resources; most are invested in health and educational facilities, followed by improvements in the secondary road network. Improved local roads increase the profitability of farming, leading to a gradual consolidation of farm properties. The average landholding increases from less than one hectare to approximately 10 hectares on the alluvial soils of the Tonle Sap. Small family farms still predominate in the foothills of the Cardamom Mountains, but larger landholdings also come into existence where tree farms and sugar cane plantations benefit the economies of scale attracting foreign investors seeking to compete in global markets. These large-scale agro industrial schemes are backed by the government because they incorporate processing facilities to transform raw feedstocks into semi-processed commodities, such as biofuels and paper.

Koh Kong, as mentioned above, has become the principal port for exporting timber, pulp and biofuels; it is also the site of a coal-fired generating facility that exports energy to Thailand and Vietnam via the regional grid. Coal is imported from Indonesia and Australia, while the abundant fresh water resources of the nearby Prek Kaoh Pao estuary make it an ideal location for pulp and paper mills. Meanwhile, the pre-existing port facilities at Sihanoukville have become the logistical center for the onshore and offshore petroleum facilities, with the Cambodia Petroleum Company locating its refinery and petrochemical complex on the east coast of the Kampong Som Bay, which is also the terminus for the pipelines that transports hydrocarbons from the Tonle Sap oil fields. The industrial and petrochemical facilities situated to the north of Sihanoukville have been kept separate from tourist facilities, which are now squeezed into the narrow coastal plain between Sihanoukville and Kep. Known as the Gold Coast of the Orient, the Cambodian Coast has developed into a booming tourist destination based on beaches, cruise ships, casinos, and recreational activities, including a Disney Indochina that receives approximately 5 million tourists annually, most of whom come from South Korea and China.

Concern for the environment increases as the country becomes more prosperous, but the decision to monetize the timber assets of the Cardamom Mountains and to maximize the cultivation of tree plantations has transformed the landscape. Although Cambodia has kept its commitment to maintain forest cover at around 60 percent, it has defined "forest" to include monoculture tree plantations. Two large wildlife sanctuaries, Phnom Samkos and Phnom Aural, have been reconfigured and reduced in size to exclude flat topography suitable for food crop production. A similar review of land use and *de facto* occupancy on the coast has led to reductions in the national parks of Phnom Bokor, Kirirom, and Botum Sakor. The Central Cardamom Protected Forest has been maintained in its natural state, but the southern reserve was rezoned as a production forest and converted, along with other state lands, to tree plantations that are harvested on seven to fifteen-year cycles. Increased urbanization has lessened the demand for land which, together with improved governance, has enabled Cambodia to consolidate the national protected area system and eliminate illegal logging and hunting. Nonetheless, the greatly circumscribed geographic area and the now-transformed production forests that separate individual units of the protected area system has isolated most populations, leading to the gradual loss of endangered and endemic species due to genetic erosion and the cumulative impact of climate change.

These changes have all required the institutionalization of a relatively efficient legal system, administrative decentralization, and improved transparency, all of which has allowed for the emergence of stable political environment needed to attract foreign investment. Prosperity and improved human welfare has slowed demographic growth, and the national population has stabilized at approximately 30 million. Poverty has been reduced to nominal levels, and Cambodia's economy surpasses all adjacent countries with much higher base populations and more rigid political systems. Cambodia has been transformed into Southeast Asia's youngest and most successful tiger economy.

The Utopian Scenario

In this scenario, Cambodia's natural resources are used to create a diversified economy, but rather than focusing on monetizing the timber assets of the Cardamom Mountains and investing those revenues in plantation forestry, the emphasis is placed on conserving natural forest ecosystems to ensure the production of goods and services over the long-term. This approach likewise supports a modern economy linked to global markets, but one in which the business models are focused on adding value to commodities and exporting finished goods, such as bamboo furniture and silk clothing. Tourism continues to play an important role, but a more diversified tourism model evolves with enterprises in ecotourism and cultural tourism, as well as traditional forms of beach tourism, all of which take advantage of the natural beauty that characterizes the Cardamom Mountains and the adjacent Cambodian Coast (Figure 1.7).

As in the previous scenario, hydropower is the centerpiece of this development strategy; however, the original design of 22 dams in nine watersheds is modified to 24 dams in seven watersheds. This modification yields the same total energy generation capacity but ensures that one river, the Stung Chhay Areng, home to one of only two *Khmer Daeum* communities, is conserved as a wild unobstructed river from its headwaters in the Central Cardamoms to its mouth in the Prek Kaoh Pao Estuary south of Koh Kong. Importantly, the Cambodian state has chosen to manage these seven strategic watersheds by conserving the native forest on their hilly terrains. This decision was a key component of the national forest conservation strategy specifically designed to reduce emissions from deforestation and forest degradation (REDD+) that eventually led to the certification of approximately 250 million tons of CO_2 emission reductions* that were monetized in international carbon markets and provided significant investment capital for companies and communities that adopted sustainable production and carbon-saving measures.

* CO_2 equivalents (see Chapter 6 for details).

Figure 1.7. In the Utopian Scenario, Cambodia pursues a low-carbon or "green" economic model that conserves and monetizes its forest resources in emerging carbon markets. Hydropower remains important, but greater emphasis is placed on fostering sustainable agriculture (top), maintaining key ecosystem services such as natural fisheries (middle), while conserving large tracts of natural forest and freshwater ecosystems (bottom).

The national forest conservation plan has slowed land-use change in the dry and seasonal forests on the north slope of the Central Cardamom Plateau, the Veal Veng Valley in the Phnom Samkos Wildlife Sanctuary and the Phnom Aural Wildlife Sanctuary. Settlement in these areas was brought under control by a combination of law enforcement and incentive agreements, which encouraged inhabitants to forego forest clearing and other practices that create carbon emissions (i.e. reliance on fuel wood) in exchange for investments in education and health facilities, improved social services, and a technologically sophisticated agricultural extension service. The incentive agreements allow local inhabitants to exploit high quality hardwood species for statuary and furniture, a major product for local artisans and businesses. Logging has been restricted to local communities and domestic firms that follow the unique harvest cycles of individual species in order to maintain viable populations in natural forest habitats. Reduced impact logging methods minimize CO_2 emissions, while silviculture treatments promote the succession and recruitment of the most valuable species. Bamboo, a fast growing and robust plant that can tolerate intensive exploitation, is promoted to support small businesses that produce flooring, furniture and other consumer goods. The combined income from carbon credits, hard wood, and bamboo products increases over time and provides income and economic opportunity to a region that is simultaneously diversifying its agricultural production by adopting new crops and technology.

In the agricultural sector, increased farm productivity leads to consolidation, but farms remain small, and profitability is increased by the formation of farmer cooperatives and the development of irrigation agriculture to allow double cropping. The irrigation model is based not on surface water, however, but on the abundant ground water resources of the Tonle Sap and Mekong flood plains.[17] Efficient micro-irrigation systems are developed to manage energy costs and the limited water flows of the Tonle Sap and Mekong aquifers.* The two-crop system includes rice grown in the traditional paddy system during the rainy season, combined with a mixture of high value cash crops for local markets and export during dry seasons, when plantings also include nitrogen-fixing legume species to improve soil fertility. Exports of these

---

*The high silt and clay content of sediments is characteristic of a back-water wetlands and leads to small pore size in water bearing underground sediments (see Chapter 2).

crops to Vietnam, Thailand and Southern China are facilitated by a completely integrated interstate highway system. This policy is matched with an agricultural extension plan to improve yields and reduce the drudgery of rice farming.

Cambodia's drive toward urbanization continues, and efforts are made to avoid the over-development of Phnom Penh. Recognizing that the flood plain of the Mekong River is inappropriate for the development of large metropolitan urban centers, the government promotes new urban and suburban centers in areas free of inundation by offering tax incentives to corporations that invest in these special economic zones. One of these special economic zones is located between Pursat and Romeas (in Kampong Chhnang Province) along a renovated rail line that has been upgraded and expanded to include a rapid-transit system, as well as a bulk commodity and other freight. This rapid-rail system serves a technological corridor that produces electronic goods and solar panels for export. The government has initiated an industrial policy that scales the technological ladder, moving from textiles to electronics. It has offered subsidies, derived in part from carbon credits, to attract high-tech electronics manufacturers who seek to improve the carbon footprint of their brands. The creation of urban corridors is also part of a deliberate national food security policy to conserve the seasonally inundated rice paddies. Food security is further enhanced by the urbanization and industrialization of the Cambodian economy, which reduces the need for rural labor, and by farm mechanization, which improves both productivity and efficiency.

The hydrocarbon and mineral sectors are developed in strict accordance with international guidelines for managing environmental and social impacts. Proposals to develop an oil refining complex are scaled down, so that the facilities supply only domestic demand, and no investments are made in a petrochemical complex. Consequently, coastal development is spared the more severe aspects of industrialization, allowing a more orderly expansion of the tourist industry, including beach hotels, cruise ships, and casinos. High-end ecotourism also flourishes, as there are no unsightly forest product processing facilities in Koh Kong or a petrochemical complex near Sihanoukville. The road network built to service the hydropower facilities is used to create a network of forest lodges in scenic locations in the Central Cardamoms, particularly at higher elevations where temperatures tend to be cooler and insects less abundant.

The development of forest-based carbon credits and the decision to use these revenues to invest in production strategies that generate wealth and improve the livelihoods of the region's inhabitants was a key step in consolidating the national protected area system in Cambodia. The realization that forest conservation brings tangible benefits to the region's residents changed the dynamic of land-use and land-speculation. Likewise, the state's decision to conserve the forests of the Cardamom Mountains led to the consolidation of a Cardamom Mountain Conservation Corridor that provided sufficient connectivity to ensure that most, if not all, species were able to maintain populations large enough to avoid extinction and improve the probability of their successfully adapting to climate change.

From an institutional perspective, creating and maintaining a transparent land registry was critical, because it enabled the government to grant incentives from carbon credits to effectively target economic actors with a vested interest in land use and, consequently, slow forest conversion throughout the Greater Cardamom Landscape, but particularly in areas open to colonization prssure. As in the utilitarian scenario, prosperity and improved human welfare slowed demographic growth, so the national population levels off at approximately 30 million inhabitants. Poverty has been reduced to nominal levels, surpassing all adjacent countries with much higher base populations and more rigid political systems. Cambodia has been transformed into Southeast Asia's greenest and most innovative tiger economy.

Business as Usual Scenario

In this scenario, the improved governance that makes the two previous scenarios plausible does not materialize, and Cambodia experiences a chaotic development trajectory in which influential individuals focus on short-term profits. The Cambodian economy grows at a rate that lags behind demographic growth, leading to the persistence of widespread poverty. Forest management remains precarious and informal, and the technological assistance and credit that farmers need to improve yields do not materialize. Industrial investment remains focused on low-tech industries such as textiles, with razor-thin profit margins operated by foreign companies. Raw commodities remain important, but exports decline as these commodities are depleted, or, in the case of food commodities such as rice, used to satisfy domestic consumption. Access to international carbon markets is limited to the technological sector because the Cambodian state is unable or unwilling to restrict access to forests. Tourism continues to play an important role in the national economy, but it remains linked to a single attraction,

Angkor Wat, and the opportunities to develop a world-class tourist industry slip away as the natural assets of the coast and the Cardamom Mountains are degraded (Figure 1.8).

As in the previous scenarios, the hydropower of the Cardamom watersheds is developed, but only the six largest hydropower facilities are built because the mega-hydropower facilities on the main stem of the Mekong River satisfy national energy demand. In addition, coal-fired power plants on the coast that import fuel from Indonesia flood the energy market and lessen the need for the smaller hydropower projects in the Cardamom Mountains. The road network built for the construction of the dams is used by settlers and land speculators who migrate into the highlands. In spite of a preference for flat land and rice paddy agriculture, population pressure and land speculation motivates many families to colonize areas with sloping topography and fragile soils. Settlers practice slash-and-burn agriculture, and a large percentage of cleared land is left idle at any given time, although it is cleared and cultivated again on an approximately 10-year cycle.

On the less humid northern slope of the Central Cardamom plateau and in the Veal Veng Valley in the Phnom Samkos Wildlife Sanctuary, a different type of land-use dynamic develops that does not favor small farmers or local communities. Urban investors displace the original settlers in order to consolidate larger properties near major roads, while displaced settlers move further into the forest. The lack of technical assistance forces small farmers to practice subsistence agriculture; families that elect not to sell their land-holding subdivide it among younger generations, limiting their ability to produce a surplus and participate in the national economy. Many are forced to work off the farm for low wages or migrate to urban areas where their limited education and lack of support networks relegate them to the margins of Cambodian society. The failure to establish a modern land registry system contributes to land speculation and as unscrupulous individuals appropriate land from unsophisticated small landholders. Land-related violence increases because the legal system is inefficient and plagued by political patronage. The proliferation of economic land concessions on state lands, combined with land consolidation in settlement areas, leads to a two-tier agricultural system, where small farmers practice subsistence agriculture on increasingly smaller family farms, while larger urban investors produce nonfood commodities for global markets.

The exploitation of oil and gas provides the Cambodian state with a windfall in revenues, which is used to expand public spending and to subsidize consumption by importing food. This policy discourages

Figure 1.8. In the Business as Usual scenario, poor governance leads to a development trajectory driven by individuals and industries seeking short-term profits. The conversion of both upland forest (top) and mangrove swamps (middle) continue unabated, as does the senseless harvest of wildlife parts for export to overseas exotic culinary consumers and sham medical treatments (bottom).

local agricultural producers, leading to increased rural poverty stimulating further encroachment on forest landscapes. It also reinforces decisions by landowners to grow non-food commodities targeted at global markets. Perversely, even though Cambodia is an energy exporter of fossil fuels and provides extensive energy subsidies to its burgeoning urban populations, the marginalization of the poor leads to strong demand for firewood and charcoal, which is produced by settlers as a byproduct of the land clearing process providing an additional incentive to clear forest.

Urban growth explodes as the population expands and rural — urban migration increases. Since the government provides subsidized fossil fuels to Cambodian consumers, imports of motor vehicles explode and traffic gridlock plagues Phnom Penh. Unplanned growth radiates outward along the major transportation arteries, particularly those connecting Phnom Penh to the port cities of Sihanoukville and Koh Kong, but also extending northward towards Battambang and Siem Reap where rice paddies are converted for commercial development, displacing small farmers and putting further pressure on the agricultural frontier. Efforts to improve rice production or implement microirrigation systems have limited success because of the disincentive to farmers caused by the subsidized imports of food.

Coastal communities experience economic and population growth approximately double the national average, spurred mainly by investments in hydrocarbon infrastructure and tourism. Rapid growth surpasses the ability of local institutions to regulate development and unplanned construction creates a chaotic mosaic of industrial plants, residential slums, and tourist facilities. Industrial development is dominated by the hydrocarbon sector, but most of the better paying jobs are held by foreign nationals, including mid-level technical positions occupied by professionals from Latin America and South Asia. Chinese construction companies that dominate the infrastructure sector rely on contract labor from Cambodia but preserve upper and middle management positions for Chinese nationals. The disparity in income and the large reservoir of urban poor contribute to the continuation of sex-based tourism, which damages Cambodia's international reputation and impedes investment in high quality tourist enterprises.

The degradation of the Cardamom Mountains rainforest has also discouraged high-end ecotourism, while the unplanned development of coastal islands has lowered the attractiveness of coastal tourism. The poor planning and subsequent degradation of coastal habitat has led astute developers to acquire pristine areas to convert to private beaches catering to wealthy foreigners. However, most of the tourist facilities are targeted at the mass market dominated by economy-minded tourists from East Asia. Environmental degradation associated with deforestation and coastal development is widespread and irreversible. The loss of natural forest cover in the Cardamom Mountains increases erosion, which reduces the energy output of hydropower facilities by as much as 50 percent in the dry season and during periodic droughts associated with the *El Niño* phenomenon. The conversion of mangrove swamps to shrimp farms and other forms of coastal development damages the near-shore fishing industry, displacing small fishermen and their families who move to urban centers. The offshore coral reefs suffer degradation from global warming and ocean acidification, but pollution from raw sewage contributes to a massive die back that damages what was a key tourist attraction and a key component of the coastal fisheries upon which thousands of local fishermen depend.

This scenario is a story of lost opportunities. The potential to generate revenues from forest conservation and carbon markets is foregone because of the failure to control land speculation and deforestation. Royalty revenues from hydrocarbons subsidize consumption to ensure popular support for the government. Populist economic policies impede the development of productive activities that create jobs via long-term, sustainable economic growth. Cambodia's population grows to more than 50 million and is characterized by entrenched poverty and economic inequality, which leads to social unrest, urban violence, and political instability that further impede investment. The failure to take advantage of the opportunities during the second decade of the twenty-first century makes Cambodia Southeast Asia's wounded and endangered tiger economy.

The Cardamom Conundrum as a Global Conundrum

Cambodia has demonstrated that it has the capacity to evolve into a stable political and economic system after decades of violence. It has shown the world that reconstruction is attainable, even after one of the most difficult and violent episodes of recent human history. These attributes will serve it well. It now has another opportunity to show the world a new pathway to sustainable development. The scenarios outlined above — while they may differ in particulars from nation to nation — will not be unfamiliar to inhabitants of Africa, Latin America, Oceania and other Asian nations. Almost every developing country in the world faces similar challenges and must make equally difficult choices about how to organize its society and manage its natural resources. Many countries suffer from what is known as the "natural resource curse," where an abundance of some global commodity — typically hydrocarbons but also minerals, timber, or agricultural commodities — can create an overreliance on revenues from these resources, distort the nation's economy and corrupt its political systems. These are typically stagnant nations, mired in poverty, characterized by social inequality and suffering from environmental degradation.

Cambodia has avoided this pathway thus far, in part because of its unfortunate history, which has impeded the exploitation of its natural resources. It also has other factors working in its favor: the cultural legacy of Angkor, and the willingness of developed nations to open their doors to Cambodia's textile manufactures for a limited but critical period during the nation's reconstruction phase. These factors created a sort of "grace period" for Cambodian development. But the natural resource curse has not disappeared and, as Cambodia moves forward into the next stage of its social and economic development, it would be easy to fall victim to this insidious curse. The secret — although it is hardly a secret — is to invest the revenue from its natural resources in the Cambodian people and communities via initiatives that are sustainable and that create long-term wealth and economic opportunity. Cambodia has already learned that exporting value-added goods is a better strategy than producing raw commodities. It has the opportunity to forge a novel development pathway based on the monetization of its natural assets, which produce revenue flows that are used as investment capital in sustainable production models. Cambodia not only has the potential to be one of the strongest economies in Southeast Asian economy, but also one of its most innovative in the management of its natural resources. Cambodia can be a global leader that demonstrates to its neighbors and other regions the possibility of having a society where conservation and economic development are part of an integrated philosophy for national development and the foundation of a low-carbon green economy.

CHAPTER 2

Natural Landscapes

The Greater Cardamom Region is perhaps the most dynamic region of Cambodia. It covers approximately one third of the country's land surface and spans nine of its 24 provinces, including Koh Kong, Pursat, Battambang, Pailin, Sihanouk, Kampong Speu, Kampong Chhang, Kampot, and Kep. The natural habitats of this region form a complex network of interconnected landscapes and aquatic systems that provide strategically important ecosystem services to the Cambodian economy. The region includes not only the mountainous landscapes of the Central Cardamoms, but also the foothills and coastal plain of the Cambodian Coast. The northern boundary is defined by the Boeung Tonle Sap,* or "Great Lake," which is surrounded by an alluvial plain that extends eastward to the Mekong River and westward to the hill country on the border with Thailand. The southern slopes of the Cardamoms are linked by fast-flowing rivers to the marshes, estuaries, beaches and tidal flats of the Cambodian Coast. Offshore, some 14 islands ranging in size from just a few hundred square meters to the more than 250 km^2 are located within a complex of coral reefs and sea grass beds that are integral components of the pelagic and benthic habitats of the Gulf of Thailand and the foundation of the region's in-shore and off-shore fisheries.

The Cardamom Mountains are the largest wilderness area in mainland Southeast Asia and home to some of Cambodia's most spectacular and endangered biodiversity. However, the landscapes that surround

* Tonle is the phonetic pronunciation for the Khmer word for a large river, Boeng or Bung is the word for lake, thus Boeng Tonle Sap literally refers to the "lake of the big river," but Tonle Sap is generally translated simply as "great lake."

that wilderness are home to approximately four million of Cambodia's 14 million inhabitants and the region is experiencing rapid population growth due to its abundant natural resources and the strategic importance of the Cambodian Coast. The biophysical features that make up the Greater Cardamom Region — climate, geology and biodiversity — create a complex web of interlocking habitats where changes in one landscape can reverberate through linked ecosystems that can lead to unforeseen impacts. These connections, as well as the characteristics of individual habitats, must be understood if one is to fully value the ecosystem services that the region provides and create a development strategy that successfully harnesses these resources via production models and exploitation systems that are inherently sustainable.

Geology

The most prominent geological features of the Cardamom region are the different mountain ridges and highlands of southwest Cambodia, particularly the sandstone formations of the Central Cardamom Plateau, Phnom Samkos and Damrei (Elephant) Mountains, and the granite massif of Phnom Aural, which is also the highest mountain in Cambodia (see Figure 1.2). More interesting from a mineralogical (and economic) perspective is the diverse mosaic of igneous and metaphorphic rocks near the Thai border in Pailin and Battambang provinces, which has been a source of rubies and sapphires for centuries (see Figure 2.1). Other interesting geological formations include the ancient lava (Basalt) flows near the Arang Valley in the Central Cardamoms and the metamorphic rock formations located along the margins of Phnom Aural.[1]

On the northern boundary of the Cardamom region, the Tonle Sap depression is a strategically important geological feature that was created by the same tectonic forces that led to the uplift of the Cardamom Mountains — a phenomenon that geologists refer to as a syncline. Thus, the same sedimentary rocks that form the Central Cardamom Plateau are also situated deep beneath the Tonle Sap Lake. At various times in the geological past, the Tonle Sap has existed as a shallow ocean bay rather than a fresh water lake; today, water levels at lakeside are a mere 10 meters above sea level during the low water period and only 25–30 meters above sea level at maximum flood stage. The lake is connected via waterways with the Mekong River, from which it has received sediments for over millions of years to form a low plain with the Tonle Sap at its center. Most of the superficial sediments were deposited during the Pleistocene and Holocene eras and are composed

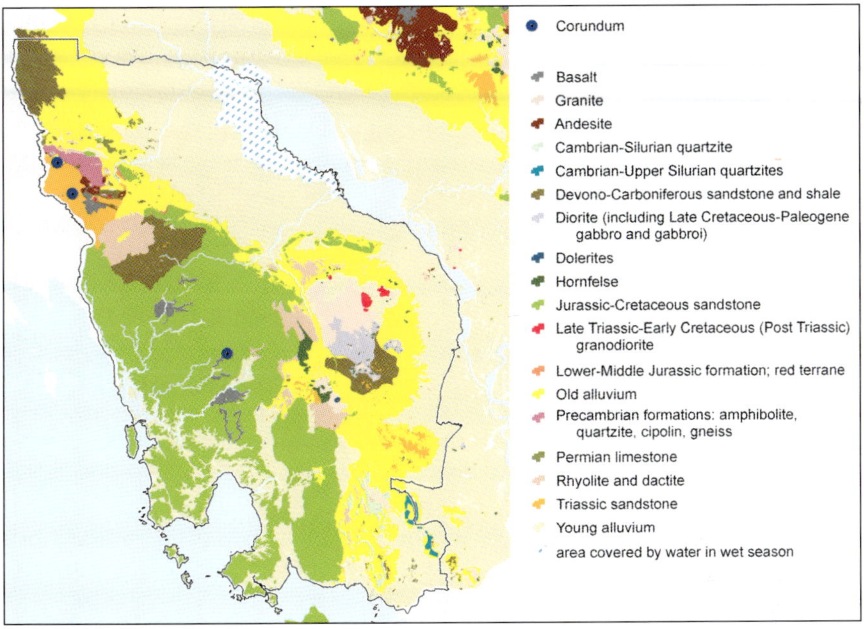

Figure 2.1. The surface geology of the Cardamom Mountains is composed largely of sandstone that dates from the Jurassic and Cretaceous, with scattered basalt lava flows. The northern and eastern sectors have more complex geology composed of igneous and metamorphic rocks, as well as shale and old alluvium. Corundum deposits in Pailin are renowned for their rubies and sapphires.

of sand, silt and unconsolidated gravels that house an important, but largely unexploited aquifer.[2] The geological eras of the Jurassic and Cretaceous during which the sandstone formations located beneath the Tonle Sap were formed are often associated with the formation of petroleum. Seismic surveys have been conducted over several decades by the Cambodian state, and the region is considered a potential area for the production of oil and gas.[3]

Situated to the south of the Cambodian Coast, the Gulf of Thailand is a shallow sea with a continental shelf that is seldom more than 40 meters deep and which extends across the Gulf to the Malay Peninsula and to the islands of Sumatra, Borneo and Java. Known as the Sunda Shelf, this submarine land form was exposed during the glacial epochs of the late Tertiary and Pleistocene, creating a land bridge between insular and mainland Southeast Asia. This land bridge influenced the distribution of many mammalian groups, including elephants, primates, cats, and prehistoric humans. Prior to the glacial epoch of

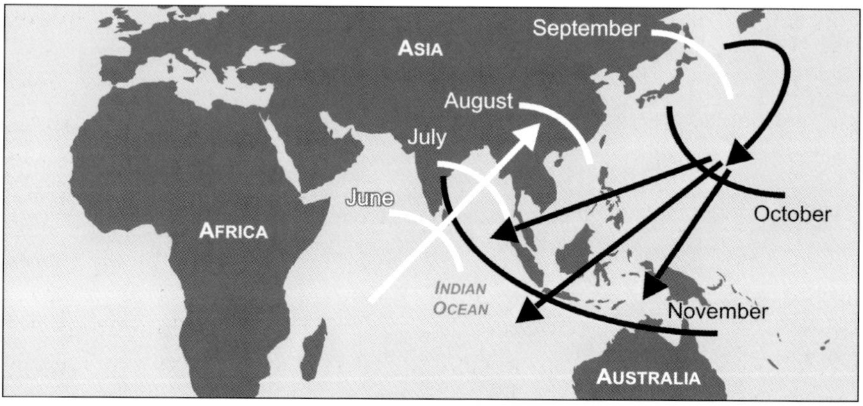

Figure 2.2. The Asian Monsoon is the principal climatic feature of Cambodia and is characterized by humid air flows that transport water from the Indian Ocean to the Asian landmass. It is expressed as a belt of rain that proceeds northward with the onset of the polar summer. The retreat of the warm winds southward around October/November ushers in the dry season and prevailing winds reverse direction.

the mid Tertiary, the Sunda Shelf was a shallow marine lagoon that received trillions of tons of sediment from the Chatheo Phraya River of Thailand in conditions that were ideal for the formation of hydrocarbons, particularly natural gas.[4] Today, the large expanse of shallow water places the nutrient rich ocean floor in relative proximity with the high light environments of surface waters, creating the productive natural fisheries that characterize the Gulf of Thailand.

Climate and Monsoon

The Asian Monsoon is the principal climatic feature of Cambodia, and understanding its patterns and influences is crucial to comprehending the impact of future climate change in the region. The Asian Monsoon is characterized by humid air flows that transport water from the Indian Ocean and the South China Sea to the Asian landmass. The monsoon dominates the climate across a broad geographic region from Pakistan to eastern China and is expressed as a belt of rain that proceeds northward, starting in early May in the southern coastal regions and arriving in mid-July in northern India and Eastern China (Figure 2.2). The monsoon is driven by differences in air temperature that occur as a natural consequence of the changing of the seasons and the lengthening of days during the polar summer. The relative positions of the Asian landmass (north of the Equator) and the Indian Ocean (south of the Equator) are important for understanding how the temperature gradients are created and how subsequent energy flows drive the Asian Monsoon: Air heats up more rapidly over land than over the ocean, which is an extremely efficient temperature sink. Since warm air rises, a pressure gradient is established across the subtropics, and moist air from the ocean is drawn into the continent. As the warm air over the land rises, it also cools and looses water-holding capacity, producing rain. This unique combination of moist air flows from the ocean and rising air over hot land surfaces is what drives the sudden onset of the Asian monsoon.[5]

The dry season is essentially the retreat of the monsoon and occurs for exactly the same reason as the onset. The band of moist warm air that surrounds the globe over the tropics and that had shifted to the north during the polar summer shifts back to the south for the austral summer. Known as the Inter-Tropical Convergence Zone (ITCZ), this band of moist warm air defines the equatorial tropics, which are always wet and humid, as well as the seasonal tropics and subtropics that receive water from the ITCZ during their respective summer seasons.[*] The term "monsoon" was originally coined to describe the onset of the summer rains in Asia, but it has been adopted by the scientific community to describe the transfer of water from the ITCZ to continental weather systems in all parts of the world.

The Asian Monsoon has many regional variations depending on ocean currents and wind systems and has been stratified by climatologists into several subcategories. The monsoon over Cambodia is located at the transition between two of these distinct subsystems: the South Asian Monsoon and the East Asian Monsoon. Consequently, Cambodia receives rain from both the South China Sea and the Indian Ocean. The onset of the monsoon in Cambodia is usually early May and is essentially a northward extension of tropical thunderstorms that originate in the seas around the islands of Sumatra and the Malay Peninsula. Temperatures in Cambodia also fluctuate with the monsoon. The highest temperatures occur at the end of the dry season (April and May) when the sun is directly overhead and the absence of standing

---

[*] The ITCZ only exists over oceans or other large marine bodies, but it interacts with continental land masses in idiosyncratic ways depending on the topographical configuration of each continent.

water and low soil moisture levels limit evaporative cooling that can otherwise ameliorate high temperatures. Maximum temperatures can exceed 40° Celsius and are a key factor in driving the onset of the monsoon. Minimum temperatures occur in January near the winter solstice when skies are clear, allowing for the radiation of heat during the night.

The other major climatic phenomenon that impacts Southeast Asia is the *El Niño – Southern Oscillation* (ENSO). The term *El Niño* describes a cyclical weather phenomenon during which the eastern tropical Pacific has warmer-than-average sea surface temperatures; it is associated with heavy rains and floods on the Peruvian coast of South America.* These conditions contrast simultaneously with cooler-than-average sea surface temperatures in the western tropical Pacific, which brings drought conditions to much of insular Southeast Asia. These meteorological conditions "oscillate" over time, with the conditions that lead to *El Niño* "events" tending to develop on average about once every seven years. The oscillation occurs in both directions; when one region experiences drought-like conditions, the other has abundant rains, and vice versa.[6] The contrasting condition is referred to as *La Niña*, which leads to floods and heavier-than-average monsoonal rains in Southeast Asia.

El Niño events are usually relatively short-lived phenomena that typically start in December and persist for several months. In Cambodia, periodic decadal-scale droughts are a common phenomenon coinciding with *El Niño* years, for example in 1986–87, 1993–94, 1997–98 and 2004–05. ENSO-related drought has also been documented in both Thailand and Vietnam.[7] The onset of *El Niño* usually coincides with the dry season in Southeast Asia and the more common "weak" events have only a limited impact on Cambodia. Stronger *El Niño* events that last more than a year can have a devastating impact on the country by reducing rainfall that limits rice production and economic growth. In contrast, the more severe *La Niña* events are characterized by humid winds that originate over the South China Sea and

* The term *El Niño* comes from Spanish and refers to the Christ child. Since the unusual heavy periodic rains would start in late December, the natives made the psychological and religious connection that these rains arrived with the Christian deity whose birth is celebrated on the 25th of December. The term *La Niña* (the girl) was coined by scientists as contrasting term to describe the opposite of *El Niño* (the boy).

converge over Cambodia with water-laden air masses from the Indian Ocean, leading to heavy monsoons with greater than average precipitation and flooding.[8]

The topography of the Cardamoms Mountains also has a large impact on the regional distribution of rain. Since the monsoon comes from the southwest, the moist air must rise over the highlands, causing heavy rains on the windward (coastal) side and a rain shadow on the leeward side. National precipitation maps are somewhat inaccurate, because they interpolate the difference between the high precipitation records of 4,000 mm per year from the coastal stations with the low records from urban stations around the *Tonle Sap* of about 1,700 mm per year.[9] The precipitation at mid-elevation of these coastal watersheds is probably considerably higher, particularly on the southern facing steep slopes where convection and wind flows combine to stimulate orographic rain. Indeed, regional precipitation maps show that the Cardamom's Mountains experience the highest precipitation of all of mainland Southeast Asia (Figure 2.3).

The Cardamom highlands are further distinguished from the surrounding landscapes because they enjoy cooler temperatures and experience greater cloud cover throughout the year, both of which are conditions that lead to the formation of unique microclimates and habitats found nowhere else in the region. These landscape features also impact the intensity of wind: The prevailing winds originate over the Gulf of Thailand and flow inland over the highlands reaching the highest mean wind velocities in the country on the southwestern crests of the Damrie (Elephant) Mountains, followed by Phnom Sankos, Phnom Aural and the Cardamom Mountains.[10]

Hydrology

Water is one of the most abundant and certainly the most valuable resource that the Greater Cardamom Region provides to Cambodia. The region has two major watersheds: To the north and east, multiple small streams and two medium sized rivers feed into the Tonle Sap and Mekong River, while to the south, seven watersheds drain into the Gulf of Thailand through the Koh Kong Estuary and the mangrove swamps of the Pream Krasop Wildlife Sanctuary or the Dong Peng Multiple Use Area on the Chhak Kampong Som Bay north of Sihanoukville (Figures 1.4 and 2.4). These watersheds are strategically important for national development, because of their hydropower potential and their contribution to freshwater and coastal fisheries. Water flows in

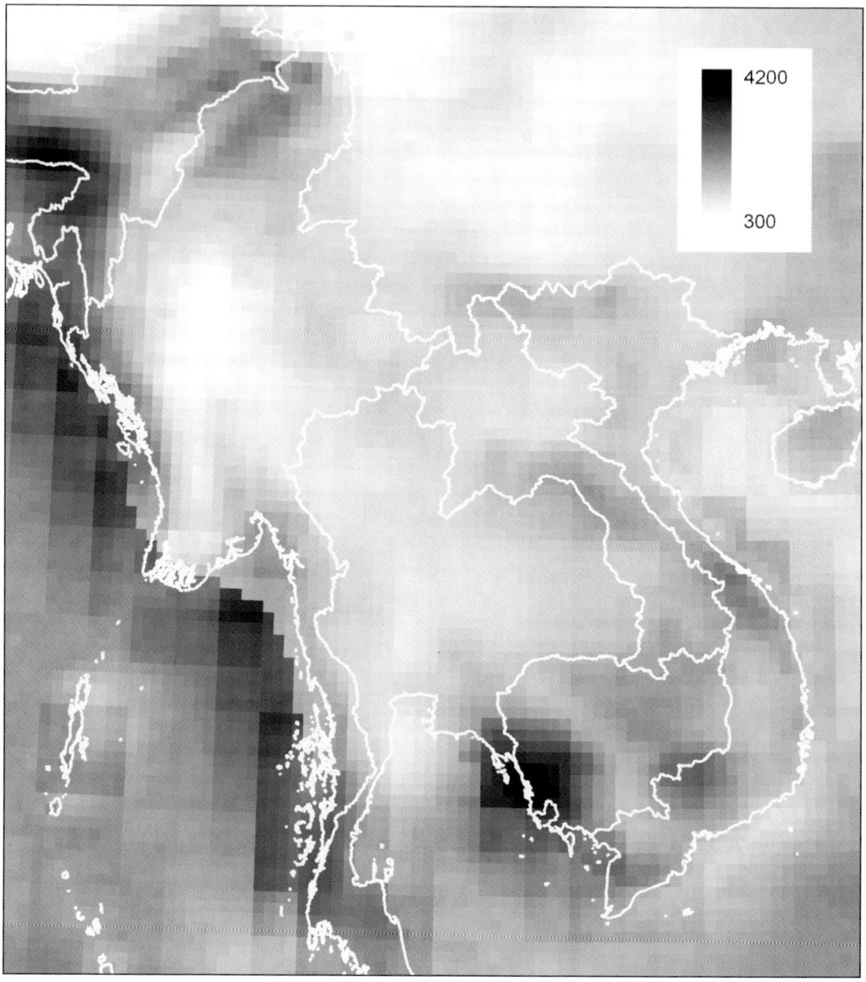

Figure 2.3. The topography and monsoonal airflows from the gulf of Thailand combine to create a local "wet spot" or high precipitation zone with almost three times the rainfall of surrounding areas. (Source: NASA TRMM satellite data for the years between 1997 and 2007)

the rivers on the windward (coastal) side of the Cardamom Mountains are much greater in volume due to the higher precipitation and have higher energy generation potential because of the rapid fall in elevation over a relatively short horizontal distance. Thanks to their dense forest cover, these coastal watersheds have low sediment loads, and provide important fresh water and nutrients into the highly productive estuarine and mangrove habitats of the Cambodian Coast.

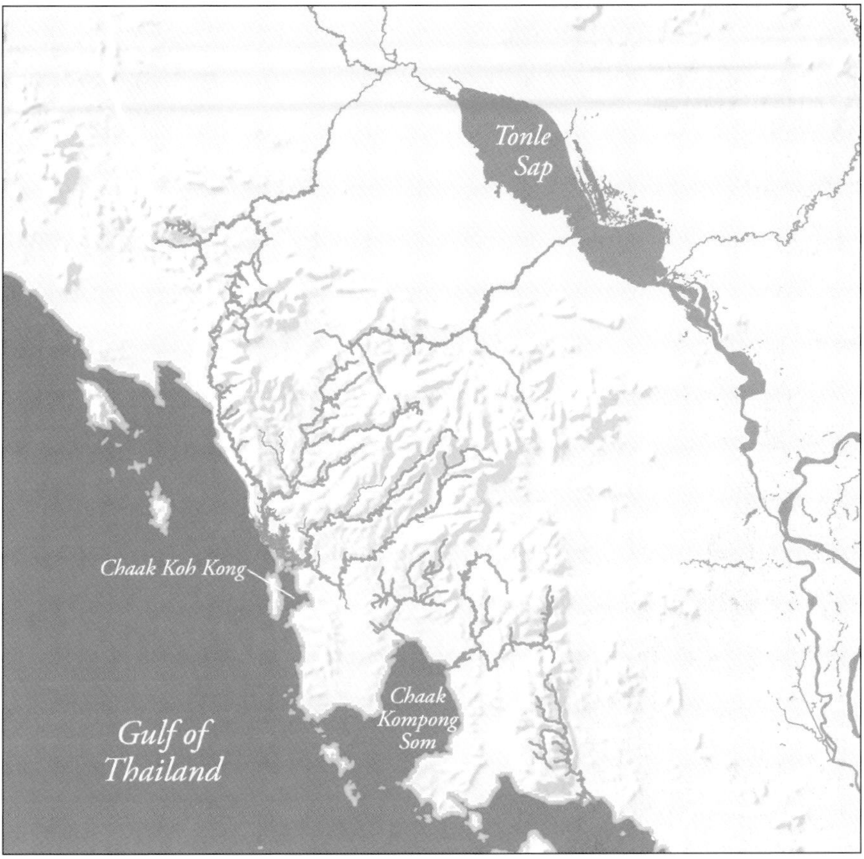

Figure 2.4. There are nine watersheds of the Greater Cardamom Region: Seven rivers flow from the high rainfall southern slopes of Cardamom and Elephant Mountains into the estuaries and mangrove swamps of the Cambodian Coast and the Gulf of Thailand, while two smaller rivers flow north into the Tonle Sap and Mekong River system.

The primary feature of the northern watersheds is the hydrological link between the Mekong River and the Tonle Sap Lake, and the pronounced seasonality mediated by the monsoonal climate of Southeast Asia.[11] Approximately 80 percent of the total precipitation that falls within the Mekong River Basin occurs during the six-month rainy season. This enormous mass of water flows down the Mekong from tributary watersheds that originate in China, Myanmar, Thailand, Vietnam and Laos and causes the reversal of water flows into and out of the Tonle Sap. At the midpoint of the rainy season when the Mekong

River approaches high water flows, the low-lying lake receives water and sediments from the river. Conversely, as the dry season progresses and water level fall in the main stem of the Mekong River, water flow changes direction and the water stored in the Tonle Sap empties out into the Mekong River.[12] This unique hydrological cycle characterizes the ecology of the Tonle Sap and underpins the extraordinary productivity of its freshwater fisheries.

Less well appreciated is the volume of water that the Tonle Sap watershed contributes to the total Mekong River discharge. Although less than the large water flows from the upstream section of the Mekong, the tributaries of the Tonle Sap and the rainfall accumulation in this very large lake account for about 30 percent of the total water flow of the Mekong and a similar volume of the sediments that are a critical part of the Mekong River — Tonle Sap ecosystem.[13] Approximately half of this water and sediment flow (about 15 percent of the Mekong total) comes from the Cardamom Mountains via the streams and rivers that pass through upland forest habitats into the wetlands surrounding the Tonle Sap, all of which are associated with the complex network of canals, fishponds and rice paddy fields, which dominate the alluvial plains of Central Cambodia.

One attribute of the Tonle Sap that has not been much discussed is the aquifer located in the alluvial deposits that overlay the sandstone bedrock of the Tonle Sap depression. The dimensions of this aquifer are poorly documented, but it is probably large, shallow and easily exploitable. Since the annual precipitation is high and the aquifer is superficial, it is most likely replenished at a high rate and could be sustainably used to create as an irrigation resource during the long and severe dry season.[14] Data from well-drilling programs has revealed that the alluvium has a high clay content, which limits the aquifer's suitability for large-scale irrigation, as water yield would be limited to a few liters per second. However, the aquifer's capacity may support family gardens or industrial scale production using super-efficient drip irrigation systems that have been developed in Israel and elsewhere.

Terrestrial Ecosystems

Based on recent forest cover maps, about 55 percent of the total Greater Cardamom Region is covered by some type of native upland habitat, with an additional 5 percent identified as native wetland habitat. The remainder is anthropogenic land cover, mostly paddy rice on the alluvial plain south of the Tonle Sap Lake, or on the coast along Kampong Som Bay (Figure 2.5). The forests of the Greater Cardamom Region

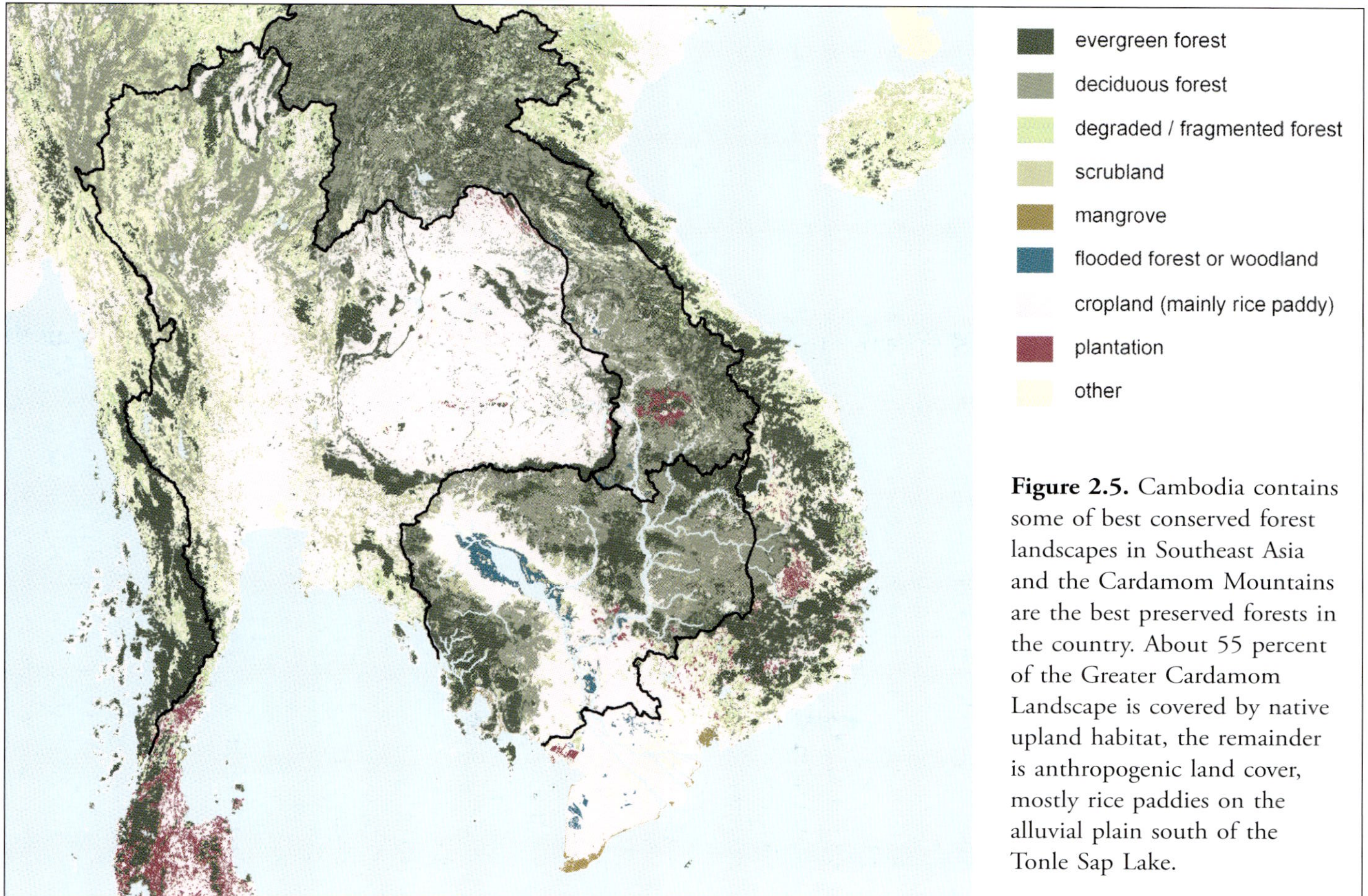

Figure 2.5. Cambodia contains some of best conserved forest landscapes in Southeast Asia and the Cardamom Mountains are the best preserved forests in the country. About 55 percent of the Greater Cardamom Landscape is covered by native upland habitat, the remainder is anthropogenic land cover, mostly rice paddies on the alluvial plain south of the Tonle Sap Lake.

have been classified into different forest types based on plant physiognomic traits:* Cloud forests occupy ridge tops, while rainforest occur on the southern slopes overlooking the Gulf of Thailand and seasonally humid moist forest predominate the northern and eastern perimeters, while dry deciduous forests share landscapes with open woodland and grassland habitats that eventually give way to the densely populated agricultural plain of the Tonle Sap and Mekong River.

Native vegetation on upland landscapes is largely correlated with precipitation — evergreen rainforest intergrades with humid forest, which transitions into semi-deciduous and deciduous forests and, eventually, into open woodland and savanna. The floristic composition of these forests has not been studied in any systematic way since the French produced a *Flora Générale de l'Indochine* in the early twentieth century[15,16] and a botanical survey is long overdue. Nonetheless, our knowledge of the natural history of the Cardamom Mountains indicates that the region contains great species diversity, including a fair number of endemic plant species. This expectation largely rests on the correlation between precipitation and species diversity — in both the Neotropics and the Paleotropics, plant species diversity increases with rainfall[17] — and on the region's geographic isolation from other rain forest and montane forest ecosystems. It is a basic tenant of evolutionary biology that isolation fosters the development of distinct local genetic variants that evolve over time into separate species. Since the end of the Pleistocene, and perhaps much longer, the rain forests of the Cardamoms Mountains have been geographically separated from similar forest types in Southeast Asia. Consequently, there is ample reason to assume that the Cardamom forests are a center of both species diversity and endemism. About 2,300 plant species are currently known to exist in Cambodia, but estimates of the eventual total range from between 3,000 to 8,000 species, and 10 percent of these are expected to be endemic.[18] It is not unreasonable to assume that as many of half of these endemic species probably will occur in the Cardamom Mountains.

A few observations can be made about the distribution of functional plant types and species richness and endemism in the Cardamom Mountains based on patterns documented for other tropical forest ecosystems in Southeast Asia.[19] For example, tree species diversity is

* Physiognomy refers to easily observed attributes, such as predominant life form (grass, shrubs, trees), structure (stem density and height), phenology (evergreen vs. deciduous) and other conspicuous features (thorns, succulent, epiphytic).

positively correlated with precipitation and negatively correlated with seasonality; cloudiness, which is a function of local topography and wind flows, is a good predictor of species diversity for epiphytes, such as orchids, ferns and aroids. Lianas are abundant where there is frequent disturbance and are more diverse where drought provides them with a competitive advantage over trees. Hardwood trees are more common in dry forests, where slower growth rates lead to high wood densities, while secondary and pioneer forests are characterized by fast growing soft wood species that are usually widely distributed over large geographic areas.

Most of the forests of the Greater Cardamom Region are dominated by genera and species of the *Dipterocarpacea*, an unusual family of trees that dominates almost all of the lowland forest formations of Southeast Asia.[20] The abundance of diptercarps over such a wide geographic area and range of tropical forest types has generated considerable speculation among scientists about the characteristics that give it such an evolutionary advantage.[21] Dipterocarops are characterized by very large boles and immense stature, traits that allow individuals to become "emergents" in the forest canopy and, thus, dominant in terms of biomass. Their size, abundance, and ecological importance explain why logging practices in Southeast Asia are more intensive than in Latin America and Africa. Simply put, this family of trees has very good timber and is naturally abundant in tropical Asian forests.

At least 6 genera and 30 species of dipterocarp trees are known to exist in Cambodia, 18 of which are on the IUCN list of endangered species.[22] This includes the dominant canopy trees on the lower slopes of the Cardamoms Mountains, including Phdiek (*Anisoptera costata*), Chheutal bangkuoy (*Dipterocarpus costatu*), Koki khsach (*Hopea pierrei*), Koki msav (*Hopea odorata*) and Koki Phnom (*Shorea hypochra*).[23] Other important timber species subject to overharvesting include *Parkia streptocar*, *Heritiera javanica*, *Swintonia pierrei*, and *Syzygium cinereum*, of which only *Swintonia pierrei* (known in Khmer as *Svay Kandol*) is listed on a recently compiled list of important tree species of Cambodia.[24]

Forests at higher elevations are home to tree species adapted to the cooler temperatures and mist of the cloud forests habitats. Of particular note are *Fagaceae* (beeches and oaks), a family of trees renowned for their abundance in the temperate forests of the northern hemisphere and a review of the literature provides the following list of Fabaceae found in the Cardamoms: *Lithocarpus cambodienseis*, *L. guinieri*, *L. farinulenta*, *L. harmandii*, and *Castanopsis cambodiana*, with the last

Figure 2.6. The Cardamom Mountains conserve a broad range of forest habitats, including pine forests on ridge tops (top), rainforest on the southern slopes overlooking the Gulf of Thailand (middle) and deciduous dry forest on the northern slopes adjacent to the Tonle Sap plain (bottom). (Middle photograph provided by A. Michaud)

species being endemic to the Elephant Mountains, though it has not been collected since 1919. Also of interest is the *Podocarpaceae*, a southern hemisphere family of gymnosperms that are particularly diverse throughout insular Southeast Asia. Other endemic plants from the Cardamom region collected during the colonial period include a ginger *Globba graminifolia* collected near Pursat, the grass *Phacelurus cambogiensis* from near Compong Chuang, a sedge *Carex cambodiensis* from the Elephant Mountains, and type of wild cinnamon *Cinnamomum elephantinum* without a known locality, but evidently also from the Elephant Mountains.

One of the most interesting habitat types in the region is the pine savanna that is found on ridge tops in the Central Cardamom and the Elephant Mountains. These savannas are maintained by naturally occurring wildfires caused by lightning strikes on the higher peaks (>1,000 m) and without the periodic occurrence of wildfire this community of plants, which consists of open grassland and widely spaced *Pinus mekussi*, would have been displaced long ago by fire intolerant rain forest species that occur at the similar elevations throughout the Cardamoms. The scattered populations of *Pinus mekussi* are examples of "relict populations" — species that were more widely distributed at lower elevations during cooler and drier periods of the Pleistocene. Similar isolated populations of pine are known to exist on high mountains in Sumatra and the Philippines, and all of these species are related to *Pinus latteri*, a more widespread species found at lower elevations in Vietnam, Laos and the Yunnan Province of China.

The dry forests that surround the Cardamom Mountains are part of the Central Indochina Dry Forests Ecoregion, which is an assembledge of species that covers most of Thailand and extends eastward to the Mondilkiri Province on the border between Cambodia and Vietnam. Like tropical deciduous forests throughout the world, rainfall varies between 1,000–1,500 mm and is punctuated by a pronounced dry season. Forest structure ranges from a nearly closed canopy forest with trees reaching up to 10 meters, to open woodland savanna with scattered trees and a dense grass sward. Like the nearby humid forests, dry forests are dominated by the Dipterocarpaceae, with the four most abundant trees being Meranti (*Shorea siamensis*), Phchoek (*Shorea obtuse*), Tbeng (*Dipterocarpus obtusifolius*), and Khlong (*Dipterocarpus tuberculatus*). Other common species include deciduous legumes, such as *Pterocarpus macrocarpus*, *Sindora siamensis* and *Xylia xylocarpal* and two conspicuous cycad species (*Cycas siamensis* and *Cycas clivicola*). Unlike the dominant timber species of the humid forest habitat, these species are all widespread and abundant throughout mainland Southeast Asia and are not considered to be at risk of extinction, despite extensive conversion of their native dry forest landscapes.

The extensive fires of the dry forest and savanna landscape are at least partially responsible for the distribution of two distinct vegetation types in an area with the same climate; many species display morphological traits characteristic of plant communities maintained by fire. Studies also demonstrate that soil texture, particularly the relative content of clay and sand, can influence the ecological processes that determine vegetation type.[25] The ongoing dependence of Cambodian society on fuel wood is likely another reason why open grassland is more abundant on the edge of the forest block that spread across the Greater Cardamom Region, which is dominated by humid forest species on the windward (coastal) side and dry forests on the leeward (Tonle Sap) side of all of the major mountain ridges.

Bamboo colonies are another ecologically important habitat type in the Cardamom region. One of the Cardamom Mountain's larger rivers is called Stung Russey, or bamboo river, testifying to the abundance of large native colonies of this conspicuous plant. Bamboos are woody arboreal grasses with many of the morphological attributes that make herbaceous grasses one of the most successful life forms on the planet. For example, almost all bamboos are rhizomatous perennials they can rapidly colonize open spaces, but because they are both woody and colonial they can persist over decades. One of the most well known population of bamboos in Cambodia is located near the village of Samlaut and this population has been harvested commercially over several decades, but is now being cleared for intensive agriculture. Nonetheless, there are other large bamboo populations in the Cardamom Mountains, particularly along river courses and on steep hillsides that suffer periodic disturbances from flash floods and landslides. Bamboos are characterized by extraordinarily high productivity and have a unique reproductive life cycle; colonies will subsist for years (even decades) in the vegetative state, then flower en masse across the entire population. Because of this prolonged vegetative state, bamboos are usually not collected during standard botanical inventories; even when they are, their identification is problematic due the lack of reproductive organs. Consequently, the diversity, distribution and status of most bamboos are often poorly documented. Bamboo is economically important throughout Southeast Asia, and in Cambodia it is broadly grouped into two categories: (1) Construction bamboo, which is valued for its strong stems for scaffolding, and (2) Paper and kitchen bamboo

Table 2.1. The Cardamom Region is home to a rich diversity of species, many of them found nowhere else on the planet, and many of them in danger of extinction both in Cambodia and globally.

Taxa	*Number of Species*	*Number of Species listed as Rare or Endangered in Cambodia*	*Number of Globally Threatened Species*
Mammals	66	17	20
Birds	174	12	6
Reptiles	74	4	9
Amphibians	32	–	1
Fish	44	–	1

that is used as a source of fiber to create mats and kitchen utensils. The former group includes Russey Khley (*Bambusa bambos*), Russey Srok Chin (*B. burmanica*), Russey Srok (*B. flexuosa*), Russey Koa (*B. vulgaris*) and Russey Ping Pong (*Oxytenanthera densa*), while the second group includes Russey Prech (*Arundinaria ciliate* or *Arundinaria pusilla*) and Russey Rlek (*Bambusa blumeana*).[26]

Rattan palms are another important plant resource of the Cardamom region's humid forest habitats. All rattans belong to the genus *Calamus* and are differentiated from other palms by their vine-like habit and spines that allow them to attach to and "climb" trees. Their tough flexible stems are an ideal raw material for the woven furniture that has been prized by global commerce for centuries. Several commercially important species referred to either as Piidau or Ropeak, are found in Cambodia, including *Calamus giganteus*, *C. dioicus*, *C. viminalis* and *C. salicifolius*. Unlike bamboo, rattans do not grow rapidly and are seldom cultivated. The global center of rattan production is Indonesia, but it is also produced in Malaysia and Thailand; like many lianas, rattans grow well in disturbed and mountainous habitats. However, over-exploitation and deforestation have reduced rattan populations in the three main producing countries over the last several decades and research is currently underway to develop sustainable management strategies.[27]

As for Cambodia's animal biodiversity, the wildlife populations of the Greater Cardamoms are among the most intact and healthy in all of Indochina; thanks to their relative isolation, these forests landscapes contain some of Asia's most unique and endangered wildlife species (Table 2.1). Among the most endangered are charismatic or "flagship"

Figure 2.7. The Cardamom Mountain and Tonle Sap are home to numerous rare and threatened species, including the Binturong (*Arctictis binturong*), the Pileated Gibbon (*Hylobates pileatus*), the smooth-coated otter (*Lutrogale perspicillata*), the Fishing Cat (*Prionailurus viverrinus*), the impressed tortoise (*Manouria impressa*), the Grey-headed Fishing-Eagle (*Ichthyophaga ichthyaetus*) and the Arowana or dragon fish (*Arapaima gigas*).

species such as the Asian Elephant (*Elephas maximus indicus*), the Indochinese Tiger (*Panthera tigris corbettii*), and the Asiatic Black Bear (*Ursus thibetanus*). Unfortunately, the presence of tigers is based only on secondary evidence (dubious sightings) and there have been no actual confirmed sightings in at least 10 years. If they are not already locally extinct, they soon will be. The size of the elephant herd is somewhat more heartening, with perhaps as many as 300 individuals that range over a wide area known as the "Elephant Corridor," which includes Botum Sakor and Bokor National Parks, the Central and Southern Cardamom Protected Forests and Phnom Aural Wildlife Sanctuary (see Figure 1.4). The traditional migratory routes of these wild elephants often bring them into conflict with communities, as they destroy crops and damage farm property. Fortunately, the inspirational work of dedicated park rangers have helped communities manage this conflict by encouraging farmers to use hot chilies, fences, and loud noises to ward off elephants. Simultaneously, authorities have used the Cambodians' cultural attachment to elephants, which are Buddhist symbols and play a prominent role in many Khmer legends and historical accounts, to encourage their conservation.*

The Cardamoms also have wild populations of two other critically endangered species: the Asiatic Wild Dog or Dhong (*Cuon alpinus*) and a type of wild cattle known as the Banteng (*Bos javanicus*), both of these mammals are extinct throughout most of their original range and now largely exist in zoos and wildlife parks. Less threatened, but still included within the IUCN list of vulnerable mammals are two primates: the Pileated Gibbon (*Hylobates pileatus*) and the Slow Loris (*Nycticebus coucang*); a wild goat: Southern Serow (*Naemorbedus sumatraensis*); the world's largest species of cattle Gaur (*Bos gaurus*); the Malayan Sun Bear (*Helarctos malayanus*); two species of cats: the Marbled Cat (*Pardofelis marmorata*) and the Asian Golden Cat (*Pardofelis temminckii*); and, finally, the Sunda Pangolin (*Manis javanica*).

The Cardamom region also contains 450 known species of birds, two of which are endemics restricted to Thailand and Cambodia: the Chestnut-headed Partridge (*Arborophila cambodiana*) and the Siamese Partridge (*Arborophila diversa*), the latter known from Bokor National Park. Other endangered birds include the Great Hornbill (*Buceros bicornis*), which attracts tourists because of its large size, attractive

---

* The 2010 Goldman Environmental Prize was awarded to Tuy Sereivathana, a park ranger who worked to mitigate human elephant conflict in Cambodia <http://www.goldmanprize.org/2010/asia> [accessed 15 June 2011].

plumage and enormous specialized beak, and the Green Peafowl (*Pavo muticus*), a close relative of the domesticated peacock, which has been reported in the Elephant Mountains. There are at least three threatened terrestrial reptiles: the Impressed Tortoise (*Manouria impressa*), the Elongated Tortoise (*Indotestudo elongate*), and the Asiatic Black Leaf Turtle (*Geoemyda spengleri*); all three are relatively widespread, but are threatened by the international trade of live reptiles.[28] Evidence that the region is important in terms of endemism, as well as worthy of the ongoing efforts to document its biodiversity, was seen in the recent discovery and description of the Cardamom Wolf Snake (*Lycodon cardamomensis*), which is known from a single specimen that was collected in the Cardamom Mountains in the year 2000.[29]

Freshwater Systems

Aquatic habitats and wetlands are enormously important components of the Greater Cardamom Region's landscape mosaic. This includes not only the purely aquatic habitats of the region's permanent lakes, streams and rivers, but also the much larger area subject to seasonal inundation and the areas where the border between terrestrial and aquatic is neither distinct nor permanent. The extreme seasonality of a monsoonal climate leads to an abundance of these transitional habitats, known as wetlands, which are important ecologically because they mediate water flows between the terrestrial and aquatic ecosystems and play an important role in providing critical ecosystem services, particularly by supporting the food web of an extraordinarily productive natural fishery.

Like all ecosystems, physical factors determine the biological composition and structure of wetland and aquatic habitats. For example, the distribution of plant species in flooded forests is generally determined by the depth and duration of flooding, because some plants are more tolerant of flooding than others. Prolonged, permanent or deep flooding impedes the establishment and survival of many vascular plants, especially those that are rooted in the lakebed substrate. This includes commonly recognized "aquatic plants" such as water lilies and bull rushes.* In contrast, seasonal or periodic flooding allows for the

---

* Ecologists use the term "macrophytes" to distinguish aquatic vascular plants from algae. Although algae play an important role in aquatic ecosystems, the structure and function of wetland habitats is largely dependent on the abundance, distribution and biomass of large multi-cellular vascular plants.

Figure 2.8. The Greater Cardamom Area also has broad range of wetland habitat types, including the flooded marshland that surround the Tonle Sap Lake (top), seasonally wet grassland in isolated mountain valleys on the Cardamom Plateau (middle), and fast flowing rivers that flow out of the mountains into the estuaries near Koh Kong (bottom).

establishment of woody and herbaceous species with physiological and morphological adaptations that permit them to tolerate waterlogged soils or submerged conditions for an extended period of time, but which must have at least some period without being submerged. Key to the lifecycle of almost all aquatic plants is an extended period of time with exposed soil, so that seeds can germinate and become established as seedlings or young plants; once established, they can survive the next period of prolonged inundation.

The severity of the dry season drought likewise greatly influences the distribution of individual species. Almost all wetland ecosystems, but especially those in monsoonal climates, are essentially cyclical in nature, with alternating wet and dry periods being an essential component of their functionality. The sediment load in the water column also influences both fertility and turbidity, both of which affect the primary productivity of aquatic habitats, because plants need both light and nutrients to photosynthesize. Aquatic habitats are further influenced by the type and amount of detritus washed into the system from nearby terrestrial ecosystems; not surprisingly, these inputs from adjacent upland habitats will vary according to climate, topography and hydrology. Overall, the combination of primary productivity from photosynthesis and decomposition of exogenous inputs form the foundation of the trophic web that determines an aquatic system's productive capacity.

Aquatic plants have evolved a number of adaptations that allow them to survive the unique conditions of wetland ecosystems: some are firmly rooted in the lakebed substrate, while others float freely; leaves can be submerged, floating or emerge from the surface of the water. In turbid, nutrient-rich systems, such as the Tonle Sap, robust plants with emergent vegetation predominate. In freshwater systems where water is clear and nutrient poor, such as the streams of the Cardamom Mountains, species with submerged leaves can successfully compete.

The seasonally inundated forests of the Tonle Sap floodplain are dominated by two tree species: Rheang (*Barringtonia acutangula*) and Phtoul (*Diospyros cambodiana*), and a ubiquitous vine species Taoor (*Terminalia cambodiana*) amid an otherwise moderately diverse plant community that varies in stature from just a few meters tall to as much as 15 meters in height. These plants form a habitat mosaic across the floodplain which is characterized by dense scrub interspersed with open grasslands. This forest has a unique phenological response that is modulated by the timing of the floods. Many species shed their foliage during the period of maximum inundation when branches are submerged and new leaves are produced during the dry season when the waters recede. Plants flower during the dry season and fruits are ready for dispersal during times of maximum inundation.[30]

The spatial heterogeneity of vegetation in the Tonle Sap ecosystem is further enhanced by disturbance from villagers who exploit the forest for fuel wood or clear patches to cultivate floating rice during the high water season. By some estimates, as much as 90 percent of the extant woody vegetation in the floodplain is secondary regeneration, which has regenerated after land clearing, and is part of what is essentially a slash-and-burn agricultural production system. Open vegetation is dominated by robust emergent herbaceous plants that have the capacity to root in the soil during periods of low water; plants become dislodged from the substrate as waters rise, however, and form floating mats of thick intertwined vegetation dominated by floating grasses, sedges, and other robust herbs, as well as the occasional small tree or shrub. Only about 200 species of flowering plants have been documented on the Tonle Sap floodplain and most are common aquatic species found throughout the wetlands of tropical Asia. In spite of the low levels of species diversity, these wetlands harbor an impressive amount of habitat diversity, because the constituent species have a great deal of structural plasticity, which allows them to form structurally distinct habitat types.

The habitat diversity of the open lake and the structural heterogeneity of the flooded forest has led to the evolution of a correspondingly high level of fish diversity, with more than 145 fish species documented to occur in the lake.[31] These span a broad range of higher taxonomic groups (36 families) and include species with a diverse variety of feeding habits and reproduction strategies that have evolved to exploit the multiple niches within the river, lake, and adjacent floodplain. The largest family both in number of species (60 species) and biomass (39 percent) is the *Cyprinidae*, a large, diverse family that includes both minnows and carp. The next most important taxonomic group is the catfish with more than 30 documented species from one of three families (*Bagridae*, *Siluridae* and *Pangasiidae*), including the Giant Mekong Catfish (*Pangasianodon gigas*), which is the largest catfish in the world. Another economically important species is the Giant Snakehead (*Channa micropeltes*), which has evolved organs that supplement gas exchange — a significant adaptive advantage in an environment that dries up for an extended period of time each year.

The very large spatial extent of the annual flood, its periodic regularity, and the biodiversity that underlies a complex trophic web are all responsible for the Tonle Sap's remarkable productivity. By some estimates, as much as 180,000 to 250,000 tons of fish are harvested annually from the lake, representing between 40 and 70 percent of Cambodia's protein consumption, depending on whether one includes the fish take from the Mekong River whose fish populations spend at least part of the year within the Tonle Sap ecosystem. The lake is renowned for its massive fish migrations both into and out of the lake, but there are less conspicuous fish migrations within the lake system itself as fish populations move into the seasonally flooded forest to take advantage of its complex trophic web during high water periods.

The Tonle Sap is apparently fairly resilient to overfishing, as the most abundant smaller species have high growth rates and rapid reproductive strategies to take advantage of the lake's high nutrient levels and the seasonal productivity characteristic of a seasonally flooded environment. As waters recede, vast numbers of fish would perish as they are crowded into an ever-diminishing volume of water. From this perspective, man is just one of many predators that gorge on this abundant seasonal resource and fishing removes a large part of what is essentially a renewable annual resource. Apparently, even the remarkably intense fishing practices that are evident to a casual observer have only a limited impact on the subsequent year's fish catch due to the reproductive capacity of natural populations. The overfishing of large predatory fish, which are typically the most desirable in commercial terms, is compensated somewhat in social terms, because populations of smaller fish increase in abundance due to the absence of predation. Since the Cambodian consumer is accustomed to smaller species, the overall catch is unaffected, or at least has been to date. This phenomenon, known as "fishing down the food chain," is, eventually, undesirable and unsustainable because the absence of predators will have long term negative impact on the fitness of prey species. If fishing practices become so intensive they damage the ability of the small fish at the base of the food chain to recover, the whole system will eventually collapse. Fortunately, this does not seem to be occurring under current fishing practices and concerted efforts are being undertaken to improve the sustainability and management of the native fisheries on the lake (see Chapter 8).

Although fish biodiversity is strategically important for the food security of Cambodia, it is also an important component of the nation's conservation strategy, because a great many wildlife species rely on the food resources produced by the freshwater habitats of the Tonle Sap. At least five species of aquatic turtles, around 16 species of snakes, and

33 known species of mollusk and large crustaceans, not to mention vast numbers of smaller invertebrates and insects, sustain themselves on this habitat.[32] There are also 46 species of mammals, including two species of otters that are highly dependent on fish that are viewed as "competitors" by local fishermen. Both the Smooth Otter (*Lutra perspicillata*) and the Hairy-nosed Otter (*Lutra sumatrana*) have been classified as endangered species by the IUCN. The area is particularly well known as a haven for storks, cranes, egrets and other "waders," which share what is essentially a massive "rookery" in the Prek Toal Bird Sanctuary on the south side of the lake. This sanctuary is the most important nesting site for the Greater Adjutant (*Leptoptilos dubius*), a massive bird with a wingspan exceeding 1.5 meters, and the closely related Lesser Adjutant (*Leptoptilos javanicus*), as well as the Black-Headed Ibis (*Threskiornis melanocephalus*), the Painted Stork (*Mycteria leucocephala*), the Milky Stork (*Mycteria cinerea*) and the Spot Billed Pelican (*Pelacanus philippensis*). Populations congregate in rookeries and breed during the dry season when their prey, small fish, is confined to small pools. Once the monsoon arrives, birds spread out across the vast seasonally inundated rice plains of Southeast Asia.

Because the high human population density of Southeast Asia impedes the ability of many of these birds to form and maintain viable colonies, the large areas of flooded forest in the Cardamom region are especially important for their conservation. Loss of the breeding grounds of these endangered species would have a much wider geographic impact, since there are very few surviving rookeries for many of these large charismatic birds.

The aquatic systems of the coastal watersheds have less biodiversity than the Tonle Sap, because they are predominantly clear water streams, which are both cooler and carry fewer nutrients. Moreover, these "oligotrophic" systems are naturally fragmented by numerous waterfalls and rapids, which limit the ability of fish to migrate upstream to spawn. These characteristics have driven an evolutionary process that has led to a relatively high number of endemic species of fish, amphibians and aquatic turtles. Researchers recently discovered the Samkos Bush Frog (*Chiromantis samkosensis*), a unique amphibian with green blood and blue bones whose unusual color is a defense mechanism to warn potential predators that it is bad-tasting. Similar studies have identified other new species of frogs, including the Cardamom Bush Frog (*Philautus cardamonus*), Smith's Frog (*Rana faber*), and the Aural Horned Frog (*Megophrys auralensis*).[33]

Perhaps, the most significant threatened species in the freshwater habitats of the Cardamom highlands is the Siamese Crocodile (*Crocodylus siamensis*), a critically endangered species that is now largely extinct throughout much of its range. This species has been bred in captivity, but many of the individuals that were once thought to be Siamese Crocodile have been subsequently revealed to be hybrids with the salt-water crocodile. This discovery has placed increased emphasis on proper management of captive populations, as well as the importance of protecting all viable wild populations. Perhaps the most high-profile endangered species of the Cardamom freshwater habitats is the Asian Arowana (*Scleropages formosus*); better known as the Dragon Fish, it is prized by collectors throughout Asia because it is deemed to bring luck to its owner. The name dragon comes from its scales, which are thought to resemble those of a Chinese dragon. Its popularity as an ornamental fish places this species in a particularly vulnerable position, because collectors and breeders are known to occasionally seek new genetic stock from wild populations.

Estuaries and Mangroves

The Cambodian Coast is composed of a series of landforms and coastal features, including large bays and small coves, estuaries and mudflats, as well as rocky headlands intercalated with long sandy beaches. Off-shore, several small islands are covered by dense forest and, essentially, represent an extension of the Cardamom Mountains in the Gulf of Thailand. One of the most critical part of this coastal ecosystems are its estuaries, which include open water channels, low islands, mud flats and mangrove swamps. Estuaries constitute a key functional link between terrestrial and coastal ecosystems, because this is where fresh-water and marine biomes meet and interact. Named for a unique type of tree adapted to the brackish environments of intertidal coastal zones, mangrove swamps are a particularly productive and critical coastal habitat. Estuaries and mangrove swamps enjoy a constant input of nutrients from inland landscapes, either directly by the mineralization of coastal soils or indirectly as runoff from rivers and streams. Because of the density of woody roots and stems, mangrove swamps accumulate sediment and contribute to long-term coastal stability. Mangrove swamps are bathed by the warm waters of the tropical seas, which creates a stable microenvironment characterized by fast growth rates throughout the year. This extremely high level of productivity is a key component of the coastal fisheries and enables mangrove swamps to accumulate biomass in quantities similar to tropical forest.[34] Moreover, the anoxic conditions of the muddy intertidal substrates allows for some of this carbon to be sequestered as part of long-term carbon pool, effectively creating a "carbon sink." This has raised interest in the conservation of mangrove swamps as a means to mitigate the impact of climate change.

In Cambodia there are approximately 85,000 hectares of mangrove swamp, but it is estimated that as much as 20,000 hectares may have been lost to conversion over the last twenty years, either from the expansion of the commercial shrimp farming industry or for the production of charcoal for urban markets. Mangrove swamps are particularly abundant within the estuaries around Koh Kong City and receive freshwater runoff from the northern sector of the Cardamom Mountains via the Stung Meetok and Stung Chum Russey rivers. To the south, the mangroves and mudflats of the Pream Krasop Wildlife Sanctuary are nourished by waters from the Central Cardamoms via the Stung Kep and Stung Areng rivers, which empty their nutrient-rich waters into the calm shallow waters of Chaak Koh Kong,* the body of water located between Koh Kong Island and the mainland. Pream Krasop is Cambodia's largest mangrove swamp and, consequently, has been designated by the Cambodian government as a wetland of international importance under the Ramsar Convention.[35] Other important estuarine and mangrove complexes situated along the coast include the estuaries associated with Stung Piphot and Preak Kampong Som rivers, both of which flow from the Southern Cardamoms into the Chaak Kampong Som and enjoy protection within the Dong Peng Multiple Use Area (Figure 1.4). Other important, if less well protected, mangrove swamps are located on the Boeng Ta Trav and Kampot rivers that originate in the Elephant Mountains and empty directly into the Gulf of Thailand.

The mangrove — the tree that gives these swamps their name — is a classic example of what ecologists call a "keystone species," meaning it has a morphological or physiological trait essential for maintaining the functional integrity of its ecosystem. The loss of a keystone species almost inevitably leads to a radical change in the function and structure of its ecosystem. The ecosystem either collapses or ceases to provide the services that defined it as a unique ecosystem. Mangrove trees are considered by ecologists to represent a "functional plant type" rather than a "taxonomic group," because the key adaptive traits that have

* Chaak is the phonetic spelling for the Khmer word meaning bay or harbor.

evolved independently in different plant families via different adaptive strategies. For example, species in the red mangrove genus (*Rhizophora mangle*) have *pneumatophores*, a specialized root that grows skyward until its tip extends above the maximum level of the low tide in order to absorb air, which is then transferred to the submerged root system. Alternatively, the Black Mangrove (*Avicenna germinans*) is characterized by large multiple stilt roots that have well developed, specialized pores known as lenticels that facilitate gas exchange during periods of low tide. These two genera are the largest and most diverse of the "true mangroves;" however, they are situated far apart on the evolutionary tree: *Rhizophora* is the only genus in the *Rhizophoraceae*, while *Avicenna* is recognized as genus in the *Avicennaceae*, or as a subfamily in the *Acanthaceae*, neither of which are remotely related phylogenetically to the *Rhizophoraceae*.

In addition to the root systems described above, mangroves have evolved a series of other traits that overcome the physiological barriers that prevent most vascular plants from growing in brackish habitats — particularly in the poorly oxygenated and saline soils of the intertidal wetlands that are subject to periodic inundation. These include large waxy leaves that minimize the loss of water from plant tissues. Paradoxically, fresh water is a sparse commodity for mangroves, which must expend physiological resources (energy) to pump water from a high salt, external environment into what is essentially a freshwater environment within its internal tissues. Some mangroves store the excess salt in old leaves, while others secrete it via specialized glands. Another important attribute of mangrove trees is the ability of mature fruits to germinate while on the tree, thus allowing new individuals to establish themselves on tidal mudflats with their physiological defense mechanisms already up and running. This combination of traits enables mangroves to colonize mudflats and other intertidal zones, which constitutes an important ecosystem service that protects against erosion.

Like most of Cambodia's plant communities, the floristic composition of mangrove swamps has not been studied in great detail; nonetheless, biological inventories in neighboring Thailand provide a glimpse into the diversity and ecological importance of this extremely productive habitat. In one mangrove swamp in Southern Thailand, biologists have documented 87 species of plants, including at least 20 species of true mangroves.[36] Different morphological and physiological attributes confer distinct adaptive advantages to individual mangrove species, which influences their distribution within the different sectors of the intertidal zone. A recent study in the Koh Kong estuary, for example, found that exposed areas tended to be dominated by *Avicennia* and *Sonneratia*, while *Rhizophora* species were more abundant along rivers and creeks. Further inland, where waters are less saline, the overall floristic composition was more diverse, and the forest canopy was dominated by *Melaleuca leucadendron*,[37] the source of "cajuput oil" of traditional eastern medicine.

Just as the plants of mangroves swamps are unique compared to other terrestrial and freshwater habitats, it is no surprise that the animals that live in this ecosystem are likewise special and interesting. Bird watchers, for example, are keen on visiting mangrove swamps because they can observe birds that they will not encounter in any other habitat, including not only resident species, but also long-distance visitors that arrive during the winter migration. In the mangrove habitats of the Cambodian Coast, numerous threatened birds have been documented, including Nordmanns' Greenshank (*Tringa Guttifer*), the Asian Dowitcher (*Limnodromus semipalmatus*), the endangered White-Winged Duck (*Cairina scutulata*), the Masked Finfoot (*Heliopais personatus*), and the critically-endangered White-Shouldered Ibis (*Pseudibis davisoni*).[38] There are also reptiles, including the critically endangered Giant River Terrapin (*Batagur baska*), which was recently discovered in the Preak Kampong Som river and is the only remnant population in all of former Indochina. Among mammals, there are three threatened otter species, including the threatened Asian Small-Clawed Otter (*Aonyx cinerea*) and the Smooth-Coated Otter (*Lutrogale perspicillata*), as well as the critically endangered Hairy Nosed Otter (*Lutra sumatrana*). The Hairy Nosed Otter is also present in the Tonle Sap, but its presence on the coast is particularly noteworthy and encouraging because of its preference for *Melaleuca* forests, which are still relatively abundant in the region. This increases the probability for the survival of what is considered to be the most threatened otter species in the world.[39]

The biodiversity of mangrove swamps and estuaries also includes a multitude of sea creatures that use intertidal habitats for a well-defined period of their lifecycle. Wild shrimp and many crustaceans, for example, use mangrove swamps as both a nesting and nursery habitat, while oysters and other shellfish colonize the stems and roots of mangrove trees and many commercially important fish species routinely migrate into and out of estuaries. Consequently, the traditional economies of coastal communities depend heavily on mangroves, either directly because villagers use natural products from these habitats, or indirectly because the productivity of the coastal fisheries is linked to mangroves swamps. The ecological importance of Cambodia's mangrove swamps has increased due to the even more aggressive clearing of mangrove swamps for the expansion of the commercial shrimp industry in neighboring Thailand.

Mangrove swamps are usually always considered to be public lands. In addition to providing habitats for fish that feed local communities, they also provide other ecosystem services that benefit more distant populations, such as carbon sequestration, which is a global ecosystem service, and protection against coastal erosion, which can benefit an entire nation from unforeseen natural disasters. The ability of mangrove swamps to stabilize coastlines is leading some nations to restore lost or degraded habitats to combat rising sea levels from climate change and to buffer the impact of violent storms. Because it takes decades for a mangrove swamp to regenerate after its conversion and abandonment, their destruction is a classic example of a short-term benefit for an individual that deprives a larger group of an important public service. This loss could be particularly critical for Cambodia and its coastal community, because the conversion of mangroves swamps is rarely sustainable and converted areas are almost always abandoned.

Coral Reefs, Sea Grass Beds and Offshore Fisheries

The Gulf of Thailand is a semi-enclosed sea that separates the Malay Peninsula from mainland Indochina. Its ecology is heavily influenced by the monsoonal climate, particularly the large influx of fresh water during the rainy season and the tendency for the currents to flow clockwise during the southeast monsoon (June–September) and counterclockwise during the northeast monsoon (October–November). Another major physical attribute is its shallowness: Its depths rarely exceeds 40 meters, which restricts the interchange of water with the South China Sea thus contributing to its reduced salinity. Because its waters are slightly less salty and nutrient-rich from the influx of five major river systems, as well as close to the ocean floor, the Gulf of Thailand is a highly productive marine ecosystem compared to most other tropical seas. This unusually high productivity has been documented from space by remote sensing instruments, which measure the photosynthetic activity of phytoplankton, and is also evident in the Gulf's historically large production of fish and seafood.[40]

According to the Cambodian Department of Fisheries, 474 species of fish from 105 families have been identified in Cambodia's offshore marine waters, an impressive number for a country with a relatively small coast. In addition, the Gulf is home to a multitude of the other sea creatures, including cephalopods (squids and octopus), crustaceans (crabs, lobsters, shrimps and prawns), mollusks (snails, sea slugs, biva-valves and osteobranchs), cnidarians (jellyfish, sea anemones and corals) echinoderms (star fish, sea cucumbers and sea urchins), seahorses, sponges, and seaworms, to name some of the more commonly known life forms. Cambodia's commercial fishing industry harvests virtually all types of fish and seafood. The most heavily sought after species are mackerel, anchovies, sardines and tuna, which occupy the pelagic zone of the offshore fisheries.* In terms of tonnage, however, the greatest amount of fish come from the benthic zone near the ocean floor and include well-known species such as stingrays, flounders and flatfish, as well as numerous other commercial species and the so-called "trash fish" that are caught as the by-catch of the fishing industry. Fishing is particularly important for Cambodians within the littoral zone, where a large group of subsistence and artisanal fishermen rely on shellfish and crustaceans as the principal catch, which is a mainstay of their livelihoods.

Fish stocks in the Gulf of Thailand have been overexploited for at least two decades, if not longer; catches reached their maximum in the mid-1980s when they were more than double the current mean harvest of around 1.1 million tons. Overfishing has impacted not just the volume of the harvest, but also the quality of fish caught, with an increase in the percentage of small, short-lived species. Overfishing is blamed largely on commercial trawlers based in Thailand and Vietnam, many of which operate within the Cambodian Exclusive Economic Zone.† Much of the overfishing and destruction is done by bottom trawlers that create havoc on the fragile benthic habitats of the ocean floor where the weighted lines used on their nets kill many sessile life

* The ocean is stratified by marine biologists into different zones according to proximity to the shore, the seafloor and the ocean surface: The *littoral zone* refers to the area that is in contact with the shore, such as beaches and mangroves, as well as the shallow waters immediately offshore. The *benthic zone* refers to the lowest ocean level and includes the ocean floor sediments as well as the waters immediately above the seafloor, which are also sometimes referred to as the *demersal zone*. The *pelagic zone* refers to the part of the ocean that is neither close to the bottom nor the shore — in other words, the great mass of ocean. It includes the highly illuminated waters just below the water surface that are sometimes referred to as the *epipelagic* zone.

† The United Nations Convention on the Law of the Sea (UNCLOS) defines the EEZ as a zone beyond and adjacent to the territorial sea in which a coastal state has sovereign rights; the outer limit of the EEZ shall not exceed 200 nautical miles from the shore. Cambodia's EEZ overlaps those of its neighbors, Thailand and Vietnam, and remains subject to negotiation.

forms and mix the sediments that define the physical environment of the ocean floor. Since the Gulf of Thailand is a particularly shallow sea, damage to its benthic habitats is especially alarming and exacerbates the overall impact of fishing; nonetheless, since the Gulf's seafloor is composed almost entirely of loose sediments, these habitats and their productive capacity would recover once this destructive practice were ended.[41]

One of the most important marine habitats of the Cambodia Coast is the seagrass beds, which are a key benthic habitat in the littoral zone. In Cambodia, seagrass habitats can be divided into two main types: extensive seagrass meadows along the mainland between Kep and Kampot, and patches of seagrass intermingled with coral around islands. Inshore seagrass beds usually occur as mixed stands of at last eight different species, whereas offshore, the large seagrass *Enhalus acoroides* occurs in extensive meadows.[42] Not surprisingly, the main threat to seagrass comes from bottom trawling, which can uproot plants and disturb sediment.

Seagrasses, like mangroves, are extremely unique plants. They are the only group of flowering plants that has become completely adapted to the marine environment. Although firmly rooted to the ocean floor and completely submerged for their entire life, they are not algae and, in spite of their name, they are also not related to terrestrial grasses. Instead, they belong to a group of closely related families in the monocot order known as the Alismatales (*Hydrocharitaceae*, *Cymodoceacerae*, *Zosteraceae*). These families are closely related to other freshwater aquatic plant families and probably evolved from a plant that was first adapted to the brackish environment of a mangrove swamp. Like mangroves, seagrasses have physiological and anatomical features that allow them to thrive in a saline environment; unlike mangroves, however, seagrasses do not "combat" the marine environment by secreting excess salt or developing specialized organs to facilitate gas exchange. Instead, they are totally adapted to the ocean and are completely submerged in the marine environment.

Individual plants look very much like grass with erect elongate leaves (another common name is tape-grass) and a buried horizontal stem known as a "rhizome." Like many grasses, seagrasses reproduce asexually via rhizomes; consequently, seagrass meadows are probably composed of hundreds of genetically identical plants derived from a single individual. In spite of being dependent on vegetative reproduction for the expansion and maintenance of colonies, seagrasses can reproduce sexually by transferring pollen from the male to female flowers, which then produce fruits and seeds. Sexual reproduction by a flowering plant is a complicated process to stage in a marine environment and demonstrates the remarkable capacity of the evolutionary process to create novel life forms adapted to every imaginable environment.

Seagrass meadows provide several ecosystem services, including the primary production of biomass, an important contribution to the capacity of fisheries in coastal waters, while trapping sediments and protecting shorelines from erosion. Their three dimensional structure provides habitat cover to a multitude of fish, as well as protection and sustenance to numerous invertebrates, including mollusks, sea urchins, crabs, bristle worms and nematodes, as well as providing sessile algae with a substrate for growth. Seagrass beds are a critical habitat for dugongs (*Dugong dugon*), a threatened marine mammal related to the manatee, as well as two other species in the order *Sirenia*, all of which are appropriately referred to as sea cows, since they all graze on seagrass. The other major taxa of endangered vertebrates that depends on seagrass habitat are marine turtles: five species have been reported for Cambodia, including the Hawksbill (*Eretmochelys imbricata*), Green (*Chelonia mydas*), Olive Ridley (*Lepodochelys olivacea*), Logger head (*Caretta caretta*) and Leatherback (*Dermochelys coriacia*) turtles; only the hawksbill and green turtle appear to be common.[43]

There are many sightings of dolphins in the coastal waters, probably of several widespread species (*Dephinus delphis*, *Tursiops truncates*, *Stenela logirostris*); however, the critically endangered Irrawaddy River Dolphin (*Orcaella brevirostris*) is known to inhabit the shallow coastal waters between the mainland and Koh Kong Island. In reality, this shy creature is neither a true dolphin (Subfamily *Delphininae*), nor particularly fond of rivers. These cetaceans are more closely related to killer whales (Subfamily *Orcininae*) and prefer the estuaries and coastal waters habitats of Southeast Asia. Once common, the species now survives only in a few scattered and isolated populations ranging from Papua New Guinea to India, with a single large robust population in Bangladesh. In Cambodia, the best known population occupies a series of the deep pools on the Mekong River upstream from Phnom Penh near the river town of Kratie.* It is one of only two freshwater populations, and recent reports indicate that its reproductive success is severely limited. The number of Irrawaddy Dolphins on the Cambodian Coast remains

---

* Despite their name and an existing population in the Mekong River, this species is (or was) more common to coastal waters.

unknown and, apparently, they have been observed only in the Chaak Koh Kong channel between the mainland and Koh Kong Island. Considering their low numbers in other populations and the highly fragmented situation of extant populations, their long-term survival remains very much in doubt, especially if the mangrove and estuarine habitats of Coastal Cambodia continue to be degraded or converted by development.

In terms of biodiversity, coral reefs are without a doubt the most important habitat of the Cambodian Coast. The extent of Cambodia's coral reefs is not large and the total area has been estimated at approximately 2,800 hectares distributed among 45 offshore islands where occur as "fringe reefs," colonizing the rocky shores of the small offshore islands; in addition, there is a single, degraded mainland site at the foot of Bokor Mountain halfway between Kampot and Kep. The largest and most diverse coral reefs are situated in the waters off two relatively large islands near Sihanoukville (Koh Rung and Koh Rung Samloem) and around a group of small islands in Kao Sdach commune located in Koh Kong Province.[44]

Coral are filter-feeding colonial cnidarians, a jellyfish-like organism that forms colonies composed of thousands of minute, individual "polyps." The most well-known corals are all "hard corals" that secrete a stony external skeleton composed of calcium carbonate ($CaCO_3$) and grow on the skeletal remains of previous generations. These accumulate over centuries to produce the elaborate and distinctive structures that most individuals assume are a single organism. The names of the major reef-building corals are emblematic of this structural diversity, with terms such as brain, staghorn, fan, star, finger and lettuce, to name a few. Corals are also "symbionts," a term used to describe two species that coexist within the same body and depend upon each other for their mutual survival. For coral, this is a golden-brown alga known as *zooxanthellae*[*] that coexists within the cnidarian polyp. These single-celled photosynthetically active protozoans provide the polyp with energy in the form of simple sugar molecules, while the polyp provides the *zooxanthallae* with a stable micro-environment and supplies

[*] The term *zooxanthellae* is the common name for all types of dinoflagellate algae in the genus *Symbiodinium*, which may contain either a single variable species or dozens of distinct species [see Takishita, Kiyotaka (2003-07-16)]. "Comparison of phylogenies based on nuclear-encoded SSU rDNA". *Phycologia* 42(3) (19 June 2003): 285–91.

nutrients for growth and reproduction. Since the algae must have sufficient light to carry on photosynthesis, this symbiotic relationship limits the vertical distribution of most coral colonies to relatively shallow waters, starting from just below low tide to a maximum of about 60 meters; the greatest biomass and diversity of coral is usually located within 10 meters of the ocean's surface.[45]

A reef is topographically complex and composed of myriad nooks and crannies, the result of the corals' morphological variability and multiple distinctive structures. This structural diversity also creates numerous microhabitats that support a wide range of other marine organisms. Coral reef microhabitats also vary due to the inherent vertical stratification characteristic of marine environments, as well as the location of individual corals in relation to prevailing currents and waves, which make the inshore zones considerably more tranquil than offshore reefs. A multitude of fish, clams, crabs, starfish, sea urchins, seaworms and other marine organisms colonize these various microhabitats, which is one reason why coral reefs have such extraordinarily high levels of biodiversity. Coral reef ecosystems also have very high levels of primary productivity, partly because of their symbiotic nature, but also because coral reef communities tend to recycle nutrients and biomass internally among the different components of the food web. Adding to the complexity of reef systems are their ecological connections with nearby mangroves swamps and seagrass beds; dozens of fish species and invertebrate larvae move between these three ecosystems during different times of the year, or use a specific microhabitat in one of these three ecosystems at critical stages during their reproductive lifecycle.[46]

Paradoxically, despite their biological diversity and inherent productivity, coral reefs are also very fragile and are being degraded by global, regional and local phenomena — all of which are a direct consequence of human action. Worldwide, coral reefs are suffering from "coral bleaching," which occurs when the algal symbiont is expelled from the interior of the coral polyp. Bleaching may have a variety of causes; one major cause is believed to be increased water temperatures due to climate change, although other regional factors also may play a role, particularly increased sedimentation and pollution, which can stress coral reefs and make them more susceptible to disease pathogens that attack either the algae or the polyp. Another aspect of climate change, "ocean acidification," is also affecting the reefs. Ocean acidification occurs because higher concentrations of CO_2 in the atmosphere lead to higher concentrations of dissolved CO_2 in the ocean, which then convert to carbolic acid via an equilibrium reaction that has been slowly lowering the pH of the ocean.[47] Increased acidity limits the ability of marine organisms to absorb $CaCO_3$ and build the skeletal framework upon which corals and most marine invertebrates — crustaceans (crabs and lobsters) and mollusks (shellfish) — depend. The coral reefs of the Cambodian Coast also face multiple local threats, including the use of cyanide and dynamite by local fisherman, damage from anchors, and, to a lesser extent, harvesting as ornaments and curios, either by tourists or by local inhabitants seeking to sell coral curios to tourists.

Recent inventories and monitoring reports have shown that Cambodia's coral and coral reef communities are still in fair-to-good condition. Inventories have identified 111 species of hard coral and 17 species of soft corals, as well as nine species of sea fans, a soft coral that produces an elaborate and attractive branching fan.[48] The most common coral in the Koh Rung reefs is an unidentified massive rock coral (*Porites* sp.) that represents about 50 percent of live coral biomass; less abundant but still common are the double star coral (*Diploastrea heliopora*), leaf coral (*Pavona decussate*), cauliflower coral (*Pocillopora* sp.) and several types of brain coral (family *Faviidae*). The most frequently observed fish are species that graze on corals, such as Butterflyfish (*Chaetodontidae*) and Parrotfish (*Scaridae*); predator species, such as snappers and groupers, are also relatively common, indicating that fishing pressure is not extreme.[49] In 2003, total cover of live coral ranged from 29.3 percent in Koh Sdach to 58.1 percent in Koh Takiev, a group of islands situated further offshore out of reach of many of the artisanal fishermen.[50] The presence of dead coral was highest at Koh Rong, accounting for as much 44.9 percent of the total reef area; nonetheless, bleached coral was low at all survey sites, with a mean average of only 0.5 percent. Most observed damage was attributed to inappropriate fishing practices, and the most degraded sites were located near Sihanoukville.[51]

Currently, the only protected area for Cambodia's coral reefs is Ream National Park, which encompasses several islands near Sihanoukville, in addition to the extensive mangrove swamp and coastal forest for which it is primarily known (see Figure 1.4). More recently, a group of marine scientists and ecotourism entrepreneurs have created an NGO dedicated to conserving the marine ecosystems of the Cambodian Coast and are working with the Fisheries Administration to improve management for a four-kilometer buffer zone around the islands of Koh Rung and Koh Rung Samloem.[52] They hope to consolidate an offshore marine protected area that would protect Cambodia's reef habitats and promote them as an attraction for the tourists who visit Sihanoukville.

Integrating the Management of a Nation's Natural Assets

A thorough survey and understanding of the Greater Cardamom Region's natural resources and landscapes is the first step in solving the Cardamom Conundrum. Conservation biologists have long argued that it is necessary to know the biodiversity of a region in order to properly evaluate its economic potential. This concept has driven national governments around the world in their policies to map the mineral resources, soil type, and water resources, as well as the other natural assets of their countries. Knowing the biodiversity of a region also leads to more sophisticated and strategic conservation measures and is essential for integrating conservation and economic development agendas. Although we have a general understanding of the Greater Cardamom Region's natural resources, the ecosystem services they provide, and the biodiversity they contain, the isolation that has protected Cambodia's biodiversity from development has also left gaps in our specific knowledge. The recent discovery of new species such as the frog found in the Cardamom Mountains and the lack of species inventories of some of Cambodia's richest biodiversity regions highlight the need for more extensive biological inventories of Cambodia.

Cambodia in general, and the Cardamom mountains in particular, were isolated during much of the latter half of the twentieth century when taxonomists in Europe and North America organized the most recent efforts to advance scientific knowledge of the planet's biodiversity. That era is now coming to an end, as traditional taxonomy programs in the developed world are downsized to make room for molecular biology and other more technologically advanced branches of evolutionary biology. In addition, many developing countries have established restrictive regulations on allowing foreign scientists to collect — and potentially exploit economically — what is now recognized as their national heritage. Fortunately, the scientific collecting expeditions organized from the developed world have been replaced by similar expeditions organized and manned by the growing cadre of scientists in developing countries.

Even more critical is how Cambodia organizes and uses the knowledge that it does have about its biodiversity. There is still a logical tendency to segregate biological and resource inventories by thematic sector, dealing with the mountains, for example, separately from the coastal region. As this chapter has attempted to demonstrate, however, the management of a forest in the Cardamom Mountains or the development of the hydropower resources in these mountains will impact the fisheries of both the Tonle Sap and the Koh Kong Estuary. Understanding the interconnectedness of the Greater Cardamom Region's physical, social, economic and environmental assets is essential for resolving the "Cardamom Conundrum." Recognition of the linkages between the region's distinct landscapes and specialized habitats is necessary for envisioning an integrated development strategy that responds to the needs of the Cambodian people. This recognition is behind the "Ridge to Reef" conservation model that has been promoted by a wide variety of both public and private institutions, which explicitly links the full range of terrestrial, aquatic and marine ecosystems characteristic of coastal regions around the world.

Coastal nations throughout the world have similarly linked ecosystems, where deforestation or unsustainable agricultural practices inland compromise the quality of the streams and rivers that flow through estuaries and into coastal habitats. Where Cambodia is unique, however, and where it could provide an interesting model to other nations is that both its coastal and its highland regions are undeveloped or at early stages of development, whereas in other nations the development process has already advanced. So Cambodia has a unique opportunity to develop a ridge to reef conservation strategy that has already slipped past other coastal nations.

Decisions to develop or exploit the natural assets of a region are not determined solely by conservation criteria, but by multiple factors, particularly those related to economic growth and investment. Just as the conservation of the mountain and the coastal habitats of Cambodia's Cardamom region are inextricably linked, so too is the need to use or exploit the natural resources in those two regions. In the next chapter, the human piece of the Cardamom Conundrum — perhaps the most vital and dynamic piece of the puzzle — is examined and described so it too can be factored into the solution for resolving the Cardamom Conundrum.

CHAPTER 3

Human Landscapes

A fundamentally important piece of the Cardamom Conundrum is the wellbeing of human residents of the region, whose needs in Cambodia's post-conflict era continue to be the creation of economic opportunity to escape poverty and the development of acceptable levels of health and educational services. Cambodia is rebuilding its economy after decades of civil conflict and is making remarkable progress towards improving the quality of life for its citizens. However, the most important productive activities — those that generate revenues and support the livelihoods for the overwhelming majority of Cambodians — are based on the direct exploitation of natural resources, particularly fisheries. In many instances, this involves investments in systems that produce goods and services from nature, such as the cultivation of paddy rice or the tourist industry. Building infrastructure to expand or support these two activities can also have a large impact on the nation's natural resource base, whether it is from the impact of hydropower development on watersheds or the building of roads to connect farms with urban markets. These investments may be sustainable, or not, depending on the management of the available resources or the effectiveness of efforts to avoid or mitigate secondary impacts that usually accompany infrastructure investments.

The paradigm of sustainable development rests on the assumption that conservation and development are mutually dependent, particularly when referring to renewable natural resources, which include soil and water, fish and tree stocks, and the combination of features that makes a landscape an attractive tourist destination. Reconciling conservation and development in Cambodia will therefore depend on the ability to engage individuals, families, communities, and corporations in activities that are simultaneously productive and yet promote the sustainable use of the region's natural resources. In most cases, humans will need to modify their behavior, especially those activities related to "making a living." For those linked to a modern economy, this involves activities related to generating income, but for many rural families in Cambodia this means acquiring or producing enough food to survive or, more importantly, for their children to thrive. Perhaps the best way to seek consensus on sustainable development is to recognize that the economic activities of individuals and businesses should become less exploitive and, instead, be transformed into investments in future productivity. Environmental advocates must realize, and accept, that fundamental human needs will drive the development process in Cambodia and these needs must be realistically and strategically accounted for in any and all conservation strategies. Poverty stems from a complex set of societal, historical and environmental factors; it is essential to understand these in all of their complexity and identify the linkages among these factors, in order to identify sustainable pathways out of poverty.

Cambodia has the distinction of being one of the least densely populated countries in South East Asia, a fact that is somewhat deceptive, because more than 90 percent of the population is concentrated on less than 30 percent of the landscape.[1] Moreover, it is a rural society in contrast to its neighbors, Vietnam and Thailand, which have large urban centers; consequently, the population density on agricultural landscapes in Cambodia is among the highest in the world. The current population of the nine provinces of the Greater Cardamom Region and its surrounding areas is about 3.7 million inhabitants and is predicted to increase to more than 5.2 million by the year 2020.[2] This projection is based largely on the internal growth rates of individual provinces and does not account for continued in-migration, which is likely to continue to bring hundreds of thousands of people to the region, particularly to the fast growing coastal provinces.*

The vast majority of inhabitants of the Greater Cardamom Region are ethnic Khmer, although a few other minorities also make the region their home. The second most numerous group are the *Cham*, who are descended from the medieval kingdom of *Champa* that covered much of South Vietnam before the fifteenth century. The *Cham* converted

---

* This is based on a mean annual growth rate of 3.3 percent, which is greater than the national average of 2.7 percent, reflecting the variation among provinces with respect to their birth rates, mean life span, and other factors.

to Islam in the seventeenth century as a result of contacts with Malay traders and originally settled in fishing villages in the Kampong Cham and Kampong Chhnang provinces along the Tonle Sap and Mekong rivers just north of Phnom Penh. The *Cham* were targeted for persecution during the rule of the Khmer Rouge and approximately half of their population was exterminated.[3] Like other Cambodians, many *Cham* have now migrated to the Cardamom Coast, and their mosques are a common sight in Koh Kong and Sihanoukville.

Other minority groups include the *Soch* in Kampot Province, the *Suoy* in Kampong Chhnang Province, the *Por*, located in Pursat Province and the *Khmer Daeum*, who reside in the heart of the Cardamoms Mountains in Koh Kong Province.[4] The *Khmer Daeum*, who are discussed briefly in Chapter 1, used to live in scattered villages throughout the Cardamom Mountains prior to the civil war, but who now largely reside in two villages on the edges of the Central Cardamom Protected Forest Reserve: Ou Som, which is situated on the upper reaches of the Stung Atay river and Thma Bang, which is located on the Stung Chhay Areng river. In both instances, these communities were greatly impacted by the long civil war, and according to reports were forced to abandon ancestral villages as part of the tactics to deny the Khmer Rouge material support. The *Khmer Daeum* have returned to their villages and now represent an important asset for conservation initiatives seeking to organize a sustainable management strategy for the forest reserve. Villagers participate in multiple aspects of that strategy, which recognizes their rights to use forest and freshwater resources. Their traditional subsistence lifestyle now faces new challenges because both villages are located in watersheds that will be impacted by planned or ongoing hydropower development (see Chapter 5).

Poverty

Cambodia's unique and unfortunate recent history has made poverty reduction the government's most important policy objective, one that is shared by its partners in the international development community. Cambodian society has made impressive strides over the last 20 years in reducing poverty, as evidenced by its progress towards meeting its Millennium Development Goals — a framework for development and conservation sponsored by the United Nations that sets clearly defined and measurable targets for such things as human welfare, social justice, environmental sustainability and economic growth (Table 3.1). There are still many obstacles to overcome. An unfortunately large number of

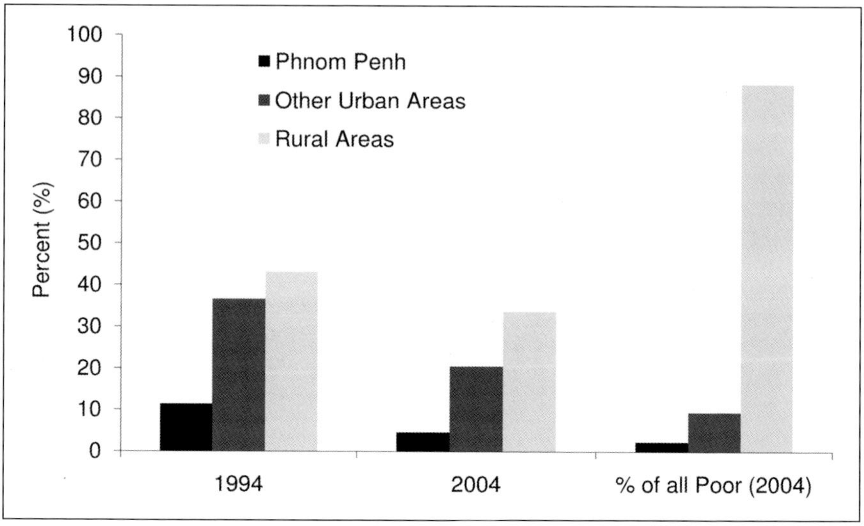

Figure 3.1. Poverty in Cambodia is an overwhelmingly rural phenomenon; only 75,000 residents of Phnom Penh live below the official poverty line, whereas more than 3.5 million people, or one in every four of Cambodia's rural inhabitants, are considered to be impoverished, earning less than $1 per day.

families still live in poverty, while many more struggle to consolidate the progress they have made in recent years.

Poverty is largely a rural phenomenon in Cambodia. That is not to say that there are no poor people in Phnom Penh and other urban centers, but the overwhelming majority of Cambodians are rural residents and a very large number of them are poor (Figure 3.1). Mean annual income is reported to be about $500 per year per person, but estimates vary widely from a low of about $200 per year for rural residents (around 55 cents per day) to almost $800 for Phnom Penh residents. Lack of money is not the only way to define poverty, which can be ameliorated by access to natural resources or exacerbated by feelings of unfairness or the lack of opportunities. One of the most debilitating and punishing aspects of rural poverty in Cambodia, or anywhere for that matter, is the feeling of helplessness in the face of adversity; a random event beyond their control, such as illness or crop failure, can cause a poor rural family to lose everything. In Cambodia there is no safety net beyond the family and a rural family can easily feel trapped by poverty.

Identifying paths out of poverty requires an understanding of the nature, extent, and manifestations of poverty and its complex societal, historical and environmental factors. A person living in a cash-based economy with insufficient income to purchase basic necessities — food, clothing, and health care — is most definitely poor. However, if that person has direct access to the basic resources necessary for his well-being, the lack of a cash income does not necessarily condemn him or her to a life of abject poverty. For example, many rural residents can grow their own food, catch their own fish, and build their own shelter, all of which are important sources of well-being. Other aspects have to do with things money cannot buy: health and self-esteem, freedom from violence, equity, and access to justice, as well as other intangible factors that are not necessarily a consequence of having, or lacking, monetary resources. This is why most poverty alleviation programs organized by nongovernment organizations (NGOs) focus on increasing production on family farms or improving the quality of health care and educational facilities, as these actions can directly benefit the poor regardless of the economic conditions that govern the monetary economy.

The nine provinces of the Greater Cardamom Region have a mean poverty rate of 28 percent, which translates into about 900,000 men, women and children. Nonetheless, poverty can vary greatly within the region (Figure 3.2). Only about 10 percent of the residents of the coastal port city of Koh Kong live below the poverty line, compared to more than 95 percent of the inhabitants of Pursat Province. In remote areas of Greater Cardamom Region, particularly along the Thai border, economic recovery was delayed after the end of hostilities due to the presence of land mines, which had to be removed by specially trained teams of individuals. The Pailin enclave and the adjacent Samlaut commune is populated by former refugees who lead a precarious existence as subsistence farmers on a deforested landscape that has suffered additional damage from decades of artisanal mining for gemstones. In contrast, equally remote areas in Pursat Province (e.g., Veal Vang and Ou Som) are home to ethnic *Por* and *Khmer Daeum* communities that are also isolated from the national economy, but whose poverty, while real, is ameliorated by living in a natural forest landscape with relatively abundant soil, water and wildlife.

The communities that occupy the agricultural plains surrounding the Cardamoms have a traditional Khmer lifestyle based on the cultivation of rice complemented by animal husbandry and subsistence fishing. These northern and eastern plains are conventional agricultural regions where families grow rice for subsistence and sell excess production for cash income. Poverty here is the result of not owning enough

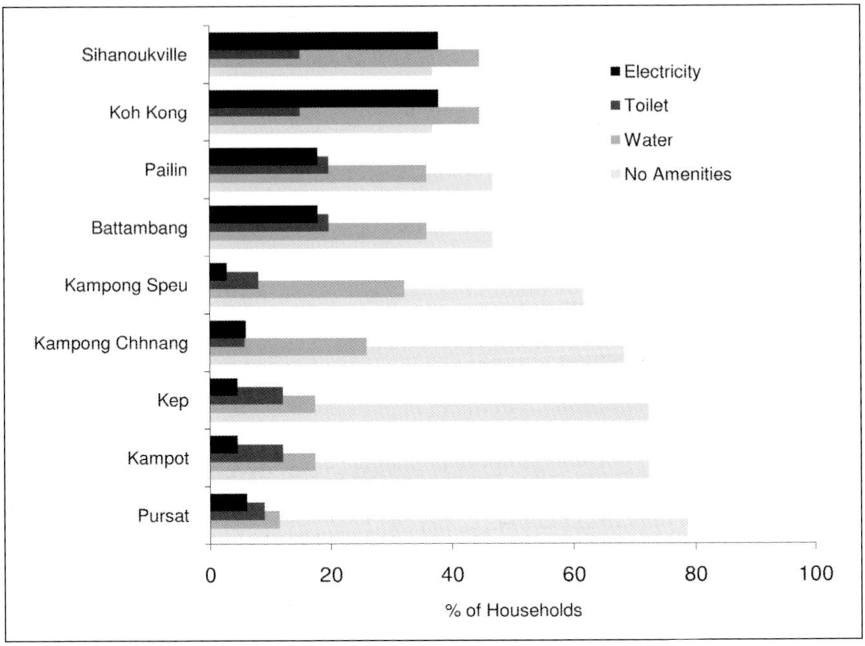

Figure 3.2. Access to services is greater in the more developed coastal provinces of Koh Kong and Sihanoukville than in rural areas, but even in these coastal areas only about half of the populace have access to secure water supplies. (Source: Ministry of Planning (2006), *A Poverty Profile of Cambodia*)

land to feed your family or lacking sufficient excess production to purchase the necessities of life, including additional food products, clothes and health care. Although the social and economic stratification in these regions is more complex, a large minority of the population — as many as 370,000 individuals, or 16 percent of the rural population — lives below the poverty line. Almost all of these families are either farmers or farm workers, who lack sufficient income or production to meet their daily dietary requirements.[5]

In comparison, the Cambodian Coast has the lowest poverty in the country after Phnom Penh, the result of a diversified economy that includes a robust coastal fishery, rice paddy agriculture, plantation forestry, and a rapidly growing tourist industry. Economic growth in this region will probably outpace the rest of the country by several percentage points each year, thanks to development opportunities provided by its ports, beach tourism and the predominance of relatively large land holdings that allow commercial farmers to benefit from economies of scale. Nonetheless, there are pockets of extreme poverty on the outskirts of Sihanoukville and, ironically, in the once posh colonial resort town of Kep;[6] many of the coastal poor are subsistence fishermen who depend on the natural fisheries of coastal waters.

Mountainous areas in Cambodia suffer the highest rates of poverty because they are isolated from urban centers and major transportation routes; neither are they integrated into the rice and fishing economy of the Tonle Sap plain, nor the more diversified economy of the coast. Efforts to alleviate poverty in these areas usually focus on improving subsistence-level production and providing basic services such as health care and education. Development initiatives, such as tourism or hydropower are bringing change to these villages, but because they lack the skills needed in a modern economy, most residents of these isolated mountain villages will not benefit from these investments. Much to the contrary, infrastructure development will negatively impact many previously isolated communities, unless special measures are taken to protect their rights and livelihoods.

Poverty in Cambodia is often juxtaposed against the wealth of the country's small business and political elite; statistics collated by the Ministry of Planning indicate that 50 percent of total national consumption is concentrated in the richest 20 percent of the population.[7] Development economists measure income inequality using a statistic known as the *gini* coefficient. A *gini* value of zero corresponds to perfect equality, where every individual has exactly equal income; in contrast, a *gini* value of one corresponds to perfect inequality, where one person has all the income and everyone else has none. Cambodia has a national *gini* score of 0.40, which is higher than Indonesia (0.36) and Vietnam (0.37), but lower than Malaysia (0.43), Thailand (0.46) and China (0.47).[8] In other words, although the distribution of wealth in Cambodia may be unequal, it is comparable to its neighboring countries.[*]

Economic growth is essential for eradicating poverty, and the twin processes of urbanization and industrialization will make large inroads on poverty over the medium term. Over the short term, however, efforts to reduce poverty in Cambodia must focus on rural development and making incremental improvements in the lives of poor

[*] Income distribution in Asia is considerably more equal than in Latin American and African nations, where *Gini* scores are usually greater than 0.50 and where economic growth and poverty reduction have consistently lagged behind Asian nations.

farmers, so that they or their children can eventually transition into less vulnerable livelihoods with greater opportunities. Opportunity is one of the key intangibles required for eliminating poverty. If individuals perceive they have the opportunity to improve their standard of living — they have hope — which is an important attribute of well-being. Although Cambodia's older inhabitants have suffered enormously, the past two decades have brought peace, stability and economic growth, which has methodically and consistently reduced poverty by almost any measure and, in the process, generated an atmosphere of hope. This positive forward looking attitude is prevalent in both rural and urban populations and is, perhaps, Cambodia's most important human resource.

Health and Education

Investments in health and education yield the most obvious positive results of any development strategy. Better-educated families tend to be healthier, more affluent and, consequently, able to invest more in health and education. In contrast, families living on the edge of poverty are forced to make tough choices based on short-term needs and often forego investments with long-term benefits. Put simply, the need to eat today precedes the need to eat better tomorrow. These choices are particularly tragic when they involve the future of children, many of whom must work to provide income for the family instead of attending school, and who lack the basic health care to offset debilitating diseases which stunts their intellectual development. Health care and education are the two social services that have the greatest impact on reducing poverty over the short term and pay enormous dividends over the long term. In the short term, better-educated mothers take better care of their children, and healthy children tend to be better students. In the long term, better-educated children have a higher probability of climbing the economic ladder and escaping poverty, thus creating the skilled labor essential to a modern economy.

For centuries Buddhist monks were the only literate people in rural Cambodian communities; the traditional educational system was based in temples where monks taught boys the basics of the Khmer language and culture, as well as the main tenets of the Buddhist faith.[9] Although relatively egalitarian in nature, Buddhist instruction was limited in scope and completely excluded women. The French introduced western concepts of education and established the nation's first primary and secondary schools; over time, educational facilities and curricula were expanded to allow enrollment for women. The nation's first university was the Royal University that was created in Phnom Penh in 1960 as part of Norodom Sihanouk's policies to modernize the recently independent nation.[10] As much as 20 percent of the national budget was dedicated to education and similar efforts were made to improve health care by constructing hospitals and training health care workers at the new university. In spite of these early efforts, most reforms and investments were centered in Phnom Penh, and to a lesser extent major provincial cities like Battambang that were targeted for industrial and agricultural development, but the vast majority of rural areas of Cambodia continued to depend on the pagoda system for educational opportunities and on traditional medicine for health care. Although a source of comfort for its recipients, traditional Khmer medicine is a mixture of superstition and herbal medicines of limited effectiveness.

Unfortunately, these early gains were completely lost during the regime of the Khmer Rouge, and Cambodia's educational system is still recovering from the near annihilation it suffered then and during the subsequent period of civil unrest that delayed reconstruction. Educated people were targeted for assassination; schools and universities were either closed or destroyed; even the traditional pagoda educational system was dismantled under the atheist regime. The nation emerged from this period not only with a largely illiterate population, but also with a severe shortage of trained teachers it needed to reconstruct its educational system. In spite of earnest efforts to provide literacy classes to the population, a rigid curriculum and the limited capacities of most teachers have impeded success. Middle class and affluent families place their children in private schools, thus further isolating urban residents from the limitations of the public school system. In the mid-1990s, more than 60 percent for the population was illiterate; although investments over the past decade have improved that statistic, the current adult literacy rate remains near 77 percent, with 85 percent for males and 71 percent for females.[11]

The government and its international partners have made improving public education a national priority — a commitment reflected in the nation's Second Millennium Development Goal, which is to achieve universal primary school education by the year 2010 (versus 2015 for the global commitment), and to provide all children with schooling through grade nine by 2015 (see Table 3.1). Additional commitments are to eliminate gender inequality by 2010 and to reduce the disparity between urban and rural areas in both secondary and university education. This commitment is backed up in the level of spending of the Cambodian government, which represents more than 17 percent

Table 3.1. Cambodian society has made impressive strides in reducing poverty, as reported for Cambodia's Millennium Development Goals, the UN framework that defines measurable targets for improvements in human welfare, environmental sustainability and economic growth based on benchmark year of 2001, midterm results for 2005, and a target date in the year 2015.*

Goal	Results
(1) Eradicate extreme poverty and hunger	Poverty rate reduced 1.5% per year from 47% to 34.7% Extreme (food) poverty fell 6% Population growth rate decreased from 2.5% to 1.8% Target is a 50% reduction in poverty Shortfall: extremely high and persistent rural poverty
(2) Achieve universal primary education	Primary School enrollment: 87% to 91% (target 95%) Secondary Education: 19% to 31% (target 50%) Poor enrollment in secondary and higher educational levels
(3) Promote gender equality and empower women	Gender imparity declining in primary, secondary and tertiary schools Elected Female Commune councilors 8.7 to 14.7% Strengthen laws against sexual trafficking Gender mainstreaming Persistent domestic violence against women (inadequate progress)
(4) Reduce child mortality	Infant mortality rate reduced from 95 to 66 per 1,000 live births 111 for mothers in rural areas 90 for mothers lacking primary education 40 for mothers with secondary education Under five mortality rates decreased from 124 to 83 Improved breastfeeding Improved immunization
(5) Improve maternal health	Maternal mortality 437–72 per 100,000 live births Women using birth spacing 18.5% to 30% (target 27%) Prenatal care for expecting mothers increased from 38% to 66% Midwife assistance 32% to 44% Shortfalls: quality health care, especially maternal health
(6) Combat HIV/AIDS, malaria and other diseases	Adult Prevalence: 3%–>1% Mother to Child Transmission: 2.6% of infected mothers enjoy ARV treatment
(7) Ensure environmental sustainability	Forest Reform (law passed in 2002) Creation of *Tonle Sap* fish reserves Fuel wood dependency in homes reduced from 92 to 86% (target 70%) Safe water (rural) 42% (target 5%) Safe water (urban) 76% (target 64%) Urban sanitation 55% (target 59%) Shortfalls: Illegal logging, wildlife trade, water degradation

Table 3.1. Continued

(8) Develop a global partnership for development	Technical Working Groups Information Communication Technology (ICT) National Strategic Development Plan (NSDP) 2006–2010 that will serve as a framework for alignment of all ODA supported activities Government-Private Sector Forum WTO membership Customs Code ILO reforms
(9) De-mining, UXO and victim assistance	Mine-free land areas increased from 30,263 ha (2005) to 35,436 ha (2006). Human casualties decreased from 875 (2005) to 440 (2006).

Note: *KOC (2007) Progress in Achieving Cambodia Millennium Development Goals: Challenges and Opportunities, Ministry of Planning, Royal Government of Cambodia, <http://www.un.org/ecosoc/docs/pdfs/Cambodia_national_rpt.pdf> [accessed 20 June 2011].

of the national budget.[12] In the most recent survey published by the Ministry of Education, Youth and Sports, primary school enrollment was 91 percent nationwide in 2005, 90 percent in rural areas and 82 percent in remote communities; attendance rates for girls have improved dramatically and now lag behind that for boys by only 1 to 2 percentage points in each category.[13] Statistics for secondary school attendance are more sobering, with a national average of only 26 percent for men and 25 percent for females, although this has increased from 16 percent and 14 percent in 2001. Secondary school enrollment in remote rural remains extremely low for the simple reason that there are no secondary schools in rural communities and parents who wish their children to continue their education must send them to live with relatives in urban areas.

Like the educational system, the public health care system had to be rebuilt from scratch. At the end of the Khmer Rouge regime, there were no doctors or nurses, no medical equipment or laboratory facilities, and hospitals and clinics were completely degraded and essentially non functional.[14] Refugees with medical training returned home from the border camps after the peace accords, but many of them had difficulty obtaining permits or licenses to practice their professions until the mid-1990s. As in the educational sector, health care was largely privatized, making it obtainable for those with economic resources, whereas it remained (and remains) substandard for the vast majority of middle class Khmer. The rural poor have no access to anything that remotely approaches modern health care. Statistics comparing Cambodia with other countries in the region show that children, infants, and woman of childbearing age in Cambodia are particularly exposed to illness, disease, and malnutrition.

In spite of these dismal statistics, ongoing reforms have made substantial improvements in recent years. Starting in 2002, the international donor community developed a comprehensive plan, coordinated by the World Bank and the Asian Development Bank (ADB), to rebuild Cambodia's health care system. Projects were channeled through NGOs who managed provincial health care facilities and programs in coordination with the Ministry of Health. The plan emphasized primary health care in rural areas and investments in provincial hospitals, as well as intensive training for health care professionals and improved drug distribution. One of the goals of the program was to increase the use of the public health care system and decrease the need for families to contract health services in the private sector, many of whom were poorly paid public servants forced to practice medicine outside of office hours.

These efforts have yielded some success, as shown by the government's interim report for its Millennium Develop Goals commitment (Table 3.1). Life expectancy at birth has increased from 57.9 years in 2003 to 61.7 years in 2008,[15] while health care statistics and the nutritional status of mothers and their children have improved overall. The incidence of HIV/AIDS has dropped from an alarmingly high rate of 3.2 percent of the adult population to less than 1 percent in 2008,[16] thanks to an aggressive campaign to promote the use of condoms by high-risk populations, such as sex workers. Currently, about 80 percent of the 25,000 individuals infected with HIV are receiving antiretroviral drugs.[17]

Malaria remains a real and present hazard on Cambodia's large inundated floodplains, as well as in the humid forest ecosystems of the Cardamom Mountains. One of the major causes of malaria in the 1990s was the large internal migration of refugees, many of whom were returning to their villages but also from families who elected to move to the forest frontier in search of economic opportunity and land. Not surprisingly, the number of "reported" cases increased following the implementation of the Health Sector Support Project (Figure 3.3), but this was because more sick people sought attention and were processed by the medical authorities. The incidence rate has fallen subsequently: from high of near 210,000 in 1997 to less than 40,000 new cases per year in 2007; the mortality rate has declined from 7 percent to 1.8 percent over the same period.[18,19] In spite of these advances, many children remain at risk; many die within days of being infected, while repeated bouts of malaria limit a child's physical and cognitive

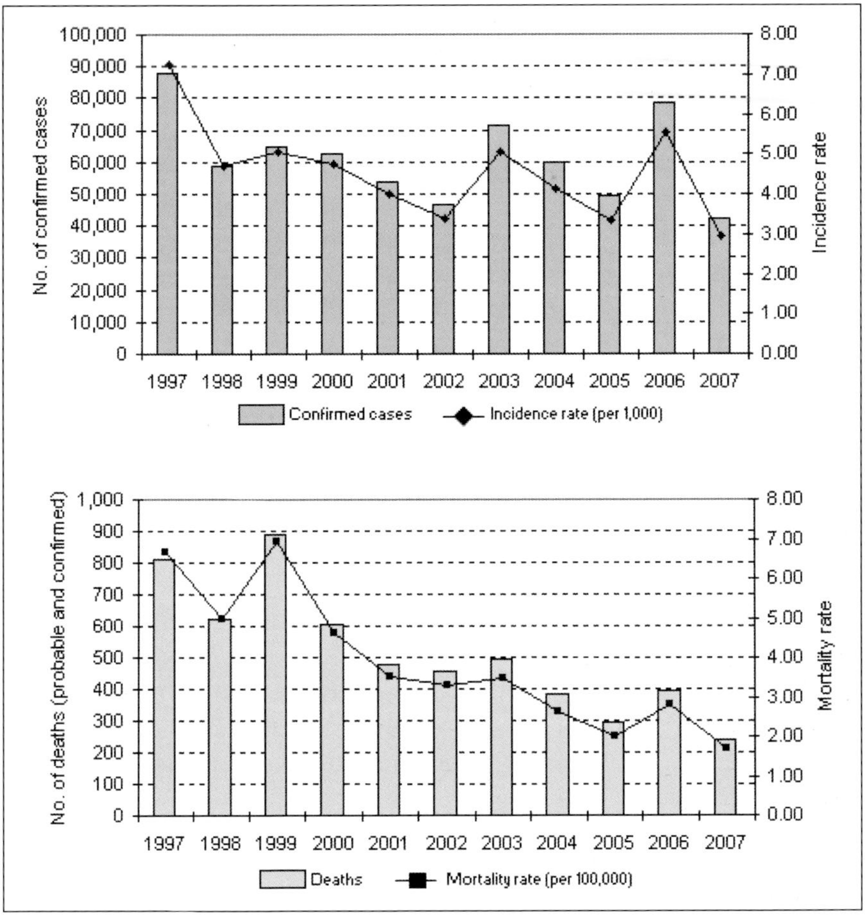

Figure 3.3. The mortality rate from malaria has been steadily decreasing over time (bottom), but the underlying incidence rate remains dangerously high (top). The misuse of prophylactic drugs could lead to the evolution of a drug-resistant strain of *Plasmodium falciparum* causing mortalities rates to return or even surpass historical levels. (Source: Cambodian National Malaria Center)

development. For children to be treated effectively, their parents must recognize the symptoms and be able to receive appropriate anti-malarial remedies. For most of the rural poor, this care must be provided at no cost, or parents will not seek it out, or they will purchase what they need from pragmatic purveyors of anti-malarial drugs.

Unfortunately, genetic variants of the *Plasmodium falciparum* parasite, which is resistant to standard drug treatments, are increasingly seen in malaria patients, particularly in western Cambodia near Pailin.

Table 3.2. Selected health statistics for the countries in mainland Southeast Asia. Compared with other countries in the region, children, infants, and women of childbearing age in Cambodia are still particularly exposed to illness, disease, and malnutrition.

	Infant mortality deaths per 1,000 births	*Life expectancy at birth*	*Fertility, births per woman*	*HIV/AIDS adult prevalence rate*	*Low birth weight (%)*	*Maternal mortality, deaths per 100,000 live births*	*Antenatal care (%)*	*Skilled personnel present at birth (%)*
Cambodia	**56.6**	**61.7**	**3.1**	**1.6**	**11**	**540**	**69**	**44**
China	20.0	73.2	1.8	0.1	2	45	90	98
Laos	79.6	56.3	4.5	0.1	14	660	27	19
Malaysia	16.4	73.0	3.0	0.4	9	62	78	98
Myanmar	74.0	62.9	1.9	1.2	15	380	76	57
Philippines	24.0	71.0	3.3	0.1	20	230	88	60
Thailand	18.2	72.8	1.6	1.5	9	24	98	97
Vietnam	15.0	74.0	1.8	0.5	7	150	91	88

Source: <http://www.unicef.org/infobycountry/index.html> [accessed July 2011].

The emergence of drug resistant strains may originally have been a consequence of the area's population, which migrated into the border area from different regions of the country. Hybridization increases the likelihood that genetic recombinants resistant to drug treatments will appear, but this may be exacerbated by poorly trained drug sellers that provide an inappropriate form of *artemisinin* drug therapy to infected patients. This is particularly unfortunate because the misuse of malarial drugs can increase the malaria parasite's tolerance to the only existing cure for *falciparum* malaria.* Because of its high cost and the danger

---

* *Artemisinin* is a group of chemicals derived from *Artemesia annua* (sweet wormwood) and is used by practitioners of traditional Chinese medicine to treat malaria. Traditional remedies based on leaf extracts provide a "cocktail" of multiple forms of the drug; in contrast, clinical trials and experimental therapies use a purified form of *artemisinin*, which is highly effective, but which loses efficacy over time as parasites become resistant to that specific chemical. Proper treatment with *artemisinin* should involve pharmaceutical cocktails with combinations of drugs that attack the parasite at different metabolic pathways and/or at different points of its life cycle. This multiple drug strategy would require the evolution of two or more mutations simultaneously in the same population, which is statistically highly improbable, and thus lengthen the expected life span of the drug treatment. <http://www.scidev.net/en/health/opinions/prudent-use-can-make-artemisinins-sustainable.html> [accessed 20 June 2011].

of misuse, drug therapy should be combined with other anti-malaria measures such as the distribution of insecticide-treated mosquito nets for all inhabitants in infected areas, and frequent malaria screening.[20] The misuse of drug therapies is also leading to the increase in drug resistant forms of tuberculosis, another major infectious disease that afflicts Cambodia.[21] Cambodia's health care challenges are a direct consequence of both substandard health care services and the inability of poor people to attain any type of health care.

In assessing the health and educational needs of the Greater Cardamom Region, it is interesting to note that the quality of education and health care services in Cardamoms is essentially a microcosm of the national situation, except that the Cardamom region suffers extreme cases because of its isolated forest frontier communities. Otherwise, the relative levels of health care and education in the nine provinces of the Greater Cardamom Region are comparable with the national average as monitored by in the Millennial Development Goal initiative (Figure 3.4). There are some exceptions: Koh Kong province lags behind the nation in education and gender equity, while authorities in Kampong Speu province have had less success in reducing child mortality and lag behind the nation in dealing with HIV/AIDS.

Demographic Growth and Migration

Prior to the regime of the Khmer Rouge, Cambodia was a typical Asian country where the family was paramount and each family's identity was closely linked to a small plot of land and village life. This way of life was radically overturned in the years of the Khmer Rouge regime and, to a lesser extent, the subsequent years of civil strife. People were displaced, the population decimated, and land and family ties uprooted. The rural to urban migration, which was at a pace in line with development in the years following independence in 1954, accelerated in the early 1970s, as war refugees flowed into Phnom Penh, when the population grew from less than 400,000 in 1963 to more than 1.2 million in 1971.[22] This migratory flow reversed direction, however, during the Khmer Rouge regime when the capital was virtually abandoned. Individuals were resettled at random into forced labor camps as part of a collective agricultural scheme that ignored familial ties. Most of the citizenry was settled into communities and collectives located on the traditional agricultural plains of the Mekong and Tonle Sap, but a few abortive efforts were also made to expand agricultural production into nontraditional areas.

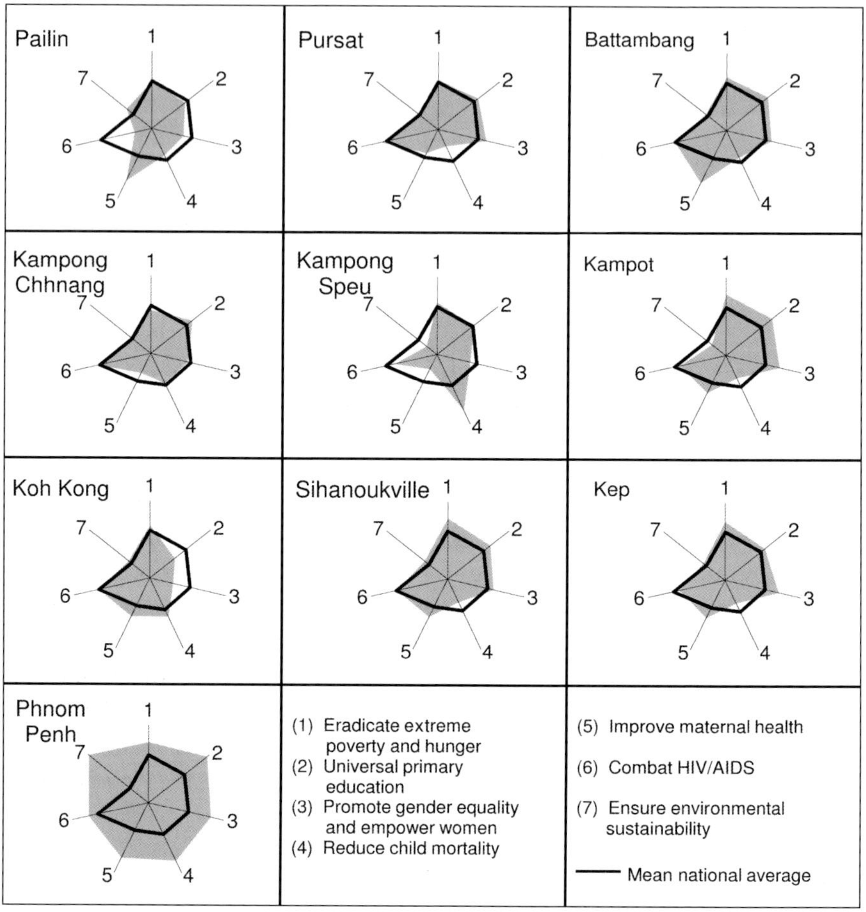

Figure 3.4. Cambodia's progress towards attaining its Millennial Development Goals for health, education, gender equality and environmental sustainability is generally considered to be on track for most categories, but varies among the different provinces depending on the level of urbanization.

The end of the Khmer Rouge regime initiated another massive migration in which hundreds of thousands of Cambodians fled to the Thai border where they were housed in refugee camps dominated by different military and political factions. Those refugees who gained access to the single UN-sponsored facility were allowed to migrate to third countries, such as the United States, but the overwhelming majority of families were repatriated eventually to Cambodia. During the 1980s, efforts were made to organize families into collective units that pooled labor and resources, but this system soon evolved into a more traditional family-based tenure and production system.[23] As normalcy returned in the 1990s, most families chose to settle down and claim the plots they had been working on during the regime of collectivized agriculture.

The armed forces assumed responsibility for organizing refugees in selected resettlements; for example, they played an important role in settling approximately 1,000 families in the the Samlaut Multiple Use Protected Area, which was part of a political agreement to disarm Khmer Rouge soldiers on the Thai border near Pailin.[24] The military also commandeered unoccupied land throughout the country, in order to distribute land among its soldiers, including the rank and file, as well as officers, a process that continues in remote areas of the Cardamom Mountains where land tenure has yet to be clearly demarcated.

Phnom Penh was repopulated during the 1980s with about 500,000 residents.[25] Since then, migration from rural communities to this and other urban centers has proceeded at pace with national development. Phnom Penh is Cambodia's largest city and is home to about 1.3 million people, followed by Battambang (200,000 inhabitants), Siem Reap (150,000) and Sihanoukville (100,000). Settlement to the coastal plain in Koh Kong province was inhibited for many years due to the continuing conflict, but as security improved in the 1990s migrants began to move into the region attracted by the relatively fertile soils on the alluvial plains around the Kompong Som bay and the opportunities from coastal fisheries. More recently, migrants have started moving into highland areas as new roads — part of a long overdue effort to connect the country via a more modern transportation infrastructure — have improved access.

While it is unlikely that migration in Cambodia will ever reach the levels seen at the end of the civil war, urbanization, industrialization, tourism, and opportunities on the agricultural frontier continue to motivate large numbers of people to seek opportunities in regions of the country with greater economic potential. Currently, about 30 to 35 percent of Cambodia's total population has passed through some sort of migratory process, of which 70 percent are rural — rural migrants, which is an extraordinarily large number for a rural society.[26] Rural migration continues at a high rate; individuals continue to seek land, because they view it as one of the surest ways to escape from poverty. About half of all landless individuals are young people who have never owned property and whose families have no surplus property. The other half are individuals who have recently lost their property due to debt or some catastrophic event, usually related to paying for

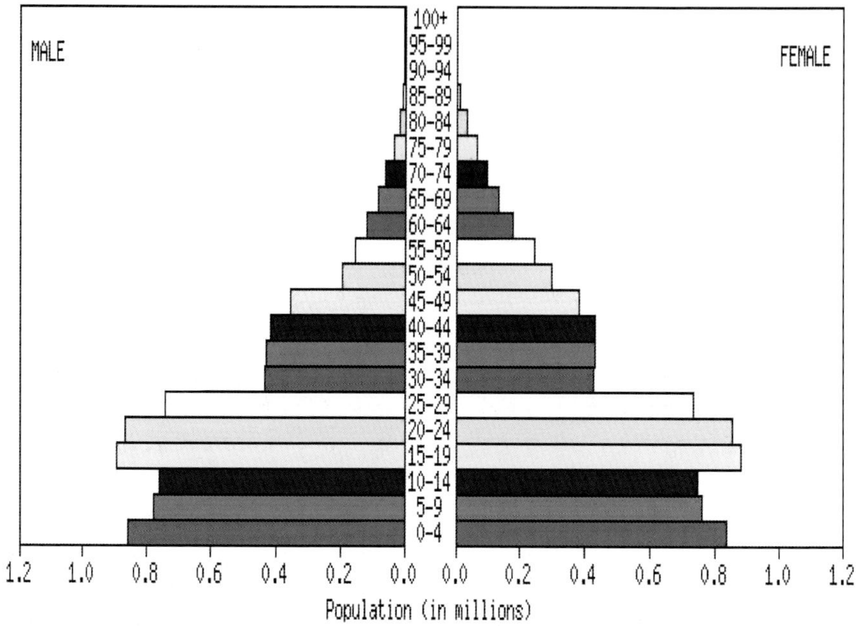

Figure 3.5. The population pyramid in 2010 reflects both the catastrophic loss of population in the 1970s, particularly adult males, as well as the depressed reproduction during this decade; this was followed by a "baby boom" in the 1980s and the first half of the 1990s; population growth has leveled off since, but is still among the highest in Asia.

health care. Most rural migrants are men, while females make up a larger proportion of urban migrants, a statistic that reflects the demand for young women to work in textile factories and the sex trade.[27]

The postwar years have also brought a period of population growth following the Khmer Rouge period when the Cambodian people suffered a massive demographic blow. Estimates vary and are subject to vigorous debate, but as many as 1 million people probably perished during the regime of the Khmer Rouge (roughly 30 percent of the population). Within this statistic is the stark reality that a disproportionate number of small children died, which evidently motivated many young adults to have fewer children. Once peace returned, the Cambodian people responded with an understandable effort to rebuild lives and replace lost children (Figure 3.5). The population growth rate shot up from 1.7 percent in 1970 to more than 2.5 percent in 2006. At this rate, the population will double in about 30 years to 28 million inhabitants. Efforts to reduce population growth, however, have reduced fertility rates from 4.5 births per women in 1995 to 3.3 in 2005. Projections show a continued reduction in fertility, assuming continued economic growth, improvements in medical care, reduced child mortality and increasing life expectancy — all factors that influence population growth. The current prediction for Cambodia is to reach a population of about 19 million by 2025, with annual population growth rates approaching 1.2 percent.[28]

The geographic distribution of this population will depend on the Cambodian government's efforts to guide urbanization and improve rural lives. As population density on the country's most productive agricultural lands (e.g., Kandal, Kampong Cham, and Prey Veng and Takeo provinces) is currently near saturation, most future population growth will either be concentrated in urban areas or displaced to the less populated areas where land is available for settlement. As mentioned in Chapter 1, an alternative scenario would be for economic growth from agro business enterprises to create jobs for rural workers who do not own land, or who own insufficient land to be self-employed farmers.

The coastal region will probably have the most diversified and fastest-growing economy of the country, which will lead to accelerated urbanization and increases in population greater than the national average. Population growth along the northern and eastern margins of the Cardamom Mountains is less easy to predict and will depend on whether small landholdings or large economic concessions predominate. While its relative isolation has impeded migration to the Cardamom Mountains in the past, the land is an obvious attraction for both individuals and businesses. Migration has already been facilitated by the vastly improved road network and a modern communication system, which allows individuals to remain in touch with family members and spread the news of new economic opportunities to their friends and relatives.

Middle Class Khmer

Although there are clearly many poor people in Cambodia, there is also evidence of economic success in the towns and countryside throughout the Greater Cardamom Region. Signs of prosperity include the abundance of motorcycles and electronic gadgets, as well as new home construction and improvements to existing houses. The standard Khmer house is built on stilts to protect it from the high water levels of the annual monsoon, but the ground floor also provides shelter for animals and storage for farm implements. The less affluent rely on housing materials from the surrounding landscapes, mainly palm leaves for roof thatches, woven mats for walls, and rough-hewn poles to support the roof. As incomes improve, sawn timber replaces rough-hewn poles and planks are used to enclose the structure from the elements; thatched roofs are replaced by corrugated metal or ceramic tiles and glass windows are incorporated into structures. At the top end in house construction, bricks, mortar and reinforced concrete predominate; air conditioning is a common appliance and evidence of affluence. Roofing materials were used by the authors of the Cambodian Atlas to estimate income distribution in the 2005 Census (Figure 3.6). Palm thatch remains the most common form of roofing in Pailin, Pursat, Kampong Speu and Kampong Chhnang, while urban buildings skew the statistics in favor of concrete roofs in Battambang and Sihanoukville.[29]

Another measure of growing wealth is the increase of home amenities, such as electricity, indoor toilets, and potable water. In 2004, more than 75 percent of all households in predominately rural provinces lacked these services, while in provinces with significant urban populations the proportion of households without these amenities was about 50 percent (Figure. 3.2). Rural electrification is still a distant goal for the Cambodian authorities, but as the national grid grows, the more affluent households on major transportation routes and around major villages have been integrated into that system. Both Pailin and Koh Kong benefit from their proximity to the Thai border where electricity is provided via cross-border agreements. Cell phones are ubiquitous and microwave towers dot the landscape as part of a national communications backbone that includes secondary roads such as Route 148 between Pursat and Veal Vang.

Even though poverty remains widespread, the increasing prosperity of Cambodia is transforming the countryside. The term "middle class" might seem out-of-place in a country where the top quintile (20 percent) of the population accounts for 50 percent of the national consumption and the bottom quintile is well below the poverty line.[30] However, the Khmer middle class represents approximately 50 percent of the population and, most importantly, own the overwhelming majority of Cambodia's most productive agricultural land. Assisting and supporting middle class farmers and the small town merchants that provide them with goods and services is critical for diversifying Cambodia's national economy. Families that have climbed out of the depths of poverty — or who soon will — are the professional and

Figure 3.6. Although rural poverty is widespread, the region is also home to numerous middle class families who are progressively improving the standard of living. Home construction materials are used to classify the relative wealth of nation's population in the national census: rustic materials and thatch roofing (top left); rough construction lumber and tin roofing (top right); brick and mortar construction material and tile roofing (bottom left); reinforced concrete building with patio roof (bottom right).

entrepreneurial class that will drive future economic expansion. As they increase their consumption of goods and services, they will drive growth in a domestic economy based on light industry, commercial goods, and basic services. Like many Asian cultures, the Khmer save a significant portion of their disposable income, and if given the appropriate incentives, they will use these savings to increase the productive capacity of their farms or educate their children, both of which are essential for Cambodia's development.

Meeting Individual Needs in a Sustainable Paradigm

The last 20 years have put the Cambodian people on a positive growth trajectory that is the result of commitments made by the government and the international aid community. Poverty remains widespread, but the Cambodian countryside is now characterized by multiple signs that the well being of the Cambodia people is improving. Official development assistance will continue to focus on the poor — and rightly so — but the private sector will have the greatest impact on the rural economy (see Chapter 5) and it is these investments that are most likely to generate the wealth needed to move poor people out of poverty and into the middle class. Without doubt, the greatest challenge will be to ensure that private sector initiatives embrace the principals that underpin sustainable production systems in both its environmental and social dimensions. If this does not happen, then the natural habitats and the ecosystems services upon which so many impoverished people depend will be diminished and degraded. This is a central part of the Cardamom Conundrum, because generating wealth is needed to reduce poverty, but if done poorly or without careful planning or without considerations of equity, it can exacerbate poverty for some, even as it reduces it for others.

The metrics that are attempting to track progress of reaching the Millennium Development Goals show that investments in "sustainability" or at least the results from those investments are seriously lagging behind efforts to reduce hunger and improve health and educational services (Figure 3.4). The government and its partners are justified, perhaps, in targeting the human development aspects of poverty, but the recognition of environmental sustainability within the Millennium Development Goals is also an explicit recognition that a healthy environment is essential to improving human well-being. Efforts to integrate environmental sustainability into national development programs will require a more innovative strategy, particularly one that reaches out to the private sector. These must include a strong participatory process that empowers communities and motivates local governments to protect the rights of the poor and the middle class, so that all members of Cambodian society benefit from economic growth.

Developing this innovative development strategy will be the responsibility of the national government and the capacity of its institutions to create an incentive and regulatory framework that rewards sustainability and penalizes short-term exploitive models. Discovering that development pathway will requiring appreciating the economic value of nature (Chapter 6) and identifying production strategies that leverage nature's assets with Cambodia's human resources (Chapter 7) to create wealth, reduce poverty and place Cambodia on a pathway to long-term prosperity in a global economy (Chapter 8).

CHAPTER 4

Institutional Landscapes

Despite current high levels of poverty, growing prosperity looms on the horizon for Cambodia and the Greater Cardamom Region will make a significant and strategically important contribution to that prosperous future. The abundance of this region's natural resources will form the cornerstone of the national economy, at least over the medium term, and its inhabitants will benefit from that development if the public and private institutions of Cambodia are able to work together to promote the wise use of these natural resources. As much as their individual needs and wants drive the development process, the inhabitants of the region do not have the luxury of contemplating how economic prosperity might come about or which pathway might be more sustainable, more equitable, and more strategic in terms of future opportunities. By the nature and the necessity of their lives, Cambodia's small farmers and rural poor are forced to seek the fastest and least risky path to meet their immediate needs for food, land, income and well-being. While this type of spontaneous development dynamic might benefit a few well-placed individuals and their families, the majority of the population will lose out in such an unorganized and chaotic free market process. Even if the majority does enjoy a net benefit from unplanned development, which is not unlikely, they will benefit much more if the development process is driven by a governance system with a clear and solid legal framework that mediates disputes among the inhabitants and other stakeholders in the region. Fairness and equity must be part of a social contract that ensures that legitimacy and legality go hand in hand, so that investors are provided the stability they need and families the opportunities they deserve.

Good governance is, perhaps, the most critical piece of the Cardamom Conundrum. The future of the Greater Cardamom Region will depend on the national government and its ability to manage Cambodia's historical legacy, its fast-changing economic profile, and its relations with the rest of the world. The goal is to foster an economy that will expand and sustain itself over time; the private sector will lead that growth, but the government must create the conditions that allow the private sector to grow and ensure that the benefits of growth are shared equitably. Cambodia has the advantage of being a late starter in comparison to its neighbors in Southeast Asia; thus, it can learn from their successes and failures. At the national level, an increasingly self confident government is encouraging foreign investment, while maintaining control over strategic sectors of the national economy — particularly those linked to the exploitation of renewable and nonrenewable natural resources such as forest products and hydrocarbons.

International assistance has provided massive financial resources and technical capacity over the past two decades and continues to influence the national development agenda. A multitude of non-governmental organizations (NGOs) bring specialized knowledge to a broad range of issues and have partnered with organizations within Cambodian civil society to promote transparency and press for an integrated approach to conservation and development. These institutions have worked well together over the past twenty years in Cambodia's reconstruction phase, but the nature of international collaboration will be substantially different during the next phase of national development. As Cambodia grows and its economy matures, development assistance will become less important in influencing government decisions or for mediating economic and social phenomena within Cambodian society.

Instead, the country's leaders will make decisions based on political and economic criteria that will largely be a function of domestic politics. An important factor in determining the rate and nature of Cambodia's growth will be its increasing integration into the regional economy and the efforts of other Southeast Asian governments to promote cross-frontier initiatives. The domestic private sector will likewise evolve and become increasingly sophisticated, while foreign investors will grow in importance; both will seek regulatory frameworks that allow them to plan and execute a broad range of investment strategies. Cambodia's development strategy will reflect the increased power and influence of Asian governments and corporations. This is due, in part, to the decreasing scale and influence of official development assistance from western countries, but is also a natural consequence of the evident success of Asian countries to compete globally and the capital flows that success stimulates in the investment market. There will undoubtedly be a continued emphasis on job creation and poverty reduction; two issues that are irreversibly linked, and which must be the primary responsibility of any Cambodian government. The growing importance of climate change will also be influential in shaping public policies agendas, at least at the international level, and this may motivate private sector investors to exploit new financial opportunities linked to climate change mitigation and adaptation or, more likely, react to market signals as consumers in industrialized countries demand sustainable products. The opportunities that will flow from a low emission development strategy, the so-called "green economy", can provide the conceptual framework to pursue the economic policies required to penetrate new markets, while increasing the resilience of the national economy and, eventually, eradicating poverty (see Chapter 7).

In order to exploit these opportunities, the Cambodia government must continue its ongoing efforts to improve administrative efficiencies and fight corruption. In the context of the Greater Cardamom Region, the potential for decentralization to improve governance is particularly important. Community participation is the essence of decentralization and is vital for improving natural resource management, but it must be formalized and made transparent, so that people feel empowered to contribute to the governance process. Coordination and cooperation among the various institutions — public, private, nongovernment, government, international, domestic — is essential, because they all have a shared interest in an environmentally healthy, socially stable and economically prosperous Cambodia.

National Agencies

Cambodia is characterized by a strong central government, which is appropriate for a small country with a relatively homogenous population. Most strategically important decisions are made by the Council of Ministers, which seeks advice from the various government ministries. Cambodia's natural resources are managed by at least eight different cabinet level ministries whose responsibilities are delegated according to logical and traditional view of development and governance (Table 4.1). This organizational structure may be logical, but it is not necessarily forward looking, nor particularly cognizant of the many cross cutting issues that characterize emerging international policy issues related to climate change and the opportunities presented by global

Table 4.1. Cambodian Federal Ministries and Departments Responsible for Sustainable Development.

National Authority	*Organizational Structure*	*Departments Related to Sustainable Development*	*Responsibilities/ Oversight*
Ministry of Planning (MOP)	General Directorate of Planning National Institute of Statistics	General Planning Economic Planning Social Planning Investment Planning International Relations	Develop & coordinate implementation of National Development Strategy Plan
Ministry of Land Management, Urban Planning and Construction (MLMUPC)	General Department of Cadastre General Department of Land Management & Urban Planning	Land Administration Land Inspection Land Registration Geographic Department Land Management Research and Regulation	Land Registry, Landuse planning, Land distribution
Ministry of Agriculture, Forestry and Fisheries (MAFF)	4 General Directorates 14 Departments, 24 Provincial Offices 2 Research Institutions 2 Universities 7 Rubber plantations	Agricultural Information and Documentation Center Agricultural Legislation Agronomy and Agricultural Land Improvement Agro-Industry Animal Health and Production; Agricultural Extension Agricultural Machinery Fisheries Administration Forestry Administration Rubber Plantations	Multiple responsibilities related to agricultural development, as well forest and fisheries management
Ministry of Water Resources and Meteorology (MOWRAM)	Director General of Technical Affairs Provincial Department of Water Resources and Meteorology	Water Resources Management & Conservation Hydrology and River Works Meteorology Irrigation Agriculture Water Supply and Sanitation Engineering	Sustainable use of water resources and facilities in land-use planning and integrated river basin management
Ministry of Rural Development (MRD)	General Directorate of Technical Affairs Provincial and Municipal Offices	Rural Roads Rural Water Supply Rural Health Care Community Development Rural Economic Development Ethnic Minority Development	Coordinating all aspects of rural development

continued overleaf

Table 4.1. Continued

National Authority	*Organizational Structure*	*Departments Related to Sustainable Development*	*Responsibilities/ Oversight*
Ministry of Mining, Industry and Energy (MIME)	General Department of Industry General Department of Mineral Resources General Department of Energy		Hydropower Mines Renewable Energy Biofuels
Ministry of Tourism (MOT)	Eight Departments	Planning Development Cultural Tourism	Planning, development and marketing of all aspects of the tourism sector Directing, controlling and maintaining natural resorts, manmade resorts and tourism development regions throughout the country; Issuing operating licenses to tourism companies
Ministry of the Environment (MOE)	Six Departments	Department of Nature Conservation and Protection Department of Environmental Impact Assessment Natural Resource Assessment and Environmental Data Management Climate Change Department	Manages the protected area system Reviews all Environmental Impact Studies Coordinates policies and actions to mitigate the impact of climate change

markets seeking sustainable products from low emission or green economies. By the same token, Cambodia must be able to respond in a timely manner to investment opportunities, especially since the country's formal institutions are still in the process of consolidation and are manned by functionaries that are still exploring the limits of their regulatory mandates.

The response of the government has been to use a "fast-track approval" process for foreign investment which is managed by a high level agency known as the Cambodian Investment Board (CIB); established in 1994, this facility has been successful in attracting and consolidating many of the country's strategic investments in energy, tourism and agribusiness.[1] The CIB has allowed investors to bypass bureaucratic pitfalls and needless delays, which undoubtedly has benefited the nation, but it has also created a not very transparent mechanism that has allowed investors to avoid environmental and social oversight. Moving forward, this agency could play a key role in moving the Cambodian economy towards a low carbon emissions strategy by including review criteria that promote sustainability or, perhaps even more effectively, broadcasting to the global investment community the adoption of a green economic model. As the nation's institutions mature, extra ministerial bodies such as the CIB will become less important for attracting investment, particularly as the legal and regulatory framework becomes a more attractive guarantor of long-term private investment.

Environmental and social oversight is by its nature a multi-thematic and inter-ministerial obligation. Consequently, the formal structure and organization of the various agencies and ministries is less important — if there is effective coordination among the entities and a willingness to adopt an integrated strategy that fully contemplates both the environmental and social complexities of development. This is particularly urgent when considering development projects that impact forest landscapes on public lands, because these are the only areas where opportunities for conserving strategic natural assets, such as natural forests and intact watersheds, still exist. Similarly, nearby agricultural frontiers should be priority landscapes for spatial planning, a key government function, in order to facilitate the implementation of sustainable agriculture and plantation forestry, both of which are production models that can be the pillars of a green economy.

Fortunately, many coordinating mechanisms would appear to be in place within the Cambodian government. The Ministry of Planning (MOP) coordinates the strategic planning of all the various ministries, while within each ministry there is also an office dedicated to planning and coordinating with partners. In addition, the government has outlined a long-term strategic, multi-sectorial development plan as a blueprint for all of its agencies. The MOP oversaw the creation of Cambodia's National Strategic Development Plan (NSDP), which is the overarching document that identifies and describes the strategies Cambodia will pursue to achieve its Millennial Development Goals.

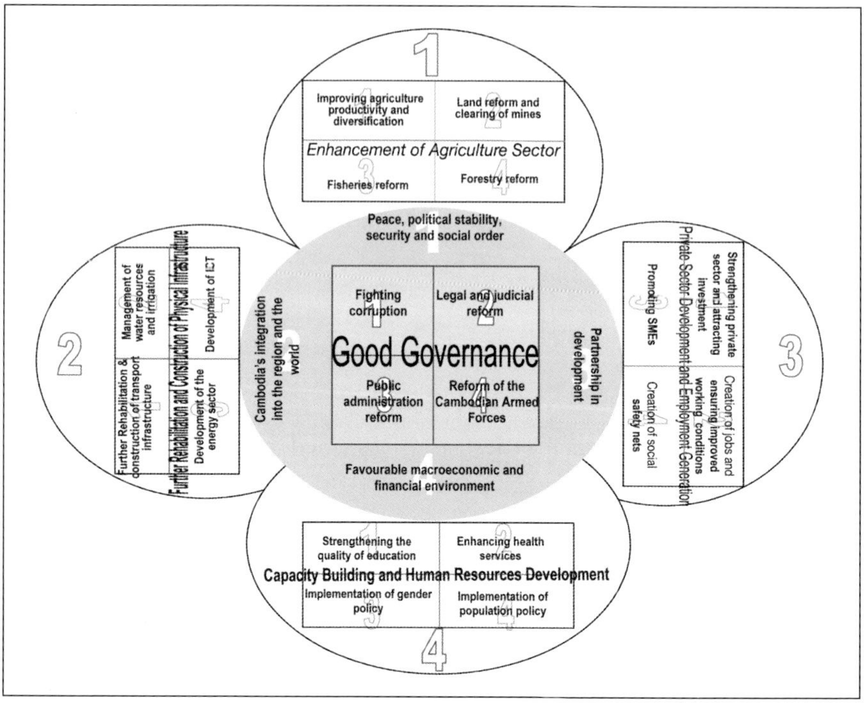

Figure 4.1. The *"Rectangular Strategy" for Growth, Employment, Equity and Efficiency* is intended to achieve Cambodia's Millennium Development Goals by improving and enhancing public institutional capacity, promoting good governance, and modernizing economic infrastructure, in order to foster economic growth, create jobs, ensure social equity, and increase public sector efficiency, as well as protect natural and cultural resources, which are vital for sustainable development and poverty reduction. (Source: National Strategic Development Plan, Update 2009–13).

The NSDP synthesizes and prioritizes the goals of the National Poverty Reduction Strategy, as well as aligns multiple different policies designed for individual sectors and planning cycles to one single common long-term vision.

Cambodia originally drafted the NSDP to cover the years between 2006 through 2010, but the document was reviewed and updated after the global financial crisis in 2008 and now covers the period from 2009 to 2013.[2] The NSDP is organized around a "rectangular strategy," which summarizes government priorities in a clever graphical representation that shows thematic links, sectorial relationships, and government responsibilities in a nested hierarchical scheme (Figure 4.1). Good governance is at the center of the NSDP and demonstrates Cambodia's recognition that at this stage of its development, elected authorities and government officials have a special responsibility to guide the nation towards a prosperous future. Like all nations and bureaucracies, special interests and human nature sometimes conspire to impede collaboration, but these conflicts also reflect legitimate but divergent views as to national priorities and the practicality of certain types of actions. Regardless of its limitations, this important document and the consultative processes used in its development, has provided the Cambodian state with an important reference point for organizing its activities and investments.

Within the strategic framework of the NSDP, each ministry is delegated certain responsibilities; as mentioned above, there are eight ministries whose portfolios address environmental issues and whose decisions impact the sustainable development of Cambodia's natural resources and the trajectory of a green economy. In the context of the Greater Cardamom Region, the most important is the Ministry of Land Management, Urban Planning and Construction (MLMUPC);[3] as its name implies, this ministry has a wide ranging portfolio that includes almost every area of the economy. Among its responsibilities is the coordination of an inter-ministerial body known as the Council for Land Policy Strategy, which recently renewed its commitment to improving "land administration, land management and land distribution" of public land.[4] Extensive areas of the Greater Cardamom Region are state owned lands and much of this land will be distributed to third parties via economic land concessions or by direct transfer to small farmers as private property. Linked to that commitment, are MLMUPC efforts to complete an inventory of state-owned land and systematizing all public and private land registration processes via a "single window."

Secure land tenure is an essential prerequisite for all investors and is the single most important aspiration for small farmers. Land tenure disputes and the subsequent displacement of small farmers are the most significant source of conflict in rural Cambodia. A consolidated and efficient land registry system would ensure the security of existing land holdings, pre-empt illegal land grabs and protect the rights of vulnerable groups. The implementation of the system is already under way, particularly in conflict-free areas where progress is most feasible and in areas inhabited by indigenous communities. Another explicit goal of the MLMUPC is to prevent improper land use and inappropriate land conversion and the ministry is seeking to accelerate land-use planning as part of an integrated process to organize an orderly, transparent and efficient allocation of land for agriculture, industry, tourism, urban, residential and reserved areas.

As part of its commitment to transparency, the government has made a commitment to regularly monitor economic land concessions for agriculture and plantation forestry; a primary goal will be to verify that concessionaires are meeting their investment commitments and are not just peculating in land values using a public asset.[5] Provincial and district governments are all expected to participate in this process and the MLMUPC maintains offices throughout the country to facilitate this liaison, as well as ensure that rural inhabitants have access to the land registry process. Although local elected officials have limited legal authority to influence the development national agenda, they do play have an important role in the distribution of land. Perhaps even more important, is their role in allocating resources to priority infrastructure investments, many of which are often essential for investors seeking to develop land for productive purposes.

Almost all of the land that is being allocated or subject to planning in processes in the Greater Cardamom Region will be dedicated, in one way or another, to either agriculture or forestry and the government agency directly in charge of those economic activities is the Ministry of Agriculture, Forestry and Fisheries (MAFF), which is one of the oldest and most respected government institutions in Cambodia. In recent years, the MAFF has provided strong leadership to improve the plight of Cambodia's rural citizens and at least some of the progress towards meeting the nation's Millennial Development Goals can be attributed to that leadership. Chief among its accomplishments are the impressive and ongoing efforts to improve the nation's rice production, most of which has been accomplished by assisting small farmers increase their productivity (see Chapter 7). The MAFF is now leading similar efforts to introduce new crops and technology, as well as helping entrepreneurs add value to Cambodia's agricultural production.[6] The ministry is large and has multiple departments dedicated to agronomy and animal husbandry, as well as extension, research, technology, and policy development. Its long-time presence in the provinces, where it maintains offices and extension services, is particularly relevant as Cambodia's renews its emphasis on decentralization.

The MAFF also develops government policies and oversees activities related to the management of the country's natural ecosystems and biodiversity. The Forest Administration, for example, manages the nation's production forests, from which Cambodians extract timber, fuel wood and other forest products that are critically important for the national economy. Many of these same landscapes are important for biodiversity conservation and the provision of key ecosystem services such as fresh water and carbon sequestration (see Chapter 6). The Fisheries Administration oversees the nation's fish resources, including which the strategically important freshwater fisheries of the Tonle Sap Lake and Mekong River, as well as the marine fisheries located off the Cambodian Coast. The MAFF has recently expanded efforts to promote aquaculture, in order to relieve pressure on the nation's natural fisheries, but also to create an export industry based on the country's abundant water resources. Aquaculture is a labor intensive production system that is highly compatible with small scale agriculture and has the potential for improving the well-being of millions of rural Cambodians. To meet its future commitments, the MAFF will require significantly greater financial and human resources; but most field programs are overly dependent on financial assistance from international assistance and the technical support provided by NGOs. As Cambodia moves into the next phase of its development trajectory, the MAFF must become more self sufficient in both income and technical capacity. Resources invested in the agricultural sector generally, and MAFF specifically, will be essential if Cambodia is to pursue a green economic model on sustainable agriculture, forestry and fisheries.

The management of Cambodia's water resources is largely the responsibility of the Ministry of Water Resources and Meteorology (MOWRAM). The wise use and management of the nation's water is extraordinarily important for any national economy and this seems even more obvious for Cambodia, where water underpins both commercial fishing industry and subsistence fishermen, while irrigation is essential for a strategic portion of agricultural production. Moreover, water use is an increasingly important component for energy generation and its safe delivery to rural communities and urban households is essential for human development. MOWRAM has been tasked with the job of ensuring that this precious resource is managed to maximize benefits for people and, incidentally, to conserve the aquatic ecosystems that underpin the overwhelmingly important natural fisheries.

Ironically, water is both Cambodia's most abundant and most limiting natural resource and managing the inevitable impacts of the monsoon represent a special challenge. These two extremes have motivated MOWRAM to develop an integrated national water management system that can be used to facilitate both irrigation and drainage, which will be part of an integrated water resources management strategy that considers all possible sources and uses of water. These include hydropower, fisheries, agriculture and urban and industrial uses, as well as the linkages between water flows, land-use, and the ecological integrity of strategically important aquatic ecosystems, such as the Tonle Sap and the estuarine habitats of the Cambodian Coast.[7] The need to coordinate all of these policy objectives with other government agencies is paramount and is an explicit part of the MOWRAMs institutional objectives. This ministry has been charged with protecting the national interest in international forums, particularly by representing Cambodia to the Mekong River Commission, a responsibility it shares with the Cambodia National Mekong Committee, an independent body chaired by the minster that reports directly to the Council of Ministers.[8] This representation is important because developments on the upstream sections of the Mekong will inevitably impact both water and fisheries resources in Cambodia. For example, the modulation of water flows that is occurring due to the operation of dams in China, Laos, and Vietnam will improve the engineering (hydraulic) feasibility of hydropower facilities on the lower Mekong in Cambodia; consequently, their development is likely to be welcomed by the Ministry Of Industry, Energy and Mines (see below). However, the physical barriers to fish migration from dams established on the main stem of the Mekong in Laos will negatively impact the ecology of the fisheries in the Tonle Sap Lake. These are complex issues and tend to stimulate passionate discussions among different stakeholder groups and highlight the need for coordination and dialogue among the various government ministries.

As its name implies, the Ministry of Rural Development (MRD) oversees development in rural areas and acts as a liaison with other ministries to implement a broad range of development initiatives in the countryside. Based on its experience and close contact with the people, this agency is among the best placed to advise the government on the most pressing needs of rural communities and report back on which solutions are bringing real improvement to the lives of rural families. Its priorities include investments in transportation infrastructure, secure water supplies, improved sanitation, and promoting small-scale enterprises by providing market research and access to micro credit.[9] A fundamental philosophy of this ministry is to promote economic self-sufficiency in rural communities by encouraging an independent attitude to solving problems and an entrepreneurial attitude so that rural residents learn to seek and identify opportunities. For example, it has encouraged communities and agro industrial entrepreneurs to form partnerships, so that the value added components of a modern agricultural economy are shared by small farmers, as well as the holders of economic land concessions. This policy and development is likewise promoted by the Council of Land-Use Policy and the MLMUPC and it is based on the assumption that an agro business sector requires economies of scale and minimum levels of feedstock supplies to justify investments in mills and processing facilities; and that current smallholder production is deemed to be insufficient or at least too risky to justify that investment. Nonetheless, the investors of the mill (and the economic land concession) are expected to source some of their supplies from neighboring small farmers. This model, while attractive in theory, requires a regulatory component to ensure its implementation and highlights again, the role different government agencies can play in its implementation. The MRD provides a forum for this type of dialogue and cooperation among various national agencies, local authorities, civil society and the private sector on issues of rural development.

One of the Ministries whose voice is becoming increasingly important in the Cardamom region is the Ministry of Tourism (MOT), which has enormous potential to support sustainable development in the Cardamom Mountains, and, even more importantly, on the Cambodian Coast. In line with this vision, the MOT has adopted the dual themes of "Culture and Nature" as the promotional banner for Cambodia's tourist attractions — a perfect fit for the future sustainable development of the Greater Cardamom Region. This ministry should be a strong advocate and partner within the government for the conservation of important natural assets, such as scenic landscapes and coral reefs; however, the tourist industry also needs good transportation infrastructure and affordable energy in order to compete with other international tourist markets. Some environmentalists view infrastructure-related components of the tourist industry agenda as a "threat" and in that respect the development of a sustainable tourist industry represents a microcosm of the Cardamom Conundrum. The MOT provides an important and influential venue for seeking consensus on ways to develop the region, because of the tourist industry's political clout, which is directly related to its contribution to the national GDP. Tourism in all of its manifestations will be one of the strongest pillars of a green economy.

Another ministry whose policies and actions can radically impact rural communities and the environment is the Ministry of Mining, Industry and Energy (MIME). The economic importance of the non-renewable (and non green) natural resources under its domain, particularly oil and natural gas and the potential for mineral exploitation to provide revenues for the government make it one of the most influential ministries in the Cambodian government. In spite of its power and influence, the MIME must still coordinate its actions and strategies with other government ministries that also have tuition over strategic natural resources that are exploited for energy. For example, energy in Cambodia is still largely based on fuel wood and electrical generation soon will be dominated by hydropower, both of which are renewable natural resources under the management responsibility of other state agencies (MAFF and MOWRAM). Liquid biofuels have the potential to supply a significant share of domestic energy needs and represent a potential export market for agroindustrial enterprises with economic land concessions, which requires policy coordination with the MAFF and land-use considerations with MLMUPC.

Surface mines are a type of land-use and traditionally, below-ground rights associated with mining concessions usually have priority over surface rights. However, the impacts and activities of concessions can significantly affect rural communities that are poorly equipped to deal with change. Dialogue with the MIME is vitally important and has not been given serious consideration by environmental advocates. The path to a green economy must also include dialogue with the representatives of the traditional economic model. For example, increases in energy efficiency are usually a cost effective measure to reduce greenhouse gas emissions, at the same time, the subsequent savings in energy consumption can reduce the need for additional energy generation, which in Cambodia might mean avoiding a hydropower facility on the main stem of the Mekong River. Energy regulation is the responsibility of the MIME and measures intended to increase energy use efficiency fall within their domain.

The government ministry which should have the most important input on the development of environmental policies is the Ministry of the Environment (MOE), which was established in 1993 and charged with environmental protection and biodiversity conservation. One of its most important departments is the Department of Nature Conservation and Protection, which manages the nation's protected area system, including the Greater Cardamom Region's two national parks, three wildlife reserves, one multiple use area and the proposed offshore marine protected area (see Figure 1.5). Another key unit of the MOE is the Cambodian Climate Change Office, which organizes research and information related to global warming and coordinates the nation's policies and actions on mitigation and adaptation within the context of the United Nation's Framework Convention on Climate Change (UNFCCC). Developing a regulatory framework that will facilitate the development of a domestic carbon market or that distributes revenues that Cambodia might be able to access via international carbon markets is part of the institutional agenda of the MOE, but the creation of those markets will involve other agencies, particularly the MAFF, which has tuition over the landscapes where the potential for carbon sequestration is greatest.

The MOE also reviews all environmental impact assessments (EIA) and monitors and regulates pollution, two responsibilities that place it in potential conflict with influential investors and other ministries seeking to fast-track industrial (MIME) and tourist projects (MOE). However, EIAs do not actually approve or reject an investment. They are usually undertaken only after projects have been approved and financed and their goal is to avoid or mitigate negative impacts. Ideally, decisions about where to build a dam, resort or highway should be done well in advance of an EIA, or well in advance of the feasibility study that determines a project's economic viability, as part of a more fundamental strategic environmental and social assessment (SESA). An SESA is a more methodological planning process that analytically compares distinct development options, so that societies and their governments can choose the options that maximize benefits over the medium to long term.[10] Many planning processes in Cambodia have adopted the precepts of an SESA methodology, at least in the sense that they bring in participation from multiple stakeholders and seek to coordinate their different needs. One of the most important tasks within an SESA is spatial planning, which is the responsibility of the Ministry of Land Planning.

Official Development Assistance

The United Nations has been leading the effort to reconstruct the Cambodian state and help its citizens recover from the devastation of decades of conflict. This effort began with the creation of a transitional authority, known as the United Nations Transitional Authority for Cambodia (UNTAC) that administered the country from October 1991 until elections in September 1993. Throughout the early period of reconstruction and continuing well into the next decade, official development assistance (ODA) has provided a disproportionate share of Cambodia's government revenue and, at the same time, has financed the activities of a large number of NGOs who contributed with specialized knowledge and expertise to the reconstruction effort. In recent years, as the Cambodian state has recovered its functional and improved its administrative capacity, assistance from multilateral agencies and bilateral assistance programs transitioned into more traditional types of programs similar to those found in other developing countries. In spite of the differences in focus and organization, the emphasis has always been on economic development, poverty reduction, improving social well-being and environmental conservation.

Assistance has covered a broad range of issues, including conflict resolution and the aftermath of civil conflict, particularly the removal of land mines and unexploded ordnance. The largest investments have been in human health and education, although efforts to increase agricultural productivity, promote industrial development and expand commerce have also been pursued as part of an integrated strategy to reconstruct the national economy. Long overdue investments were made in basic physical infrastructure for transportation, energy and telecommunication systems, as well as to improve fundamental services such as potable water and sanitation. There was also a strong focus on governmental reform and strengthening government institutions, as well as a concern for human rights, the environment and cultural issues. Efforts to promote decentralization are ongoing, with capacity training for local commune councils a key priority for fostering the sustainable management of natural resources.

The special nature and sheer volume of the assistance provided to Cambodia brought with it unique organizational and implementation challenges; from the onset donors were concerned that, without strong coordination, the aid effort would be plagued by inefficiencies, duplication of efforts and a waste of financial resources. As the Cambodian government assumed greater responsibility for its internal affairs, it too wanted to ensure that the development assistance would be in line with its priority investments and would conform to the agenda that both it and its donors had approved via a consensual process. To this end, the Cambodian government and the donor community joined together to create an oversight and coordinating body that meets annually to review financial aid strategies and map out future actions.* This annual meeting, now known as the Cambodia Development Cooperation

---

* Donor assistance has been since coordinated by a series of *ad hoc* bodies known originally as the International Committee for Rehabilitation of Cambodia (1993–5), which was later changed to the Consultative Group (1996–2007). The Consultative Group has been co-chaired by the Cambodian government since 2002 and now meets annually in Phnom Penh under the name of the "Cambodia Development Cooperation Forum (CDCF)" <http://www.cdc-crdb.gov.kh/cdc/gdcc/eighth_objectives_procedures.htm)> [accessed 1 July 2011].

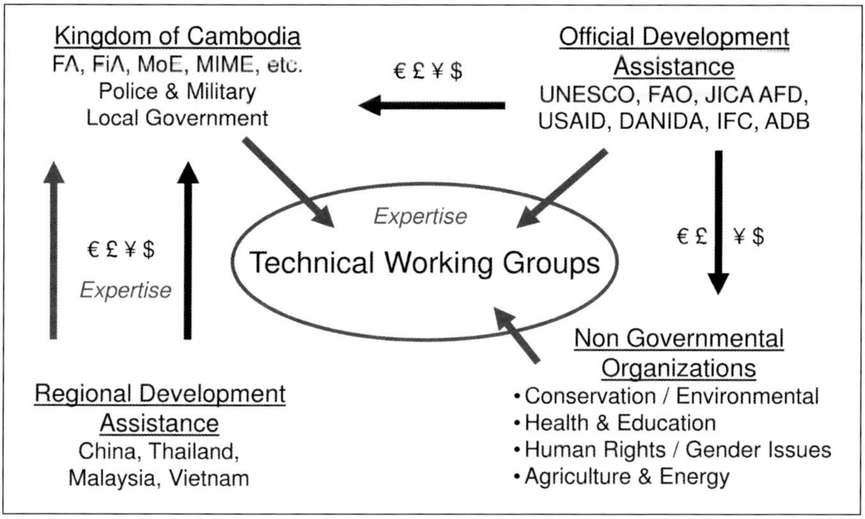

Figure 4.2. A schematic representation of the coordinating role of the Technical Working Groups, which are charged with ensuring that development assistance to NGOs are in line with the overall development strategies and goals identified by the government of Cambodia and its partners.

Forum is composed of individual Technical Working Groups (TWGs) that develop action plans to meet the objectives agreed upon by the Forum, the Cambodian government and the donor community (Figure 4.2). Each TWG is co-chaired by a senior official from the Cambodian government and a functionary from one of the major aid agencies — typically the one making major investments in the TWG's focus area (Table 4.2). Since their creation in 1999, the TWGs have evolved into an effective means of consultation, collaboration, and cooperation — precisely because they involve the major stakeholders from the government, development assistance agencies, and civil society.

The effort to improve coordination also included a commitment to transparency; to this end, a public database of all the ODA flowing into Cambodia was created.[11] Although data from the first few years of the international assistance effort is incomplete, the number and scope of projects in Cambodia's reconstruction is an impressive example of the international community's capacity to act when there is consensus. Since 1993, more than 1,400 separate ODA projects have been implemented in Cambodia for a total of more than $10 billion. Only $100 million (1 percent) of this has been dedicated to environmental issues; nonetheless, many mainstream development projects on health and rural development have also incorporated important environmental criteria into their design and execution. The leading donor agencies have been, and continue to be, the Asian Development Bank ($1.9 billion), followed by Japan ($662 million) and the United States ($628 million); Europe has provided similar amounts of assistance that include resources from the European Community ($395 million), as well as assistance from individual countries ($715 million). Significantly, more than 90 percent of the assistance to Cambodia has been made as a grant, with the remainder being "concessional loans" with low interest rates; these are usually converted to grants as part of debt forgiveness programs for heavily indebted poor countries.* On an annual basis, the amount of aid to Cambodia ranges between $480 million (2002) and $670 million (2007).[12] Approximately 30 percent of this assistance is channeled via the government and constitutes about 25 percent of its annual budget.†

China has also made important contributions to Cambodia's development, but the $234 million reported by the Chinese government is an underestimate. A large part of China's assistance is not listed on the clearing-house mechanism coordinated by the Cambodia Development Cooperation Forum because it does not qualify, strictly speaking, as "Official Development Assistance." China also provides support to Cambodia via its state-owned corporations, which arrange commercial loans to private Chinese corporations, which then invest in Cambodian infrastructure as part of for-profit ventures.[13] The total investment of Chinese corporations in Cambodia was recently estimated at $1.7 billion dollars. Similarly, South Korea has provided about $300 million in development assistance, although its total direct foreign investment in Cambodia since 1997 is estimated at $2.9 billion.[14] As in most countries, this type of investment benefits from government subsidies in the form of loans from governmental agencies, and Cambodia was one of the largest recipients of loans from the Korean government, which are then used to contract private Korean companies, who build the infrastructure that Cambodia so urgently needs.[15]

---

* Usually, the donor country or agency forgoes repayment of the loan but requires that the revenues that would have been used to pay off the debt be used to finance some specific activity, such as health, education or environmental protection.

† In 2006 the Cambodian government reported revenues of $814 million and expenditures of $973 million; foreign financing accounted for $213 million of this budget. *Bureau of East Asian and Pacific Affairs, US Department of State*, November 2008, <http://www.state.gov/r/pa/ei/bgn/2732.htm> [accessed 1 July 2011].

Table 4.2. Technical Working Groups coordinate the large amount of Official Development Assistance (ODA) Cambodia receives from the U.N., foreign governments and multilateral organizations to ensure that these resources are used in line with overarching development strategies and goals defined by Cambodia and its donors.

Legal and Judicial Reform	Mine Action
Partnership	Agriculture and Water
Public Administration Reform	Fisheries
Public Financial Management	Forestry and Environment
Land	Private Sector Development
Decentralization and Desconcentration	HIV/AIDS
Gender	Infrastructure and Regional Integration
Health	National Strategic Development Planning
Education Food Security and Nutrition	

The role of ODA in the development and conservation of the Greater Cardamom Mountains landscape has been and will continue to be decisive. For example, a major investment by the Asian Development Bank (ADB) laid out the strategy for a biodiversity corridor that would link the Cardamom Highlands with the Elephant Mountains, while the French Development Agency has made important investments to help the Forest Administration protect the biodiversity of the Central Cardamom Protected Forest. At the same time, Chinese companies will build six hydropower facilities in the region, while the ADB, Japan and Germany are financing different components of an integrated energy grid that will connect generation facilities to urban markets.[16]

These investments represent potential conflicts between development and conservation agendas, but the TWGs can play a role in seeking a consensus that will benefit Cambodian society. The Forest and Environment TWG has assumed a leadership role in this regard, but important development issues within the Cardamom region should also concern TWG charged with coordinating investments and policies related to Fisheries, Agriculture and Water, Infrastructure & Regional Integration (Table 4.2) and, in the process, engage the appropriate government ministries. The international donor community can also provide support by using its influence to ensure that crosscutting issues are brought to the attention of the appropriate TWG — particularly the most important of all TWGs, the one dedicated to National Strategic Development Planning, which is coordinated by the Ministry of Planning.

Regional Integration

As Cambodia grows and its economy matures, the Cambodian government will rely less and less on development assistance to make its decisions and more and more on political and economic criteria, both on the domestic front, and with an eye towards integration into the regional economy of Southeast Asia. Governments throughout the region are promoting cross-frontier initiatives and corporations are seeking economic opportunities in neighboring countries. The benefits of looking "over the frontier" instead of "across the ocean" have been brought home by the global financial crisis of 2008. Capitalizing on these benefits will require trade policies and administrative measures to facilitate interchange among neighboring countries. Agreements that simplify border crossings, lift passport requirements, or allow transportation companies to carry goods across multiple borders can stimulate cross-border trade. Harmonizing manufacturing standards and communication protocols can help companies expand sales in neighboring countries. One of the most ambitious but difficult, goals — the elimination of tariffs between neighboring countries — can foster economic growth and create jobs through trade. Even the most enlightened free trade policy, however, will be ineffective if there are no physical links among countries. For this reason, physical infrastructure is perhaps the most essential component in regional integration schemes — be it highways, energy grids or communication systems. Infrastructure is viewed as a threat by most environmental conservation organizations, but it is essential for economic growth and contributing to environmental impact studies or, better yet strategic environmental analysis, can identify consensus approaches for sustainable development.

Regional integration has both public and private components. In Cambodia, direct foreign investment originally focused almost exclusively on the textile and tourism industry, which created an export oriented industry targeted mainly at markets in Northern Asia, Europe and North America. More recently, investments in agriculture and real estate have increased, with capital flows that are more closely linked to regional markets and investors.* Still, it is difficult to track the extent of private sector cross border enterprise because it involves so many

* An exception to this trend was the early interest of Thai and Malaysian companies in Cambodia's timber sector, but that was largely an exploitive activity, not a long-term investment in the productive capacity of the country.

people and entities — from small businessmen and traders to global corporations — and because most business transactions in Southeast Asia are based on personal relationships. Nonetheless, some activities are coordinated and tracked though organizations such as the Cambodian Chamber of Commerce.[17]

In contrast, the activities of governments and multilateral agencies are easier to document, especially those activities financed by or coordinated by institutions dedicated to regional integration. These institutions sponsor multiple initiatives and programs, provide forums for cultural and academic exchange that involve civil society groups, and encourage regional commerce in private industry. Several of these initiatives for regional integration have particular importance for the Greater Cardamom Landscape.

The Greater Mekong Subregion is a program of the ADB that implements high priority projects in transportation, energy, and telecommunications. It also dedicates financial resources to initiatives for the environment, human resource development, tourism, and agriculture within the six-nation area of the Mekong River Basin.* Like all activities of the ADB, it is based on the "3C philosophy" — connectivity, competitiveness and community — and makes a special effort to promote investments by the private sector. The Association of Southeast Asian Nations (ASEAN) coordinates policy among member states† with the objective "to create a stable, prosperous and highly competitive region in which there is a free flow of goods, services, and investments," or in more simple terms, to create a common market. The United Nations Economic and Social Commission for Asia and the Pacific (UNESCAP) is one of five regional commissions that make regional cooperation the central component of its mission. One of its goals is to bring nations together to address issues collectively so that they can learn from one another and can resolve issues that are explicitly trans-boundary in nature.

The Mekong River Commission (MRC) is an intergovernmental body that coordinates actions and investments specifically related to the Mekong River and its tributaries among nations of the lower Mekong River Basin.‡ Originally conceived to promote a series of

* Cambodia, China, Laos, Thailand, Myanmar and Vietnam.

† Brunei Darusalam, Cambodia, Indonesia, Malaysia, Myanmar, Philippines, Laos, Singapore and Vietnam.

‡ Cambodia, Laos, Thailand, and Vietnam.

large-scale hydropower projects on the Mekong River, the MRC now largely compiles data and coordinates feasibility and environmental studies for the individual nations' numerous plans to develop the hydrological resources of the Mekong River. The centerpiece of that coordination is the Mekong Programme, with a descriptive subtitle that describes that challenge: The Regional Cooperation Programme for the Sustainable Development of Water and Related Resources in the Mekong Basin. The goal of the MRC is to achieve more effective use of water and related resources by promoting a concept known as Integrated Water Resources Management (IWRM), which purports to be a "balanced, peaceful, equitable and sustainable development process" that will mutually benefit all of the countries of the Mekong Basin. The challenges of the MRC and its adoption of the IWRM are not unlike those that define the Cardamom Conundrum, because they are attempting to reconcile the exploitation of a strategic natural resource, without endangering or damaging those aspects upon which nature or communities depend. In that capacity, the MRC seeks dialogue and consensus, so that the infrastructure investments in one part of the basin will not cause negative environmental and social impacts in other regions.

Non-Governmental Organizations

The need for technical capacity in a broad range of disciplines and the abundance of official development assistance has led to the proliferation of NGOs in Cambodia. The Cooperation Committee for Cambodia states that there are 450 NGOs operating in Cambodia, but its online database lists 950 separate entities in 17 different areas of action (Table 4.3). In fact, the exact number is not known, because even though NGOs in Cambodia are required to register with the Cambodian Foreign Ministry, there are many small organizations that are either unaware or which choose to avoid this requirement.

The list of NGOs includes large international organizations that raise financial resources from and operate in multiple nations as well as those that raise resources in a single country but operate in many regions across the globe; others are essentially bi-national in nature, obtaining funds in a single country for activities dedicated entirely to Cambodia. Many NGOs work with government agencies, but most choose to interact directly with the Cambodian people or, even more likely, to partner with a Cambodian NGO. Consequently, layered across the network of international NGOs is a parallel network of national NGOs. To attain a sustainable or effective level of action, international NGOs, or the individuals who staff them, will sometimes create a Cambodian NGO to fill a perceived need.

The staff of NGOs in Cambodia reflects the diversity of the countries that provide or have provided assistance to Cambodia; nonetheless, the vast majority of NGO staff are Cambodian nationals. Funding for NGOs comes from many sources, including the official development agencies, but many international groups bring private resources to Cambodia to pursue activities that have highly motivated constituencies, such as women's rights, indigenous groups, biodiversity conservation, and human rights. The role of private foundations from the United States is well known, as is the benevolence of religious groups from across the world. Many of these groups increasingly make investments to Cambodian NGOs and bypass the intermediation of international NGOs, although precise information about the extent of these activities is not available.

Within the Greater Cardamom Region, 268 different NGOs are listed as actively engaged in different aspects of development. Of these, 47 characterize themselves as dealing with environmental issues, although only about six can be characterized as "conservation organizations," whose primary objective is to preserve biodiversity and natural ecosystems. Three of these have active "on the ground" programs to help Cambodian authorities develop and implement conservation programs in the greater Cardamom region. The rest are dedicated largely to human development, although most of these have long realized that human wellbeing and environmental quality are intimately related. Meanwhile, environmental organizations have been promoting conservation as part of a larger political and development agenda. Consequently, conservation and development initiatives are becoming more and more integrated. Even the most narrowly focused conservation organizations now commonly engage in some type of sustainable development work and support projects that attempt to improve human livelihood in the communities that live near the biodiversity they are trying to protect. Similarly, almost all social NGOs realize that environmental conservation is essential to human well-being. This convergence of institutional agendas is part of the solution to the Cardamom Conundrum.

CHAPTER 5

The Inevitability of Change

Cambodia is changing. And this change is the result of a complex set of factors — from individuals acting locally to commodity markets influenced by policies in distant capitals. In many ways this change is positive and welcome: Cambodia experienced an annual growth rate of between 8 and 10 percent in the nine years prior to the economic downturn of 2009 (Figure 5.1). Cambodia is the most recent example of an Asian country to experience the phenomenon of rapid economic expansion, following Thailand, Malaysia and Indonesia, who have all seen their economies expand markedly over the past three decades, and who are often referred to as the "Southeast Asian Tigers.* The economic crisis of 2009 turned out to be a relatively mild market correction in much of Asia and growth has rebounded across the region. Much of the recent expansion has been linked to China, but astute economic management is also responsible for the success within Southeast Asia. Vietnam has followed China's example and liberalized its economy, which has led to high annual growth rates over the past decade. Cambodia is following the same model and it is often said in the Phnom Penh popular press that "Vietnam is the next China, and Cambodia is the next Vietnam," which reflects the country's optimism and the trend toward economic growth and the expansion of commerce across borders.

* This distinguishes them from Hong Kong, South Korea, Taiwan and Singapore, known as the "East Asian tigers."

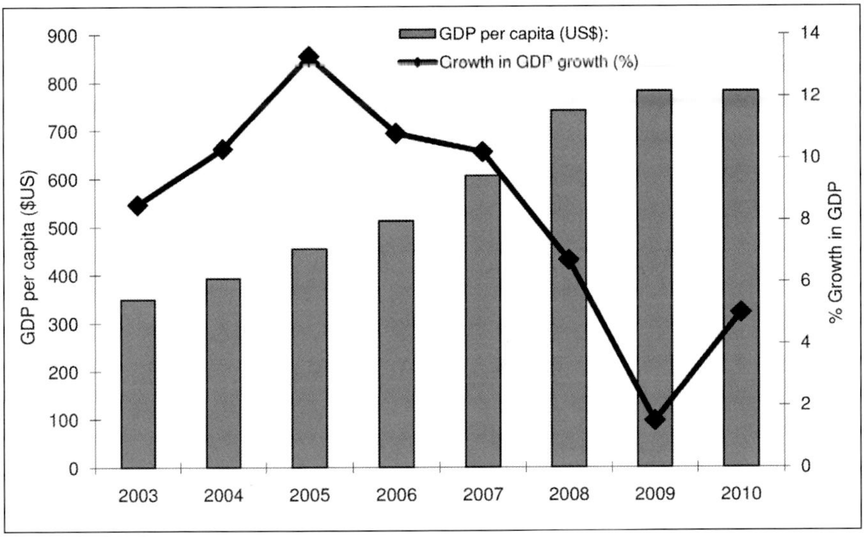

Figure 5.1. Cambodia's economy has seen an annual growth rate of between 6 and 12 percent between 2000–08, which dropped to near zero during the global recession of 2009 with modest rates of growth predicted for 2010 and 2011.

Cambodia's recent growth stems from several policies, including preferential trade agreements with the United States and the European Community, as well as the decision to join the World Trade Organization. Another feature that has fostered foreign investment is the fast-track investment approval process facilitated by the Council for the Development of Cambodia (CDC), a cabinet-level entity that approves investment projects by foreign companies and the Cambodia Special Economic Zone Board, a group that provides multiple services to investors who locate their factories in designated zones. Cambodia's openness led to a surge in direct foreign investment from South Korea ($2.36 billion), Malaysia ($1.98 billion), China ($1.58 billion) and other nations for a total of $12 billion between 1994 and 2006.[1] This pace of investment can be expected to continue over the next decade and, if reports in the popular and financial press are to be believed, will expand from a narrow emphasis on the garment industry and tourism, to focus on essential infrastructure, particularly transportation, communications and energy, which in turn will stimulate growth across the wider economy. The roster of investors will also become more diversified and include companies from Thailand, the United States, Australia, France and Russia. Reflecting the surge in investor confidence,

Standard and Poor's financial analysts granted Cambodia an investment-grade rating for government debt securities in 2007. Another sign of investor confidence is the interest of private capital funds and the decision by a Wall Street investment firm to create a fund dedicated to Cambodia, which has invested in telecommunications, hydropower and electricity transmission, banking and microfinance, the food and beverage industry, and real estate.[2]

The Cambodian economy can be roughly divided into three major sectors: agriculture, industry and services, with the last category including tourism. Each sector has contributed to past growth and will be a key component of future growth. Within the domestic economy, agriculture is by far the most important, because it provides employment to the vast majority of the population; in the future, agricultural production will expand due to increasing mechanization and technological improvement, in part due to a trend towards larger landholdings under corporate management. The exploitation of the country's natural resources will diversify from a reliance on renewable resources, such as timber, rice and fish, to include mining for industrial commodities and gold, and drilling for hydrocarbons believed to exist under the Gulf of Thailand. Cambodia's industrial sector will remain reliant on its strong textile sector, which has prospered due to special trade agreements and a "fair labor" marketing strategy that has allowed it to protect a market niche targeted at high end retailers, but these companies desperately need the country to complete its planned investments in infrastructure and energy in order to compete in a highly competitive global industry. Tourism, a fundamental pillar of the economy since the end of the civil war, is expected to continue its almost exponential growth over the next decade as it expands the number of destinations, such as the Cambodian Coast that is particularly convenient and attractive to winter season visitors from China.

The growth of these industries is essential for reducing poverty and will change the face of Cambodia, which is to be expected and welcome, but it will also bring impacts that are neither planned nor desirable. For example, poorly planned real estate development and resort projects can degrade the environmental assets upon which the tourist business ultimately depend or, at the very least, add value to the tourist experience. Highways and power facilities are essential for future growth but high priority investments require strategic planning so the impacts that accompany them do not degrade forests and wetlands that support the livelihoods of hundreds of thousands of rural families. Similarly, industries such as mining, forest extraction and shrimp farming can damage ecosystems, undermine the renewable natural resource base and further marginalize impoverished communities.

It is not necessary to slow down development to minimize or avoid these impacts; much to the contrary, if investments are accompanied by a coherent and agile planning process that identifies the full range of positive and negative impacts, economic progress will accelerate. Most negative impacts can be mitigated — if they have been identified — but this can only be guaranteed if the planning process includes the participation of all of Cambodia's stakeholders, the ones who are promoting change, as well as those who will feel the impact of that change.

Highway Infrastructure

Modern transportation infrastructure is a prerequisite for the economic growth of any country or region, but highways can also degrade natural landscapes and disrupt remote communities poorly equipped for rapid change, especially in previously remote areas such as the Cardamom Mountains and adjacent landscapes. The direct impacts of roads and highways are easy to identify — cleared land and agricultural fields that are physically adjacent to the roads themselves: the link between cause and effect is obvious and usually immediate. However, road improvements also bring secondary and cumulative impacts that occur further afield and over a longer period of time, usually because a secondary road network develops spontaneously and expands into adjacent landscapes. Environmental impact analyses (EIAs) seek to identify and manage this phenomenon and are standard requirements for all major transportation projects, but these typically are not conducted for secondary road networks, because local authorities are exempt from conducting an EIA, either by regulatory fiat or by the absence of an oversight body.

Consequently, the full impacts associated with highways are not foreseen and the land-use change they stimulate can infringe on land that was intended for — and is better suited for — other purposes. Deforestation is one of the biggest side-effects of road development when it occurs in pristine areas, but Cambodia's road network has not yet stimulated massive deforestation, at least in the more remote regions of the Greater Mountains. This is undoubtedly due to the past (and ongoing) impact of land-mines, which were put in place to protect road infrastructure during the long period of civil conflict. This practice continued into the 1990s in areas where Khmer Rouge holdouts remained active and impeded settlement along many roads, especially near the Thai border in Pailin and Battambang provinces. The removal of land mines has been a slow, but steady, process which originally focused on clearing land near schools and villages. As efforts to improve secondary and tertiary roads move forward, adjacent land will also be cleared of mines and the inevitable side effect of deforestation will occur.

Cambodia's most significant highway investment is part of a 1,500 km-long transportation corridor that links the Vietnam coast with the Andaman Sea in Myanmar (Figure 5.2a), which will be extended eventually into India and integrated into a larger continental system known as the Trans-Asian Highway. The southern leg of the Trans-Asian Highway is Cambodian Route 5 which was upgraded between 2000 and 2005 and connects Phnom Penh to the Thai border via Battambang. Route 5 transects a landscape that has been occupied by humans for millennia, so negative environmental impacts have been minimal, while the social and economic benefits are overwhelmingly positive, because it reduces travel time and lowers transportation costs, enabling inhabitants to transport their products to urban markets more easily.

The other major transportation investment in the Greater Cardamom Region is Route 48, which is a more typical of the potential positive and negative impacts associated with transportation corridors in tropical developing countries. Although it is essentially parallel to Route 5, it provides a more direct route between Phnom Penh and Bangkok, as well as a shorter route between Bangkok and Ho Chi Minh City (Figure 5.2b). This is a totally new highway throughout most of its extent and features four newly constructed bridges, including the 1,000-meter bridge that crosses the Prek Kaoh Pao estuary at Koh Kong.* Only 150 kilometers in length, Route 48 has ended the isolation of the coastal communities of Koh Kong province and the southern slope of the Cardamom Mountains, which are home to key hydropower facilities under construction. Route 48 passes through spectacular mountain scenery, opens new areas of the Cambodian Coast, and facilitates travel from Thailand for weekend tourists visiting Phnom Penh, thus bringing significant benefit to the tourist industry. It is a strategically important investment in every sense of the word.

* Financing for the project obtained from Royal Thai Government has provided a soft loan of around 800 million baht ($25 M) for road improvements and grant of around 288 million baht ($8 M) for the construction of the bridges.

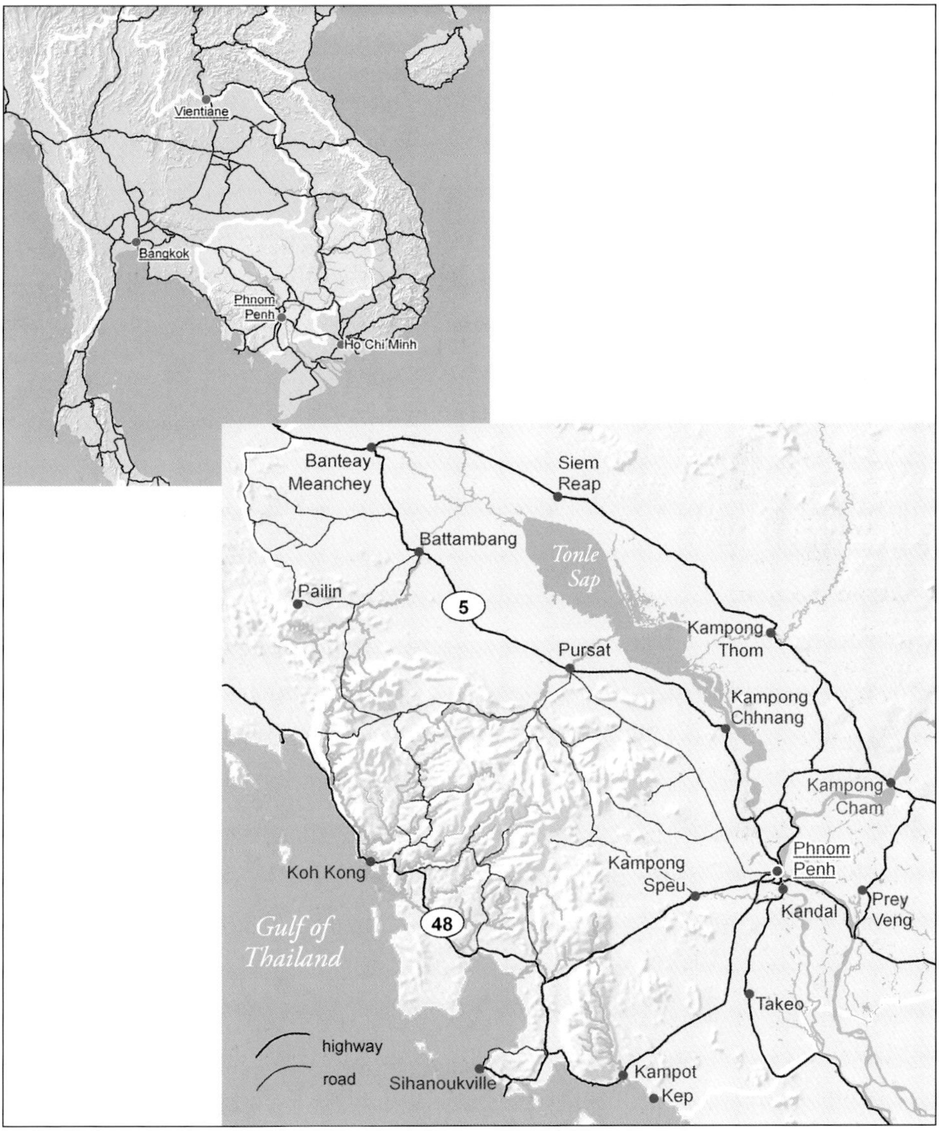

Figure 5.2. Southeast Asian nations are improving their transportation networks as part of joint strategy to further integrate their economies. Cambodian highways now connect two of the region's economic powerhouses: Bangkok and Ho Chi Minh City (top). Modern highways bracket the Cardamom Mountains, while secondary road networks are increasingly penetrating previously remote regions (bottom).

However, it also represents a threat to the natural ecosystems and human communities on the landscapes through which it passes, because it transects pristine forest landscapes, setting the stage for extensive deforestation. Fortunately, there has been surprisingly little settlement and deforestation along the route between Andung Tuek and Pream Krasop, which is situated on the border between Botum Sakor National Park (south of the highway) and the Southern Cardamoms Protected Forest (north of the highway), making it one of the few (perhaps only) examples of a major highway in a tropical country that has not led, almost immediately, to a land rush and subsequent deforestation. This is due largely to the efforts by the Cambodian government's Forest Administration and Department of Nature Conservation and Biodiversity (DNCP) of the Ministry of the Environment to control illegal activities in this region.*

Along with these two major highways, secondary road networks are expanding across the Greater Cardamom Region. Route 148, for example, connects the provincial capital of Pursat with the villages of Phnom Kravanh and Veal Veng and has brought increased settlement into what still remains a forest wilderness. Recently paved, this road was first developed as an oxcart track in the 1950s by the Sihanouk government and was extended to the Thai border during the 1980s. Following the end of hostilities in 1997, the roadbed was upgraded and a series of steel and plank bridges were built over the numerous small rivers and streams. With the improvement of Route 5, settlement and deforestation along Route 148 has increased steadily between Pursat and Veal Veng, as evidenced by satellite-based studies conducted by the Forest Administration in 2005. Small farmers have established plots along the road (Figure 5.3), while reduced travel time from Phnom Penh has fostered interest on the part of land speculators from the capital city. The eastern segment of the road runs through state forest land that was part of the old concessionaire system that ended in 1999; although still classified as a production forest, legal logging remains on hold, although the area continues to be lightly exploited for timber by local populations and illegal loggers. The western segment of the road passes through the Phnom Sankos Wildlife Sanctuary, which is administered by the Department of Nature Conservation and Biodiversity of the Ministry of the Environment. Settlement along the

* The FA manages both production forest and protected forests, while the DNCP manages national parks, wildlife reserves and multiple use protected areas.

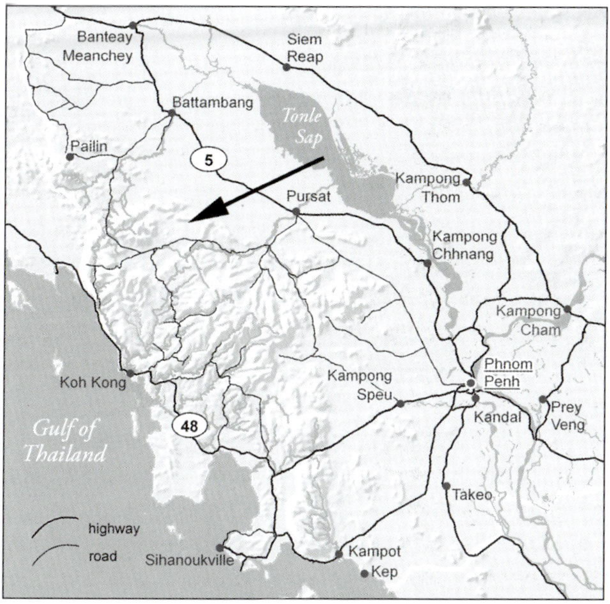

Figure 5.3. Land clearing by small farmers along Route 48 in the Phnom Sankos Wildlife Sanctuary is a typical and predictable impact of road construction in wilderness areas; the arrow shows the orientation of the view, the scale bar corresponds to 2 km and the Phnom Sankos massif is located in the upper left corner of the image, which is from 2007.

road must be approved by the government and most of the settlers claim to have been granted legal title to the land they currently occupy.

A similar process is underway on the agricultural plains situated to the north and east of the Cardamom Plateau and the Phnom Aural massive that overlooks the Tonle Sap Lake. Route 53 in Kampong Chhnang Province provides access to the northern flanks of the Phnom Aural Wildlife Sanctuary, where illegal logging and hunting are recurrent problems. To the south in the Kampong Speu province, Route 44 provides access to the flat arable soils of the southern Phnom Aural Wildlife Sanctuary and terminates just inside the Central Cardamom Protected Forest (Figure 5.2b). This region has ready access to Phnom Penh, the nation's largest urban market, which makes it particularly attractive for agricultural expansion. This area has been open to settlers since at least the late 1980s, as satellite imagery from the period shows, and has numerous well established hamlets associated with rice paddy fields along water courses. The plain has increasingly been occupied by larger farms, most of which have been cleared since 2005. The combination of small settlers and agro industrial investment make it highly unlikely that the integrity of the Phnom Aural Wildlife Sanctuary will be maintained, at least in this sector.

The most intensive recent deforestation in all of Cambodia is associated with a secondary road network in the northern part of the Greater Cardamom Region, which link several small border towns with the provincial capital of Battambang. Settlement began in the mid 1990s, but intensified in 2000 after the removal of land mines. By 2005, approximately 90 percent of the Roniem Daun Sam Wildlife Sanctuary had been deforested. A similar process is underway in the neighboring province of Pailin, where it has transformed previously forested landscapes surrounding the Samlaut Muliple-Use Protected Area (Figure 5.4).

The heart of the Central Cardamoms has been crisscrossed with a network of secondary routes that were established as access roads during the timber boom that ended in 1999. These roads provide limited but vital access to the remote communities at Ou Som and Thma Bang. Several of them are being upgraded as part of the government's investment plan for hydroelectric facilities on the Stung Russey Chrum and Stung Meetok rivers and, when that work is complete, the inland cities of Pursat and Battambang will be connected to port of Koh Kong (Figure 5.2). This connection will more completely integrate the remote regions of the Cardamom Mountains into the national economy and, simultaneously, create new opportunities for settlement. Whether these

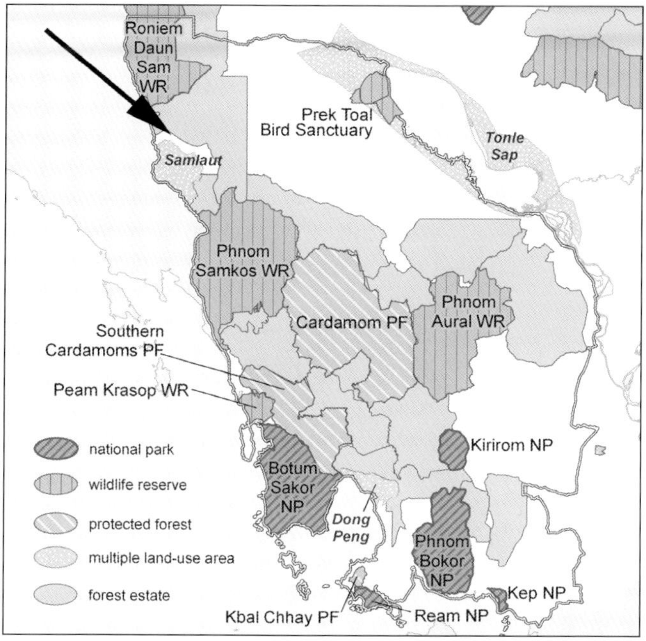

Figure 5.4. Deforestation along the Thai border near the provincial capital of Pailin is constrained by the mountainous topography, which has helped conserve the forest cover of the Samlaut Multiple Use Protected Area; the scale bar is 2 km and the satellite image is from 2006.

new roads lead to deforestation will depend on how the Cambodian state manages the environmental impacts of the hydroelectric facilities and their access roads.

Hydropower

Cambodia suffers from a severe deficit in electricity power, and the need for greater investment in generating capacity is an urgent national priority. Electricity consumption has been growing at about 12 percent per year and is estimated to reach about 1,200 MW in 2020 — approximately six times the installed capacity of the national system in 2009.[3] This deficit causes frequent power supply disruptions and has caused a proliferation of stand-alone generators, which are used by private industry and affluent individuals, as well as numerous official agencies and NGOs. Diesel-powered generators, both large and small, are inefficient and expensive, and expose Cambodia to the volatility of international petroleum markets. Moreover, the proliferation of small private generators is also fundamentally unfair, since this solution is unavailable to the overwhelming majority of the country's population.

To meet the growing need for electricity, the Ministry for Industry, Mines and Energy (MIME) has developed a strategic plan for a series of priority investments that began in 1999 and are scheduled for completion in 2016.[4] Although now outdated, this document describes the government's current policy: short-term investments in diesel-powered generation and a national transmission system that will be integrated into the Greater Mekong Subregion energy grid.* As part of a long-term drive towards energy independence, the plan prioritizes the construction of five hydropower facilities. The first of these was the renovation of an existing hydropower generation facility on the Kirirom Plateau, which was originally built in 1968 but ceased functioning in 1970; the plant resumed operations in 2002 with a modest capacity of 12 MV.[5] Now known as Kirirom I, this dam is located within the Kirirom National Park, which was created in 1993.

This first modest investment was followed in 2005 by the decision to construct a 145-meter-tall dam on the Kamchay River inside the Bokor National Park in the Elephant Mountains. This site was selected for early development because it could be easily integrated into the

* Subsequently, the government approved a coal-fired power plant in Sihanoukville to add security and flexibility to a strategy deemed to be over-reliant on hydropower.

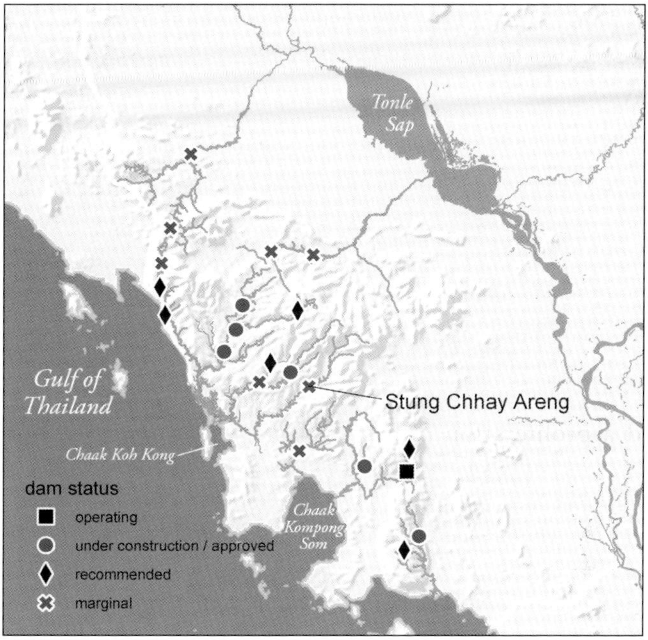

Figure 5.5. The Cambodian government has fast-tracked six hydropower projects for development in the Cardamom Mountains and has identified 15 additional sites in the Cardamom, Elephant and Kirirom Mountains. These facilities will resolve one of the most significant impediments to economic growth in Cambodia, the high cost of electric power, but will modify every river basin of the Cambodian coast and impact numerous freshwater and estuarine species.

existing distribution system and supply Phnom Penh, where demand is highest. Also, road access in the area was better than in other regions, such as the Cardamom Mountains, which have greater hydropower potential, but more rudimentary highway infrastructure, making construction more difficult. The Kamchay facility is scheduled for completion in 2010 at an estimated cost of $300 million and will have a capacity of 127 MV. Subsequently, three other sites were approved for hydropower development: two on the Central Cardamom Plateau (Lower Stung Russey Chrum and Atay rivers) and another dam in Kirirom National Park (Kirirom III). In June 2008, the Cambodian government announced two additional projects — Middle Stung Russey Chrum and the Stung Chhay Areng and the — both on the edge of the Central Cardamom Protected Forest and located just downstream from the communities of Ou Som and Thma Bang, respectively (Figure 5.5).

These sites were most likely chosen for fast-track development because of their geographic location in Cambodia's rainiest region with significant hydropower potential; additional considerations include their proximity to each other and to potential urban markets, at least when compared to other potential sites in the more remote highlands of northeastern Cambodia. The companies will be able to coordinate the construction and operation of these sites using a single road network and, more importantly, integrate the generating capacity into the national power grid via a single transmission line across the Cardamom Mountains to Pursat, where it will connect with the soon-to-be-completed Battambang–Phnom Penh 350 kV transmission line.

In addition to these priority projects, the government has identified 63 additional sites — 21 located in the Cardamom and Elephant mountains — for hydropower facilities of differing scales (Figure 5.5). The total potential of these sites would surpass 8,000 MW and allow operators to export energy to adjacent markets via the regional Southeast Asian power grid. More than 50 percent of this total would come from two large proposed projects on the main stem of the Mekong at Sambor and Stung Treng, both of which are highly controversial due to their impact on the hydrology and sedimentation patterns of the Tonle Sap and the Mekong Delta.[6] To evaluate the costs and benefits of these proposed hydropower facilities, the national energy office in MIME and the Japanese International Assistance program (JICA) contracted a study which evaluated the dams according to five different criteria: economic viability, power generation potential, technical feasibility, environmental impact and social impact (Table 5.1). The study excluded the five sites already operating (Kirirom I), under construction (Kamchay) or with approved contracts (Kirirom III, Lower Stung Russey and Stung Atay) but included the other planned and approved projects. Ultimately, the study identified ten sites as priority investments; of these, six were located on the south slope of the Cardamom Mountains. The Middle Stung Russey Chrum project was ranked as the most attractive, due to its high potential for economic return and favorable technical characteristics, as well as its acceptable scores for social and environmental impacts. In contrast, Stung Tatay did not make the top-ten list because of a low score for economic feasibility, although it did have adequate scores in other categories.[7] In summary, a total of 11 sites were selected for fast-track development, with a total of 710 MW of installed capacity producing about 4,000 GW-h of energy. This is expected to account for 60 percent of the country's power needs in 2020 (Table 5.1).

Table 5.1. The hydropower facilities in the Greater Cardamom Region: proposed or under construction.

Hydropower Facility	*Province*	*Watershed*	*Capacity (MW)*	*Annual Production (GW-h)*	*JICA Ranking*	*Year Authorized*	*Estimated Completion*
Stung Battambang I	Battambang	Tonle Sap	24	120	LEP		
Stung Battambang II	Battambang	Tonle Sap	36	187	LEP		
Bokor Plateau	Bokor	Kamchay	22	147	4		
Kamchay	Bokor	Kamchay	127	558	E	2005	2010
Kirirom I	Koh Kong	Ocham	12	63	E	1970	2002
Kirirom II	Koh Kong	Ocham	1	6.2			
Kirirom III	Koh Kong	Ocham	13	70	E	2005	
Lower Stung Russey Chrum	Koh Kong	Stung Russey Chrum	125	668	E	2005	2015
Middle Stung Russey Chrum	Koh Kong	Stung Russey Chrum	56	668	1	2008	
Upper Stung Russey Chrum	Koh Kong	Stung Russey Chrum	38	211	9		
Stung Chhay Areng	Koh Kong	Stung Chhay Areng	109	1350	11		
Stung Kep I	Koh Kong	Stung Tatay Koh	4	27			
Stung Kep II	Koh Kong	Stung Tatay Koh	114	169	8		
Stung Meetok III	Koh Kong	Stung Meetok	23	350	10		
Stung Meetok I	Koh Kong	Stung Meetok	50	350			
Stung Meetok II	Koh Kong	Stung Meetok	29	384	7		
Stung Piphot	Koh Kong	Stung Piphot	25	140			
Stung Tatay	Koh Kong	Stung Tatay	80	410	LEP	2008	2015
Stung Atay	Pursat	Stung Russey Chrum	110	588	E	2005	2010
Stung Pursat I	Pursat	Tonle Sap	75	443	LPD		
Stung Pursat II	Pursat	Tonle Sap	17	71	E		
Operating			12	63	1 percent	The Percentage of total energy supply from the complex	
Approved or Under Construction			443	2,615	41 percent		
Recommended			225	1,261	21 percent		
Marginal			237	1,264	22 percent		
Controversial			109	1,350	10 percent		
Total			1090	3,939			

Notes: E: excluded from study, LEP: low economic potential, LPD: low power density.

All of Cambodia's hydropower facilities are being constructed under a build, operate and transfer (BOT) plan, which means they will be operated as private enterprises by companies that undertake the construction and assume the financial risk. After thirty years, the facilities will be transferred to the Cambodian national electrical authority. This private sector model greatly facilitates Cambodia's ability to move rapidly despite its current technical and credit limitations. The contracting companies are all state-owned entities from the People's Republic of China with experience developing hydropower plants; the total reported investment is estimated at more than $2 billion.*

In addition to the large- and medium-scale hydropower projects, the Cambodian government has also identified opportunities for micro-hydro projects, where *micro* is defined as a facility that produces less than 100 kW.[8] Although large- and medium-scale hydroelectric installations generate most of the world's hydroelectricity and are particularly well suited to resolving Cambodia's energy deficit, small- and micro-scale hydropower facilities can make important contributions to energy security, while avoiding many of the social and environmental impacts associated with larger projects. The opportunities for micro-hydropower projects include numerous sites in the Cardamom region, mostly on the periphery of the Cardamom Mountains close to rural communities that currently have no power and are isolated from the national energy grid. China leads the world in small and micro-hydro capacity and could provide the technical assistance as part of their assistance program to Cambodia.

Many environmentalists have a visceral dislike of large-scale hydropower projects and most, if not all, of these projects have been vigorously opposed by advocacy groups.† Their opposition is based on decades of experience witnessing the negative consequences of dams that become manifest after their construction — particularly the loss

* The names of the companies cited in the Phnom Penh press are: Synohydro (Kamchay), China National Heavy Machinery Corporation (Stung Tatay), China Datang Corporation & China Yunnan Corporation for International Techno-Economic Cooperation (Stung Atay), Michelle Corporation (Middle Stung Russey Chrum), Yunnan Southeast Asia Economy and Technology Investment Industrial Co Ltd (Transmission line), China Southern Power Grid (Stung Chhay Areng), this last in spite of not being approved.

† International River Network <http://www.internationalrivers.org/>, Probe International <http://www.probeinternational.org/>, NGO Forum of Cambodia <http://www.ngoforum.org.kh/eng/core/> [accessed 15 July 2011].

of native forests and the collapse of fish populations. These concerns are true for the rivers that will be changed by the Cardamom hydro facilities, which are known for their endemic and endangered fauna, particularly the Cardamom water turtle, the Siamese crocodile and the dragon fish.[9] Unfortunately, and like many places in the world, the environmental impact assessments for these projects have been commissioned and executed by the same company that will construct and operate the facility, leaving ample reason to doubt the objectivity of those studies.

In the Cardamom region, dams raise other concerns related to watershed protection: hydropower stations are managed to maximize energy generation over 12 months of the year. Water flows are thus increased during the dry season and decreased during the wet season. As water flows are altered, the reservoir behind the dam becomes a sediment trap, changing the chemistry of the aquatic habitats below the dam; if there is significant erosion above the dam, the reservoir will eventually fill up with silt, and it will not be able to store enough water to make operating the turbines economically viable. In some cases, the reservoir will become a large shallow lake and wetland. If and when this occurs largely depends on how much erosion occurs upstream from the dam. The degree of erosion depends on how, and how intensely, the land is used and on soil management practices. Forested landscapes protect from erosion, while agriculture leads to increased sedimentation and a decreased lifespan for the hydropower facility (see discussion in Chapter 6). Since most of Cambodia's proposed dams are located on forested landscapes on the edge or within protected areas, sediment flows should remain low, and the productive life of the hydropower facilities should span many decades — *if* those protected areas are maintained.

While some argue for the environmental benefits of hydropower because it reduces the emissions of greenhouse gases caused by burning fossil fuels, this argument also has another side. When reservoirs are initially flooded and forest landscapes inundated, carbon stored in the biomass of trees is released as a pulse of CO_2 as plants die and rot, or are cleared and burned prior to flooding. The remaining biomass then settles to the bottom of the new reservoir and begins to decompose in anoxic conditions, producing methane (CH_4), a potent greenhouse gas (GHG) with approximately 20 times the warming effect of CO_2. Consequently, the alleged climate benefits from hydropower may take decades to materialize, as the savings in GHG from fossil fuels is negatively offset by emissions of CO_2 and CH_4. According to the World

Commission on Dams, hydropower projects created on forest landscapes must have a power density ratio >100 watts/m^2 of surface area of the reservoir, or else the GHG emissions from the project actually will be higher than those of a conventional thermal generation plant.[10] Although the Master Plan developed by hydrological engineers on behalf of the Japanese and Cambodian governments did not calculate the net balance of GHG emissions for the priority hydropower projects, it did calculate the energy density based on the above rule (Table 5.1). All of the projects in the Cardamoms met that benchmark with the exception of Stung Chhay Areng.[11] The impact of methane emissions can be avoided by harvesting the timber and clear cutting any forest that might exist prior to filling the reservoir.

Finally, the impacts of large hydropower plants on human populations are also a concern raised by opponents to large hydropower projects, particularly when the area is home to traditional communities with a strong cultural attachment to the landscape. Communities are typically resettled or paid monetary compensation, but they are often relocated to areas that lack the same natural resources as their original lands, and monetary reparation seldom adequately replaces the loss of a way-of-life. Fortunately, most of the dams in the Cardamoms are located in areas with low population densities, and only a few dozen households would need to be resettled; however, the Stung Chhay Areng project would impact one of the last intact communities of the *Khmer Daeam*, who were forced to leave the Areng Valley during the civil war and who have only recently returned to their ancestral village. Current plans call for the impacted families to be relocated to state lands near the proposed dam site.

Land Tenure

More than 85 percent of Cambodia's population lives in the countryside, and as many as 45 percent of these rural inhabitants live below the poverty line.[12] These residents depend overwhelmingly on agriculture — either as subsistence farmers, as small landholders or as contract labor for more affluent farmers. As in most societies, land ownership is skewed with 20–30 percent of landowners holding 70 percent of farmland, and the poorest 40 percent owning only 10 percent of farmland.[13] Poverty afflicts the 25 percent of rural families who hold properties under 0.5 hectares, which is the minimum amount of land needed to produce sufficient food for survival; worse off still are the approximately 20 percent of rural families who own no land. Women make up more than half of the agricultural workforce, and poverty is significantly higher for female-headed households. This group is most vulnerable to having their land rights ignored, in part because they lack knowledge about land rights and land title procedures. Unlike many countries characterized by rural poverty, the distribution of land in Cambodia is not the result of centuries of land tenure precedent. Rather, the distribution of land occurred after the end of the civil war in 1991 and the dissolution of the socialist state, when land was allocated among members of Cambodian society according to a poorly documented process based on prior occupancy, prestige, wealth, gender and the use of force.

The rural population of Cambodia is concentrated largely on the flood plain of the Mekong River and the alluvial plains surrounding the Tonle Sap where rice is the overwhelming crop of choice. The total cultivated area in Cambodia is approximately 4.2 million ha, of which about 70 percent is dedicated to rice cultivation. The primary goal of every household is to grow sufficient rice to feed the family for the upcoming year. Rice following rice is the principal mode of cultivation, and rudimentary farming methods provide low yields. The risk of drought is relatively low, but it still affects production at decadal scale regularity (see Chapter 2). Cambodia has been self-sufficient in rice production and has been exporting a modest surplus since 2000; improved methods have increased yields from 1.35 t/ha in 1990 to 2.03 t/ha in 2004.[14] In addition to rice, almost all lowland farmers also have a vegetable garden; sugar palm is a conspicuous component of paddy landscapes where it grows on small mounds and on dikes separating paddies. The intensity of rice cultivation varies, and areas around the edges of the plain are often dotted with small clumps of woody vegetation.

The remainder of Cambodia's agriculture is dedicated to upland crops, which are inherently more diverse than paddy rice. Cultivation on upland landscapes is notoriously less intensive than on the flood plain, due largely to the well-known limitations of tropical soils that compel low technology farmers to adopt slash-and-burn farming methods. Upland small farmers cultivate a combination of other crops including cassava, corn, soy, peanut, bean, squash, sweet potato, sugar cane, banana, mango, and other vegetables.[15] Plantation crops, mainly rubber, have increased by almost 50 percent to more than 180,000 hectares; older plantations are senile remnants from the colonial past

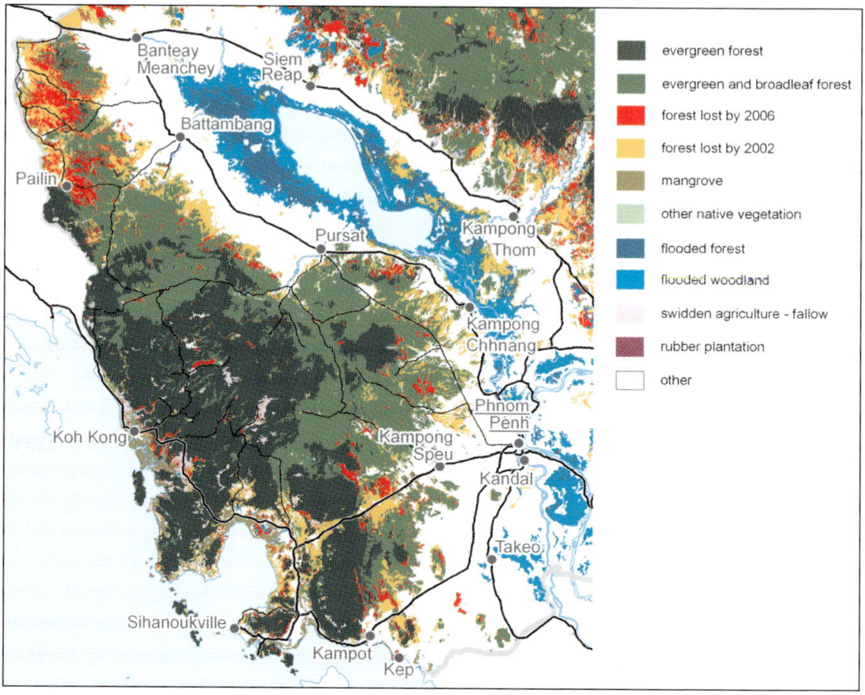

Figure 5.6. Land-use change in the Greater Cardamom Region between the mid-1990s and 2006 has largely been limited to peripheral areas on the northern and eastern slopes of the mountains that are characterized by relatively low levels of biomass and along the coastal plain and foothills where humid and mixed forests are more common. There was a surge in deforestation on the Thai border following the clearing of land mines that allowed farmers to expand production.

and are being renovated in response to global demand for rubber. Most of the recent expansion is occurring in the eastern part of the country and is financed by Vietnamese investors in industrial plantations. However, rubber is a relatively lucrative cash crop and can be stored on-farm in its crude storage, which makes it attractive to smallholder and more than 50 percent of the nation's rubber production originates from family plots.[16] Although the amount of land dedicated to rice cultivation probably has not changed appreciably over the past decade, upland areas particular those on the periphery of the alluvial plains in Pursat, Kampong Chhnang, and Kampong Speu province have seen considerable change as more a more land is converted for agriculture. Deforestation is greatest along the Thai border in Battambang and Pailin provinces, having accelerated since 2002 (Figure 5.6).

Cambodia's forest and woodland cover was recently estimated at 10.7 million hectares;[17] of this, approximately 50 percent is located on topographically flat or gently rolling upland landscapes suitable for crop or plantation agriculture — making Cambodia one of the few places on earth with appreciable amounts of unexploited arable land. Approximately 40 percent of the forest cover is located inside protected areas, while an additional 35 percent was granted as commercial forest concessions in the 1990s. These halted operations in 1999 and the Cambodian government has yet to formally implement a comprehensive policy determining land use for the vast area of public forest located outside the protected area system, except for a limited effort to promote community-based forest management. Essentially, all of this forest land is up for grabs and is at the heart of the debate about land-use and the distribution of state lands.

Considering the pervasiveness of rural poverty and the relative abundance of arable soils, it is hardly surprising that deforestation is occurring across Cambodia and is particularly abundant around the margins of the Greater Cardamom Region. The rural poor have been migrating to areas where they can obtain land or where they can find employment opportunities on newly created agricultural estates. Under Cambodian law, settlers can acquire title for up to five hectares of land after five years of peaceful occupation.[18] Occupation means clearing the land and converting it to agriculture; failure to do so is equivalent to ceding the land to another individual. For many small farmers, however, land clearing is also equivalent to subsistence and survival. Like every agricultural frontier in the world, past and present, migration and settlement is a chaotic process dominated by opportunistic and enterprising individuals acting in their own self-interest.

Not surprisingly, this scramble for land perpetuates and accentuates asymmetry in the existing land-tenure system, as Cambodia's weak institutional structure inevitably provides affluent and politically connected individuals with inherent advantages. "Land grabbing" as it is termed by the local media, takes several forms; the most common is patronage, where local politicians and military commanders use their influence to assist relatives, employees, loyal soldiers and party stalwarts obtain a small plot of land. Many, if not most, of these individuals are small farmers and grants of 2-hectare plots are a common form of political patronage.

The squatter dynamic is now also driven by land speculators, who are largely urban investors responding to several separate, but linked, global market forces. For example, the worldwide boom in agricultural commodities is fueled by increased consumption from emerging nations and the growing demand for biofuels. This has contributed to an increase in the price of land values, which has sky-rocketed over the past few years, as the increase in commodity values drives producers to put more land under cultivation. Demand therefore increases for marginal lands that could not be profitably farmed when prices were low and stimulates the expansion of the agricultural frontier into forest landscapes. Urban investors seek land to invest in agriculture; many own small farms as a form of savings and investment and sell their production into the cash economy. More successful farmers acquire land from the less successful, or less lucky. This natural market driven process displaces small farmers who migrate to remote locations and acquire another plot of land. Some settlers are also speculators in their own right and own multiple plots, preparing one for sale while setting up another as the family homestead (Figure 5.7).

In addition to this small land-holding dynamic, Cambodia is also fostering a nascent attempt to create an agro-industrial sector based on larger properties, where the economies of scale facilitate the cultivation of cash crops, such as rubber, cassava and oil palm, which require a relatively large capital investment in processing facilities. Toward this goal, the Cambodian legislature enacted a land law in 2001 and the government formulated subsequent regulations in 2005, which established procedures for granting Economic Land Concessions (ELC). A maximum of 10,000 hectares is allowed per concessionaire, where control over the surface rights is ceded to a corporate entity for a period of 99 years.[19] According to the regulations, public lands must be registered under a special category entitled "eligible for privatization" and conform to provincial or municipal land-use plans. The proposed development must complete an environmental impact assessment and no involuntary resettlements are allowed; furthermore, concessions must be validated by a public consultation that involves both territorial authorities and local residents. Finally, investors must present a coherent investment plan with a detailed time table in order to spur real development and avoid land speculation.

Following the formalization of the reform process, a comprehensive nation-wide review revealed that 60 companies had been granted economic land concessions covering a total of 987,769 hectares in 14 provinces. None of these concessions had been approved via the rigorous procedures outlined by the 2001 land law. Observations as to the

Figure 5.7. Most of Cambodia's agricultural production comes from the densely populated flood plains of the Tonle Sap and Mekong River (top), but increased clearing by smallholders is occurring on upland landscapes of the agricultural frontier (middle), and on economic land concessions granted to corporate investors, such as the industrial rice farms in Kampong Thom Province (bottom).

size of these land holdings and the lack of transparency motivated the government to conduct a comprehensive review of all existing concessions. Subsequently, the MAFF cancelled 43 concessions covering 373,545 hectares, largely because the owners had not met the investment conditions and timelines explicitly detailed in the original agreements. Nine concessions were found to be larger than 10,000 hectares and were required to undergo a renegotiation with the objective of reducing each concession to the allowed maximum area. In 2007 and 2008, the government and five firms agreed to reduce their concession areas to the 10,000 hectare maximum, while two concessions were annulled in their entirety; apparently, the status of the two remaining concession remains under negotiation.[20] Presently, there are 95 valid ELCs in the country, which cover a total of 1,012,010 hectares (~7 percent of the total land surface of Cambodia) in 17 provinces, involving investors from France, Singapore, India, China, Vietnam, Korea and the United Sates, as well as Cambodian nationals.[21] Provincial authorities also are empowered to grant economic land concessions if the area under consideration is less than 1,000 hectares; this process has led to the establishment of an additional 40 land holdings totaling 41,171 hectares.[22] Almost all of these mid-size concessions are made to urban investors and are frequently associated with allegations of the abuse of the rights of small farmers who are evicted from the land being leased to corporate entities.[23]

The urban economy in Cambodia contributes in other ways to the expansion of the agricultural frontier. Most obvious is the increased consumption by the growing middle class, which must be met by increased production or imports. This type of increased demand is overwhelmingly positive for almost all landholders in the rural economy. A more important, and subtle, link between the urban and rural economies is the super-heated urban real estate market that has allowed investors to diversify their portfolios by adding rural properties. Along with using profits from the sale of urban real estate to purchase farms, it is now common for urban investors to use equity from their urban holdings to borrow money and acquire rural property. This practice puts rural property owners at a competitive disadvantage, since banks will not accept rural properties as loan guarantees, because the land registry is essentially nonexistent and riddled with fraud. This situation reveals both the lucrative potential of rural investments, as well as the inherent risk of investing in properties with uncertain land tenure.

The financial dependence of rural investment on the urban real estate market creates an inherent injustice. Rural residents are effectively denied access to capital, because their real estate holdings — which are often substantial — are not subject to credit. Meanwhile, urban investors expand their land holdings via increased rural production. Providing rural families with access to credit was part of the impetus behind an ambitious program Cambodia initiated to create a national land (cadastral) registry within the Ministry of Land Management, Urban Planning and Construction in 2004.[24] This program hoped to register approximately 1 million land parcels by 2008 (out of an estimated 7 million total) and was to be part of an overall strategy to reduce poverty and promote sustainable development. The logic was that a transparent land registry would enhance equitable access to land, thus increasing economic opportunities for the poor. Some analysts question whether a land-titling program can reduce poverty, because risk-adverse peasant farmers would probably not seek credit, even if it were available. Nonetheless, a more transparent and efficient land market would help develop the commercial agricultural sector and strengthen the ability of local governments to formulate and levy a land tax.

Forest Exploitation

Cambodia has been generously endowed with forest resources that covered more than 60 percent of the country at the end of the twentieth century.[25] The humid rain forests of the Central Cardamoms Plateau and Phnom Sankos massive are the most productive forest ecosystems in Cambodia, because high precipitation levels translate into high growth rates and, consequently, greater timber volumes.[26] Geographical and historical factors have impeded the intensive logging of the greater part of these forests, which are relatively intact and rich in biodiversity and timber resources. Population densities in the forested areas have stayed low, as the cultural emphasis on rice cultivation and access to abundant fisheries has channeled settlement to the flood plain of the Mekong River and Tonle Sap. During the colonial period, the ethnic minorities who occupied mountainous regions relied on the forests for food and fiber, while harvesting resins and other non-timber products as the basis of regional trade.[27] Some logging occurred during the late colonial and post-colonial period, which followed the French tradition of scientific management and the economic necessity of generating revenues from its overseas colonies;[28] however, the commercial exploitation and exportation of timber was slow to start in the Greater Cardamom Region due to its relative isolation, which was exacerbated by civil conflict. Southeast Asia's major timber companies instead focused on exploiting the now-depleted forests in their home countries of Thailand and Malaysia.[29,30] The Khmer Rouge played an important role, first by hindering the exploitation of Cambodia's forest resources when they occupied the forest during their pre-government stage in the early 1970s and then by obsessively focusing development on communal agriculture during their reign of terror. However, after they retreated into the forest following the Vietnamese invasion, logging became an important source of revenue for their forces on the Thai border, even though the rudimentary infrastructure restricted their trade to a few high-value species, which left the forest intact as a functional ecosystem.

During the long period of civil strife in the 1980s, timber traffic was a major feature of Cambodian politics, involving a web of Thai companies, military officers and government authorities, as well as different factions of the Cambodian resistance who competed for control of timber resources.[31] The role of factional politics in the logging sector became even more pronounced following the Paris peace accords in 1991 with the liberalization of Cambodia's economy and the arrival of Malaysian logging companies, who entered into a head-to-head competition with Thai firms. In 1993, large areas were set aside as protected areas by Royal Decree; however, these were largely paper parks and illegal logging continued unabated in most areas. By the end of the decade, essentially all forest landscapes outside of protected areas had been granted to concessionaires, a total of about 6.5 million hectares.[32] Forest revenues were an important source of funds for armed groups and their political allies who were competing for territory and influence. The Cambodian Army was closely involved, and most of the international logging companies formed alliances with military leaders in what was still a dangerous landscape. Military personnel were used to guard the concessions, and some officers allegedly organized their soldiers into logging crews under contract with the companies. Elements of the Khmer Rouge that remained outside the peace process continued to exploit and sell timber across the border into Thailand.

The international assistance community originally viewed Cambodia's logging industry as a legitimate way to monetize a capital asset, create a domestic economy and generate the financial resources needed to reconstruct the Cambodian state. There are no reliable statistics because the forest sector was chaotic and harvest volumes were deliberately underreported, but an estimated 10 million cubic meters of timber were exported between 1990 and 2000 with an approximate value of US$2.4 billion. Only about US$120 million was paid in

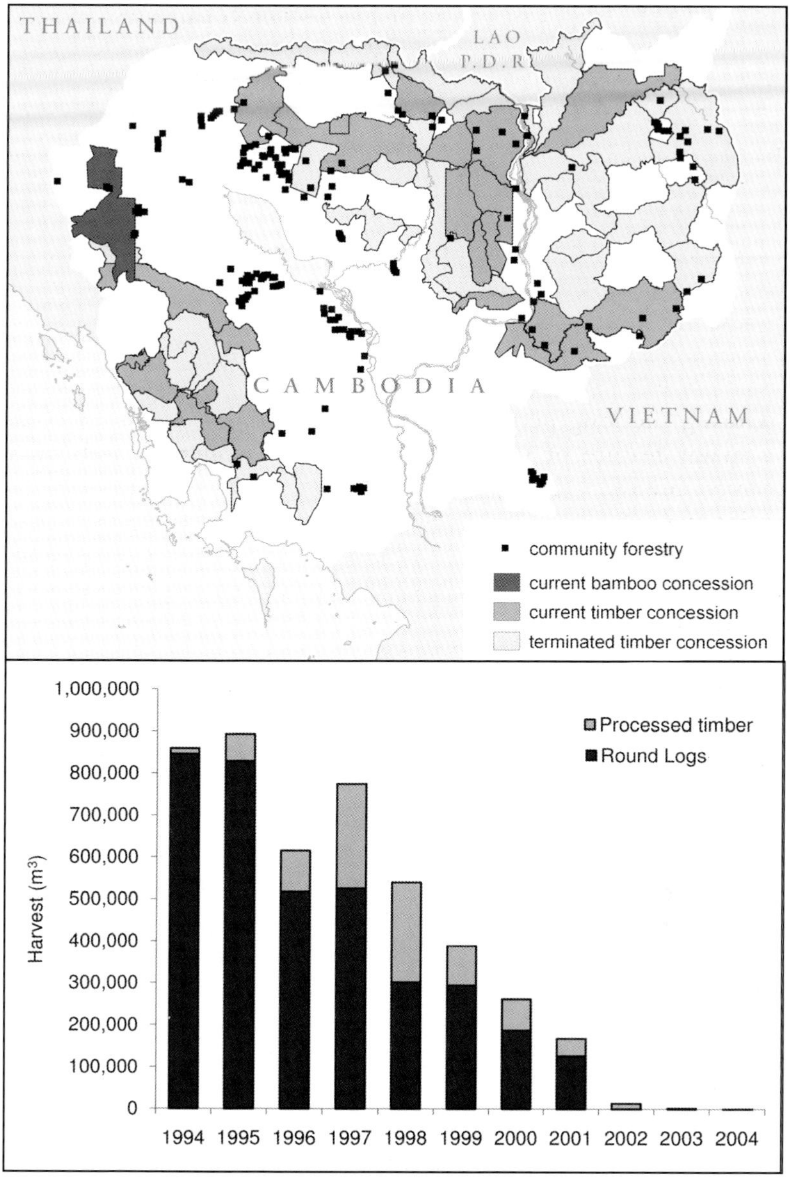

Figure 5.8. In 2001 the Cambodian government declared a moratorium on logging and timber exports and terminated most commercial forest concessions, while creating hundreds of community forests (top). This has essentially halted the legal export of timber, which were declining due to over harvest (bottom). Future forest management will be managed by the Forest Administration according to a quota system, in collaboration with local communities and regional governments.

royalties and the international donor community, which was providing the overwhelming portion of the government's revenues, eventually called for reform. The export of logs was banned, but the domestic processing of timber continued, stimulating greater illegal logging; companies purchased timber from informal operators to achieve economies of scale and improve profitability (Figure 5.7).

The reform of the forest sector became a priority for the new government elected in 1998, which took steps to improve both regulation and forest governance; a "crackdown" on illegal logging led to the closing of about 1,900 illegal sawmills and wood-processing plants over a two-year period. In 2001, the government went further and suspended the licenses of 17 companies with rights to 3.5 million hectares, as well as halting operations in 12 logging concessions covering 3.37 million hectares (Figure 5.8). The Forest Law of 2000 and subsequent regulation issued in 2002 outlined criteria for sustainable forest management, and remaining concessionaires prepared new management plans in the hopes of renewing activities.[33] Nonetheless, all concessions remain suspended and the formal timber sector in Cambodia has been greatly reduced, essentially consisting of harvesting just enough wood for the domestic market. Illegal logging has also been greatly reduced and is now largely a clandestine activity dedicated to exporting high value hardwoods to Vietnam (Figure 5.9). It continues within wildlife sanctuaries and protected areas, particularly the dry and semi-humid forests of the Phnom Aural Wildlife Sanctuary, but also in the production forests surrounding these core reserves. It is done by organized teams with mobile sawmills manned by military personnel, as well as by individual operators that sneak planks and beams out of the forest in passenger cars. Illegal logging is actively combated by the Forest Administration (MAFF) and the Department for Nature Conservation and Protection (MOE) in collaboration with international conservation organizations.

Cambodia's new forest regulations put forward three categories of forest: protection forests, production forests and conversion forests. There are four protection forests reserves that were established between 2001 and 2005, while the delineation between conversion and production forests is vague and, apparently, open to interpretation. Conversion forests are part of the forest estate, but as the name explicitly states are destined to be converted to agriculture or forest plantations. They include all land allocated via certain legal processes, such as the granting of economic concessions (land grants to businessmen), social concessions (land grants to communities), land-use change for infrastructure

Figure 5.9. Forest rangers, park guards, military police and members of local communities conduct joint patrols (top left) and have confiscated dozens of rifles and chainsaws (center). Illegal loggers target high value hardwoods species in their attempts to smuggle rough cut beams (top right) in small cars to Vietnam; dozens of cars have also been seized, providing additional incentive to respect logging regulations (bottom).

development (highways and reservoirs), and mining concessions, as well as the *de facto* processes of encroachment by small farmers. Production forests are "everything else" and include most former timber concessions; according to existing proposals, these lands will be managed directly by the central government via a contractual system. The sustainable forest management (SFM) paradigm that has been embraced by the FA will be based on a quota system fixed for geographic units (known as cantons). The proposed system will change the nature of public-private partnership in the logging sector, as the state will assume management responsibility (inventory, access, licensing, etc.), while private companies will act as contractors with annual harvest quotas distributed via an annual bidding process and fixed by the state. An ongoing review process may modify this approach and there are proposals under consideration to decentralize some of the government's management responsibility via "partnerships" with a local institution, either a commune or provincial government.[34]

Cambodia is also experimenting with another approach to forest management: community forestry, which seeks to improve forest conservation and management by providing local communities with a legal right to forest resources and empowering them to use these resources sustainably (see Figure 5.7). This approach assumes that local people will protect forests from illicit exploitation and manage them responsibly if they are given tenure over their forest lands. Community forestry has been promoted by NGOs and reflects their priorities. Some projects focus on ethnic communities whose rights are explicitly recognized by the Forest Law;[35] however, most community forests seek to redress the perceived unfairness of the concessionaire system, where large tracts of land were given to foreign firms or urban investors. More than 275 community forest projects covering about 220,000 hectares were established between 2002 and 2004,[36] along with another 400 communities who received rights to 80,000 hectares prior to 2002.[37] This total area is not great, however, representing less than 5 percent of all production forests and most of the community "forests" are actually located on landscapes dominated by agriculture. The concept and its spatial extent must be expanded dramatically to encompass areas on the forest frontier, if community forests are to successfully foster forest conservation. Community forests must also overcome many hurdles to achieve sustainability: the challenge of local governance, the volatile nature of land tenure in the Cambodian countryside, as well as the unfortunate state of the degraded forests that are bequeathed to communities.

Finally, although logging has dominated forest management discussions in Cambodia, the nation's most important forest product in terms of the national economy is fuel wood (Figure 5.9). Cambodia relies on fuel wood in its various forms for more than 84 percent of its total energy consumption.[38] Almost 96 percent of all households use wood or charcoal for cooking, while some industries depend on fuel wood for processes such as brick making and drying grain. Demand for fuel wood is strongest in rural areas, where supplies tend to be local; however, urban consume vast amounts of charcoal produced, which is often produced as a byproduct of land clearing in areas undergoing settlement. For example, the plains south of the Phnom Aural massif, within the confines of the wildlife sanctuary are a major source of charcoal for Phnom Penh, as well as the important ceramic and brick-making industry located in the city of Kampong Chhnang.

Shrimp Farming

When most people think of tropical deforestation, they envision peasant farmers practicing slash-and-burn agriculture or agro-industrial estates of oil palm, soy or pasture grass. However, the clearing of mangrove swamps to create shrimp farms is also a type of tropical deforestation (Figure 5.8). The practice of farming shrimp, like other forms of aquaculture, has exploded across the globe in the past half-century in response to increased demand in developed countries. Shrimp was once seen as an expensive delicacy, but is now a common and affordable item on many menus in the United States, Japan and Europe, as well as in the fast-growing markets of emerging nations. The demand for shrimp has been partially met by more intensive fishing practices and expanded fishing in unexploited areas, but also from a massive increase in cultivated shrimp from tropical countries, which now provide about 35 percent of all the shrimp consumed in the world.[39]

Shrimp and prawns have been cultivated in Southeast Asia for centuries. The original aquaculture system used small brackish ponds to raise shrimp with other fish species.[40] Mangroves swamps were favored because their ecological conditions were ideal for growing shrimp, and they contained a naturally abundant supply of wild juvenile shrimp. Traditional shrimp farms supplied local and national markets throughout Southeast Asia until the 1970s, when a series of technological innovations developed in Japan and Taiwan together with the assistance of multilateral development banks led to a surge in shrimp farming, starting first in mainland China and Thailand, then spreading to

Figure 5.10. Thirty-five percent of the world's mangrove swamps have been converted to shrimp farms in the last three decades. Deforestation threaten the conservation of Cambodia's 85,000 hectares of mangrove forests, such as shrimp ponds located in the Koh Kong estuary (top) and the Preah Krasop Multiple Use Protected Area (bottom). The satellite image is from the year 2000 and the scale bar is 1 km.

other countries in the Southeast Asia and eventually to Latin America, Australia and Africa. Currently, about 75 percent of the world production of farmed shrimp comes from Asia, with the three leading nations being China, Thailand and Vietnam. Thailand is the largest exporter of shrimp to North America with more than $1.2 billion in export earnings in 2006.[41]

As the nature of shrimp farming changed in the late 1970s and 1980s, it evolved into a production system that was both more intensive and extensive — intensive in that more shrimp could be raised per square meter of surface area, and extensive because producers amplified the area under cultivation, in some instances to hundreds of hectares in a single production facility.* At first glance, shrimp cultivation is an attractive economic activity for developing countries of the coastal tropics because shrimp is a commodity with a relatively high unit value and a robust international market. In both Thailand and Vietnam, small farms predominate, with individual families constructing and managing one or two small shrimp ponds. Private sector companies also play an important role by providing key downstream services such as purchasing, processing and commercialization.

Shrimp farming comes with its own unique set of business risks and the experience of all major shrimp farming countries is illustrative. Without exception, each country has experienced a boom and bust that devastated its shrimp farming sector — all because of poorly managed environmental risks. These are largely related to the improper handling of animal waste as shrimp farms, like any animal-raising operation in a confined area, contaminate the production pond with fecal waste and degrade the water resource, thereby threatening the sustainability of the system. Farmers attempt to manage the contaminants by flushing them into an adjacent river or estuarine system, but these waters become polluted very quickly in an area with many shrimp farms. Also, if groundwater is used to flush the ponds, excessive pumping can change the balance between fresh and sea water, leading to increased salinization of the coastal aquifer. Poorly managed shrimp farms suffer

* "Extensive" production systems use large semi-natural ponds and rely on natural feedstocks to yield up to 500 kg/ha; "semi-intensive" systems use smaller ponds and provision feed to yield between 1,000–5,000 kg/ha; "intensive" systems use very small ponds with manipulated aeration and specially formulated feeds to yield up to 20,000 kg/ha. From About Shrimp Farming, Shrimp News International <http://www.shrimpnews.com> [accessed 15 July 2011].

from periodic outbreaks of a variety of pathogens, including protozoa, bacteria and fungi; the most devastating infections have been viral pandemics that have led to the catastrophic failure of the shrimp farming industry in all of the major producing countries. Measures to improve the conditions in a specific pond, such as changing the water, actually spread the waterborne diseases to other producers and worsen the problem. The market demand for specific species likewise tends to narrow the genetic base of production and increase the risk from pathogens. This boom and bust cycle has brought many producers to face bankruptcy after a few years. In Thailand, farmers with ruined ponds were forced to move to other regions or become urban dwellers.[42] In Vietnam, small shrimp farmers have been impeded to switching back to rice farming because of the high salt content of the fields. In the best of circumstances, farmers would lose a year's harvest to decontaminate their fields, an option that is unattractive or simply not possible for many small farmers.

From the standpoint of biodiversity conservation, shrimp farming has been devastating; 35 percent of the world's mangrove swamps have been converted to shrimp farms in the last three decades.[43] Cambodia originally had slightly more than 85,000 hectares of mangrove forests, considered by many ecologists to be most pristine mangrove swamps in all of Southeast Asia. These mangroves, like the extensive upland forests of the Cardamom Mountains, escaped exploitation due to the delayed development trajectory of Cambodia, but expansion of shrimp farming industry continue to threaten their long-term conservation. The impacts to biodiversity from mangrove deforestation go beyond the loss of mangrove forest, because many of the vertebrate and invertebrate species native to marine and coastal ecosystem use mangrove ecosystem as nurseries, so loss of the mangroves threatens these populations as well.[44] Even fish that live most of their lives in offshore coral reefs may use mangrove swamps during the larval or juvenile stage of their lifecycle. Certain species of fish for example, have been documented as being more abundant in coral reefs with nearby mangrove swamps than in coral reefs without them.[45] Wild shrimp are also very dependent on mangrove swamps and the productivity of coastal fisheries is directly linked to the conservation of mangrove swamp forests.[46]

The socioeconomic aspects of shrimp farming also provide mixed benefits, depending on how it is organized, on the perspective of the social group, and on the timeframe when the cost and benefits are analyzed. On the positive side, a robust export market boosts the national economy and shrimp farms create a value-added production chain, requiring a series of goods and services. Some providers, for example, specialize as hatcheries and nurseries, supplying juvenile shrimp to production farms, while processing facilities require refrigeration equipment, transportation, and packaging. As a labor-intensive production system, shrimp farming thus provides economic opportunities for rural populations that are often marginalized in developing countries. Most positions in processing factories are given to young, unmarried women.[47] In Vietnam, the number of jobs created by the industry was estimated at more than 500,000 in 1998, the beginning of that country's rapid expansion of the industry. So the shrimp farming provides both micro- and the macro-economic benefits to exporting countries.

Since mangrove forests have not been categorized as farmland, they are considered to be part of the public commons and have been used by local communities who have depended on them for decades for fuel, food and other resources. But, like the public commons throughout the world, they are vulnerable to encroachment by individuals acting in their own self interest.[48] Coastal fishermen, who are usually landless, will then suffer from decreased catch of wild shrimp and other fish. Although there have been no long-term studies to examine the link between the clearing mangrove swamps and diminishing fish catches, the correlation between the extent of mangrove habitat and the diversity and the size of fish catch, and the extensive scientific knowledge on the lifecycles of marine and coastal organisms make this connection evident.

It is almost inevitable that shrimp farming as it is practiced in Thailand and Vietnam will become widely adopted in Cambodia as well, sandwiched as it is between two shrimp production giants. Hopefully, the lessons learned in the past 30 years in Thailand will allow Cambodia to avoid the worst aspects of that industry, but the short-term profits that can be obtained by converting mangrove swamps may prevail on the Cardamom Coast. Current regulations governing aquaculture prohibit encroachment on existing farms and natural mangrove forests and mandate the appropriate treatment of wastewater. Like many environmental regulations in Cambodia, however, enforcement is weak and the lure of the short-term profits is strong. Community management may provide a solution; for example, a village committee was formed in 2001 to reduce fishing conflicts and foster the sustainable management of mangrove swamps and fish resources of Peam Krasop Wildlife Sanctuary in Koh Kong Province and the adjacent Chrouy Pros Bay. Villagers also protect the local seagrass beds that are linked to the mangrove ecosystem and provide habitat for fish spawning. In the mangroves themselves, community patrols do their best to control illegal clearing and fishing practices.[49]

Coastal Tourism

Cambodia is well on its way to becoming the tourist giant of Southeast Asia. Tourist visits have grown from less than 200,000 in 1995 to more than 2 million in 2007, earning an estimated $1.7 billion dollars annually[50] and the number of tourists is expected to exceed 3 million by the year 2010 (Figure 5.11). Approximately half of all visitors go to Angkor Wat, while most other arrivals are probably business travelers, particularly if they come from Vietnam, Thailand or Malaysia, countries with important investments in Phnom Penh's booming urban economy. Angkor Wat is the cornerstone of the Cambodian tourist industry. Like Pompeii in Italy, Tikal in Guatemala, and Machu Picchu in Peru, it is an archeological and cultural treasure — a lost city that provides a rare view into a complex civilization of the past. Angkor Wat has particular appeal to Asians, because it reaffirms their sense of historical place and the technological sophistication of past Asian states.

Not surprisingly, there is a growing interest in other attractions that complement Angkor Wat and the Cambodian tourist industry is expected to grow over the next decade to develop a more diversified tourist product with a broad appeal to an expanding Asian market, as well as visitors from Europe, North America and Australia (Figure 5.11). Indeed, many visitors seek a second destination as an add-on trip to the temples and Phnom Penh is an exotic city that attracts hundreds of thousands of tourists each year. Nonetheless, tourist industry analysts are unanimous in identifying Cambodia's extensive beaches as the country's largest unexploited tourism asset.

Cambodia's coastal area is 440 kilometers in length when measured as a straight line from the Thai to Vietnamese border, but is actually about twice that length due to the bays inlets, and peninsulas that characterize the coast. There are four urban areas on the coast, all relatively modest in size; the largest city is Sihanoukville, Cambodia's major ocean port, which has grown from about 75,000 inhabitants in 1998 to more than 250,000 in 2007. The others are Kampot and Kep in the south and Koh Kong in the north. Around each of these cities, coast real estate is the focus of intense speculation driven by the anticipation of Cambodia's expanding tourist industry.

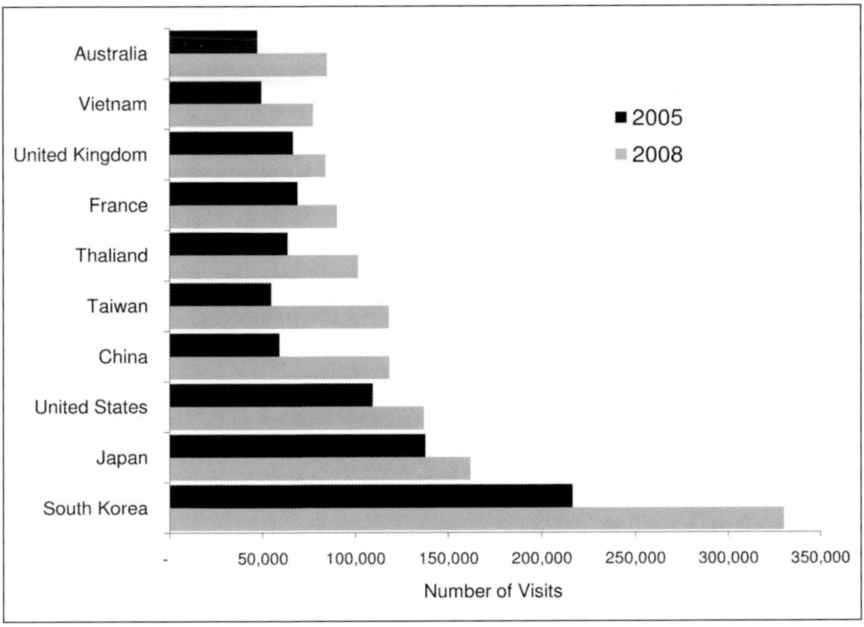

Figure 5.11. Tourism is one of the fastest growing sectors of the Cambodian economy with annual growth estimated between 10 and 20 percent over the last five years, with the greatest growth from middle class visitors from Asian countries.

This future development is linked to the rapidly expanding economies of northern Asia, specifically China and South Korea. It is assumed that as inhabitants of these expanding economies acquire more wealth, they will mimic the consumptive and recreation habits of North America and Europe and want to escape their cold climes and the drab, grey winter skies for a stretch of white sand, palm trees, and other tourist amenities. So just as the analogy of Tikal and Pompeii with Angkor Wat provide an example of the similarities between North American and East Asian tourist industries, so does the comparison of Kep with Cancún or Costa del Sol. Even if it takes decades for East Asia economies to reach the affluent consumption characteristic of Western society, its demographics will create a recreational tourist industry that will dwarf the Mediterranean and Caribbean. Cambodia will share this tourist industry with other Southeast Asian countries, but geography favors Cambodian beach resorts. The broad tranquil Bay of Thailand is protected from the brunt of most typhoons that tend to track to the northeast, while the shallow waters of the continental shelf and the prevailing water currents maintain relatively warm water temperatures throughout the year.

The speculative real estate craze has followed a predictable trajectory, but has been influenced by erratic governance and an inadequate land registry. The original owners of this coastal land essentially disappeared or were dispossessed during the civil war, while its current occupants are migrants who came to the region following the cessation of hostilities, attracted by the ability to meet their family's dietary needs by subsistence fishing. The spectacular beaches, remoteness and low cost soon attracted the "backpacker tourist," which stimulated the growth of guest houses and restaurants. This was followed by investments in more up-scale facilities, often with Western expatriates acting as investors and managers. The Internet provided a novel marketing tool that attracted a more affluent client base. By 2009, several cruise liners had added Sihanoukville as a port of call in their package tours between Vietnam and Singapore, as do the growing number of yachts plying the placid waters of the Gulf of Thailand.

As the economy in Cambodia has steadily grown over the last decade, investors have sought to diversify their portfolio; for the hotel industry, this meant leveraging ongoing investment in Siam Reap and Angkor Wat with new facilities on the coast. The real estate boom took off in 2006 when the Cambodian government started to receive bids for economic land concessions on the coast and, even more lucrative, for the unspoiled islands situated just off shore. Among the first leases granted was to a consortium of Russian investors for the 5,000-hectare Koh Puos (Snake) Island located south of Sihanoukville. The project calls for basic infrastructure, including a 700-meter causeway that will connect the island to the mainland; the planned 5-star resort will include convention center, golf courses, hotels and residential units, as well as private beaches and yacht facilities. Ground-breaking ceremonies were held in May 2008 for the bridge, which is under construction by Sinohydro, one of the Chinese companies building the hydropower dams in the Cardamoms. Executives associated with the project claim the total investment will exceed $470 million, while a government spokesman estimated the creation of about 25,000 new jobs. Other similarly ambitious investments have been announced by French and British real estate developers who have acquired properties on the mainland and entire islands backed by venture capital from the Hong Kong and Paris financial markets.[51, 52]

Figure 5.12. Investment in coastal tourism is being encouraged to diversify Cambodia's tourist industry and offer travelers a "second destination" in addition to Angkor Wat. Ongoing and planned investments include luxury hotels (top left) and beach resorts (top right). The coast provides easy access to scenic waterfalls in the Cardamom Mountains and scuba diving opportunities on offshore islands near Sihanoukville.

The real estate scramble is not limited to western and Russian entrepreneurs. Some of the most prominent real estate development projects have been organized by two of Cambodia's most successful entrepreneurs — Sok Kong, the charismatic CEO of the Sokimex energy and logistics company, and Kith Meng of the Royal Group of Cambodia. Mr. Kong has pioneered hotel investments in Cambodia via the subsidiary Sokha Hotels Ltd. and is the major force behind a visionary and ambitious effort to restore the hotel and casino complex at Bokor Hill Station on the Southern summit of the Elephant Mountains overlooking the Gulf of Thailand, a $1 billion investment. Not to be outdone, Meng has formed an alliance with the Hong Kong-based Millennium Group to develop coastal and island resorts that will require an estimated $2 billion over the next decade.[53,54] Another prominent — and controversial — investment is being promoted by the former Prime Minister of Thailand, Thaksin Shinawatra, to build a large casino and resort complex on Koh Kong Island near the international border separating Thailand and Cambodia. Thaksin envisions the construction of a model city designed according to ecological principals of sustainability, which would include residential, educational and health care facilities, as well as a financial center and a shipyard for refurbishing oceangoing vessels. Just across the straits of Koh Kong Bay, the Union Development Group of China has broken ground on a large 36,000 hectare resort complex in coastal land situated within Botum Sakor National Park.

It is difficult to predict when, or even if, all of these grandiose projects will be completed. Real estate developers are well-known for exaggerating the potential of their projects, as well as the commitments of existing investors. As the maxim states, "time will tell." The remarkable growth in Cambodian tourism over the past decade may not turn out to be a reliable guide for future growth. Angkor Wat draws people to Cambodia due to its very unique qualities while beach destinations in Cambodia are not substantially different from well established destinations in Thailand and Indonesia, as well as new beach resorts in both Vietnam and Malaysia. The presence of Angkor Wat is most definitely an advantage, and the tourist industry will exploit the well-known industry paradigm of a "second destination," where visitors combine a visit to the temples with a few days at the beach.

Regardless of the rate and dimensions of coastal tourism growth, change will come, and it will be accompanied by substantial environmental and socioeconomic impacts that will need to be identified and mitigated. Although the macro-economic benefits that come from the infusion of foreign currency will boost the national economy, unplanned development could very well lead to the deterioration of water resources and the degradation of the coastal landscapes and offshore habitats upon which the broader Cambodian Coastal economy depends and which provide Cambodian tourist operators a competitive advantage. Similarly, the exposure to people from different cultures will lead to an improvement in language skills that filters into other sectors of the economy, while the entrepreneurial spirit of the population will be enhanced as hotels outsource services to control costs. As development spreads and the population along the coast grows, residential neighborhoods will be created or spontaneously appear to provide housing workers for hotels workers; land rights could become divisive, unless steps are taken to plan growth and avoid conflict. Reports of land grabbing have accompanied several of the recent large investments, including the forced eviction of subsistence fishermen and small farmers who settled along the coast after the civil war and who, according to Cambodian law, have legal land-tenure rights if they have inhabited their plots peacefully for more than five years.[55]

Mineral Extraction and Hydrocarbons

The production, transformation and consumption of mineral resources are fundamental aspects of all national economies, and Cambodia's dependence on these mineral commodities will increase as its economy expands and diversifies. Like its neighbors, Cambodia is a net importer of mineral resources. It relies on foreign sources for virtually all of its energy, as well as for steel and other metals used in light industry, and cement for its fast growing construction sector. Mineral extraction contributed a miniscule 0.39 percent to Cambodia's gross domestic product in 2005 ($24 million dollars), employing an estimated 19,000 people. In contrast, the country imported approximately $200 million in petroleum-based fuels in 2004, $55 million in cement and limestone, and $28 million in unfinished steel stocks.[56] To close this gap, the government and the private sector are seeking opportunities to improve Cambodia's domestic capacity to supply these basic mineral commodities and, potentially, to move from being a net importer to an exporter of industrial materials and petroleum products.

The exploitation and exportation of mineral commodities provides revenue to the state, stimulates economic opportunity for business, and creates jobs, while improving a country's balance of payments. As Cambodia has improved its climate for overseas investors, there has been increased interest in exploiting its mineral resources for overseas markets. In the last five-year period, several concessions were granted to multinational corporations to explore for industrial metals, such as iron ore and bauxite, but also gold and specialized metals such as zinc and tungsten.* Cambodia imports about one million tons of refined petroleum products a year and is hoping to attract investors to develop both the upstream and downstream sectors of a nonexistent petroleum industry. Since Cambodia has a relatively small domestic market, any large discovery would allow the investor to monetize those reserves in international markets.

Perhaps the most publicized investments are those in offshore oil fields in the Gulf of Thailand, where multinational companies have been exploiting oil and gas in Thailand's territorial waters for more than three decades. Although the exploration has attracted much recent attention, oil exploration dates from the 1950s and 1960s when geologists from China and Poland conducted the first geological surveys and the first exploratory wells were drilled by western companies in the early 1970s. The offshore hydrocarbon-bearing formations are referred to as the Khmer Basin and, like all the oil-bearing formations in the Gulf of Thailand and mainland Southeast Asia, were formed by tectonic forces generated by the impact of the Indian Plate with the Eurasian Plate compounded by additional stresses from the movements of the adjacent Philippine Sea Plate and the Indian-Australian Plate. This combination of stresses led to the buckling of geological strata and the uplifting of topographic features, such as the Cardamom Mountains, as well as depressions such as the Khmer and Tonle Sap Basins. The oil-bearing strata dates from the early to mid Tertiary when alluvial and marine sediments were deposited in what is now the Gulf of Thailand.

The Cambodian National Petroleum Authority (CNPA) has divided the Khmer Basin into six blocks (designated as A through F), which have been leased to multinational companies from North America, Europe and Asia (Figure 5.13). Early discoveries in Block A are quite

* These include BHP-Billiton, Southern Gold, OZ Minerals of Australia (formerly Oxiana), Kenertec Resources, and a consortium of Chinese steelmakers (Wuhan Steel, Baosteel Group, Anshan Iron & Steel Group, Shougang Iron & Steel Group). Source: Invest in Cambodia (2009) <http://www.investincambodia.com/minerals.htm> [accessed 25 Jan. 2012].

close to production blocks in Thailand's territorial waters, where the same corporations have more than 30 years experience in exploring and extracting hydrocarbon reserve.* Seismic surveys and six exploratory wells indicated the presence of both gas and oil, which led some analysts to estimate Cambodia's total offshore oil potential at between 500 million and 700 million barrels of oil and 11 trillion cubic feet (tcf) of natural gas. Reserves of this magnitude would grant the state at least $15 billion in royalty revenues, and there was much speculation in the local and international press about Cambodia's future as an oil exporting country.† Estimates and projections must be validated, however, and after a second round of exploration in late 2007, Chevron issued a more sober assessment of Block A's offshore potential. The oil and gas reserves, although real, are dispersed through a wider area, rather than being concentrated in one or several core fields.[57] The other blocks, leased to companies from China, Singapore, and Malaysia, are at earlier stages of exploration.‡ Perhaps the greatest potential for new discoveries lies in the Overlapping Claim Area (OCA) in the disputed zone between Cambodia and Thailand, which includes the seabed located between Block A and Thailand's current producing fields. Both Cambodia and Thailand have awarded exploration rights to the OCA,

* Block A: Chevron (55 percent), Mitsui (30 percent), Caltex (15 percent). <http://www.rigzone.com/NEWS/article.asp?a_id=68777> [accessed 15 July 2011].

† To compare, Chevron's investments in upstream activities in Thailand approached $15 billion between 1962 and 2007, and about $4 billion in petroleum royalties were paid to the Royal Government of Thailand from 1981 to 2007. Much of this investment occurred during a period remarkable for "cheap oil." <http://www.chevron.com/Documents/Pdf/ThailandFactSheet.pdf> [accessed 25 Jan. 2012].

‡ Block B: Singapore Petroleum Company/Malayisian Petronas/Thai PTT. The Singaporean company and its partners have invested US$1 million and will invest US$2.5 million more when oil deposits are found in this block.

Block C: Contracted to Polytec of Hong Kong.

Block D: China Petrotech Holding (CPH) reportedly completed a three-dimensional seismic survey of this block in early 2007, which led them to estimate that Block D contains about 227 million barrels of recoverable reserves of crude petroleum and about 14 billion cubic meters (496 billion cubic feet) of natural gas. CPH announced in April 2007 that it would draft a plan for its first exploration well-drilling program for Block D (Demongeot, 2007).

Block E: Indonesian Medco International Petroleum company/Kuwait Energy/JHL Petroleum, a Swedish company. For this block, Medco of Indonesia paid US$4.5 million to a social development foundation of the Cambodian government.

Block F: China National Offshore Oil.

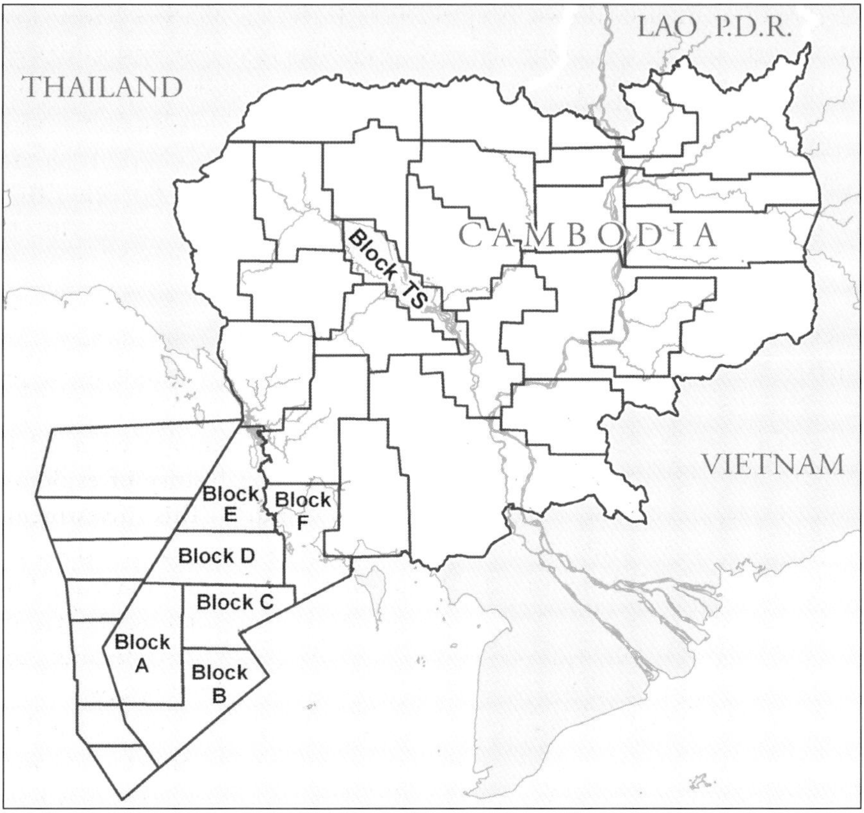

Figure 5.13. There is great interest in Cambodia's potential hydrocarbon resources in the Gulf of Thailand offshore and underneath the Tonle Sap. The Cambodian National Petroleum Authority (CNPA) has divided the Khmer Basin offshore into six blocks (designated as A through F) and leased these to different multinational companies from North America, Europe and Asia for exploration and development.

but the current tensions between the two countries have impeded exploration.*

In spite of the publicity generated by Cambodia's potential offshore reserves, there is another prospective region situated on the mainland; known as the Tonle Sap Basin, it is situated underneath the lake of the same name (Figure 5.13). In 1996, the Japan National Oil

* ConocoPhillips and Chevron, both U.S.-based/British Gas/Australia's BHP Billiton/ Japan's Idemitsu, Inpex and Moeco, Asian Times, <http://www.atimes.com/atimes/ Southeast_Asia/JG30Ae01.html> [accessed 15 July 2011].

Corporation identified two deep features in the Tonle Sap Basin that are composed of sedimentary rocks and, therefore, have potential as hydrocarbon-bearing formations. Although exploration in and around the Tonle Sap has been much slower than the offshore blocks, the area has been divided into concessions that are available for interested companies. Japanese and Chinese companies have conducted preliminary surveys, while the CNPA continues exploration using its own resources after initiating seismic surveys in 2008.[58]

Regardless of the slow pace of development and the political setbacks, many geologists consider Cambodia and Thailand to have the highest oil-producing potential in all of Southeast Asia Their development is more a question of "when," rather than "if." Evidence of that potential was recently confirmed when Chevron announced a $3.1 billion investment to expand and continue its long history of exploration, exploitation and production in Thailand.[59] This investment reflects both the business risk involved in hydrocarbon production, as well as the long lead time in developing projects and the tendency for investments to replicate as the knowledge base increases and physical assets multiply. Unless Cambodia is very unlucky, it is destined to develop a hydrocarbon industry, starting with its offshore reserves.

This development will bring major changes to the Greater Cardamom Region, impacting both marine and coastal habitats. Marine habitats will need to accommodate drilling rigs and pipelines, while coastal areas will be modified to make room for ports, storage facilities, refineries and the industrial development that almost inevitably grow to support an active hydrocarbon industry. If oil or gas is discovered in the Tonle Sap, that area will also experience change. On the positive side, it would bring new resources for the central government and some of these revenues would be funneled back to the local communities. The negative impacts, however, also are well known, particularly the risks from oil spills in a highly vulnerable and strategically important wetland fishery that is a food resource for millions of people. Export of the oil or natural gas would also require the construction of a pipeline that would transect the wildest part of Cardamom Mountains.

Due to its economic and environmental importance — both local and global — hydrocarbon exploration and investment usually dominates media coverage of the mineral sector. Nonetheless, the exploration and extraction of Cambodia's many other mineral riches has also been proceeding apace. Most of the interest has focused on the east and northeast of Cambodia where gold, iron and bauxite deposits have attracted Australian companies that dominate the mining sector, as well as Chinese companies seeking access to mineral resources essential for their continued economic growth.[60] However, the Cardamom Mountains have also hosted some important developments, including a tungsten mine in Kampong Chhnang Province on the flanks of Phnom Aural Wildlife Sanctuary, a commercially important chrome and copper discovery in Pursat Province in the Phnom Sankos Wildlife Sanctuary.

In the south, several rock quarries have been developed in the limestone hills overlooking the agricultural plain east of the Elephant Mountains along Highway 31 in the Banteay Meas and Dang Tong districts in Kampot Province. These investments are a direct response to the shortage of cement that has driven up costs and constrained growth during the recent construction boom in Phnom Penh. Domestic cement factories will replace between 2 and 5 million tons of imported cement annually at a value of approximately $50 million. The first phase of the largest factory was completed in 2008 at a cost of approximately $90 million; production in the first year was near 1 million tons and was expected to double by 2010. The plant is operated by the Kampot Cement Company, a subsidiary of one of Thailand's largest industrial conglomerates and an example of the cross-frontier investing environment that characterizes much of the recent economic growth.*

The oldest and most traditional form of mineral extraction in Cambodia is the mining of gemstones. The frontier towns of Samlot and Pailin† are famous throughout the world for their high quality sapphires and rubies. Rubies and sapphires are red and blue in color, respectively, although they can cover a broad range of tones depending on their mineralogical composition and various treatments. They belong to the corundum mineral family and were created under high pressure and temperature deep in the Earth's crust; over geological timescales,

---

* Ninety-three percent of Kampot Cement Co. is held by the SCG, a Thai conglomerate with investments in cement, chemicals, paper, and building materials; about 30 percent is owned by the Crown Property Bureau, the holding company that manages the assets of Thailand's royal family. <http://www.siamcement.com/en/index.php> [accessed 15 July 2011].

† According to legend, the presence of gemstones in the Pailin area was discovered by Burmese traders in the fourteenth century when they observed otters playing with gemstones in the rivers of Northeast Cambodia. The word "pailin" is apparently a Thai word that refers to otters (pai) playing (lin), as well as the Thai word for blue sapphire. <http://gemwanderings.googlepages.com/pailin> [accessed 15 July 2011].

they are transported to the surface by magma and released from these rocks via erosion and weathering over millennia. Prospectors seek gemstones in alluvial deposits near basalt rocks of volcanic origin[61] and this area has been exploited for centuries and was subject to intensive mining during the 1980s when gemstones were an important source of income for the Khmer Rouge factions based in and around Pailin. In the past, artisanal techniques and heavy machinery were used to extract these gemstones, under the thinking that digging deeper and moving more earth would yield new discoveries. Currently, mining in the area is limited to small scale artisanal efforts by local inhabitants. However, at least three companies have filed suit for mining concessions with the MIME and are preparing environmental impact and feasibility studies.*

The other major mining activity in Cambodia is gold mining. Although there are currently no major gold mining prospects within the Greater Cardamom Landscape, that may be due to lack of exploration rather than geology. Plate tectonics generate primary gold deposits, which are sometimes referred to as "hard-rock" gold. Gold is soluble in super-heated water and is transported via geological faults and then deposited when the water cools in igneous rocks, such as basalt and granite. Other minerals such as copper, tin and tungsten, are susceptible to the same process and are often found associated with gold. Alluvial or placer gold deposits, called "secondary gold deposits," are formed via the erosion of primary deposits over millennia. Gold is transported with other sediments via rivers and streams and will sink into trap sites due to its extreme weight. Ancient alluvial deposits have been transformed via geological process over hundreds of millions of years and are the source of gold found in sedimentary rocks, such as sandstone. The third type of gold deposit is formed from the mineral translocation of gold during prolonged weathering and is concentrated at the bottom of laterite soils, where it is deposited within iron oxides. Many of these geological attributes, including igneous rocks such as granite and basalt, are present in the Cardamom Mountains, while laterite soils are common throughout Cambodia.† Thus, it may just be a matter of time before gold is found and mined in the region.

---

* In 2006, Ultra Marine Kiri (Cambodia) Ltd. received a license to mine near Samlot; Jirech International Cambodia Construction and Import-Export Co. Ltd., and Sonuba Cham Industries Co. Ltd. received concessions in 2005 (see USGS document).

† Angkor Wat and other temples are all built using laterite bricks.

Although many developing economies depend on the export of nonrenewable natural resources to finance their government and to provide the underpinnings of the national economy, these revenues often create problems of social inequity. Developing countries that are richly endowed with oil and gas all too often become addicted to the revenue streams from royalties and use them to subsidize consumption, rather than invest in sustainable production models. Another common problem in developing countries is what has been termed "elite capture," where a small group of well-connected individuals or companies use their personal or family connections to obtain a disproportionate share of the benefits from the exploitation of a country's natural resources.

Cambodia has been subject to this criticism from individuals linked to human rights NGOs. As a response, the Cambodian government has discussed the possibility of joining the Extractive Industries Transparency Initiative (EITI), an international mechanism that promotes good governance for revenue generated from natural resources and proposes to monitor both corporate and government practices.[62] Like other initiatives that attempt to certify sustainable forestry or best practices within certain industrial sectors, the EITI involves all stakeholders — governments, investors, companies and NGOs — as well as multilateral and bilateral agencies such as the World Bank and the International Monetary Fund. Essentially, the EITI requires companies to publish what they pay to governments in royalties, taxes and other fees and, simultaneously, for governments to disclose what they receive from the companies operating within their borders. The EITI was first proposed in 2002 by the former Prime Minister of the United Kingdom, Tony Blair, and has since established its headquarters in Oslo, Norway. The world's largest energy concerns, including BG, Chevron, ConocoPhillips and BHP have joined 20 resource-rich countries to participate in the initiative; Cambodia would be the first Asian state to participate in the initiative.

Industrialization

The development strategies of emerging Asian economies are explicitly designed to foster diversified economies with a strong industrial base. The creation of an industrial sector has significant benefits for an emerging economy: Manufactured products are less susceptible to the cyclical price fluctuations of commodities, so they provide economic stability for countries that are overly dependent on their natural resources. Since it was first traveled by the English in the nineteenth century, the route to industrialization has taken many forms, but the recent success of the Asian tigers has been built on export-driven enterprises, which capitalize on the cost advantage of large populations and low wages. Another key aspect of their success has been the creation of stable political regimes over decades and the adoption of policies that promote direct foreign investment.

Cambodia under the tutorship of the UN, in conjunction with massive multilateral assistance programs and with an eye to the example set by neighboring states, has successfully applied this development model over the last 15 years. Direct foreign investment accounts for almost all investments in its industrial sector and its products, mainly textiles and footwear, are almost entirely destined for export. Although the current model is built on overseas markets and organized by foreigners, it benefited from a long tradition of silk and cotton weaving that is as old as the Khmer nation. The manufacture of textiles was expanded and mechanized during the colonial regime of the French, although these early efforts were destroyed during the civil war.

Cambodia received an opportunity to restart this textile industry during the reconstruction effort, which included a decision in 1995 to grant Cambodian special status under the Multi-Fiber Agreement (MFA) of the General Agreement on Tariffs and Trade (GATT). This agreement created a quota system to limit exports from emerging economies to developed countries* but excluded Cambodia due its special status as an impoverished state. As a consequence, investors from Hong Kong, Singapore, Taiwan, Malaysia, South Korea and Thailand rushed to create subsidiaries in Cambodia to take advantage of the preferential access to lucrative markets in the U.S. and Europe. The textile industry grew exponentially between 1995 and 2000, from less than $4 million in 1994 to more than $2 billion in exports in 2004.[63] The MFA was only intended to last 10 years, and it was obvious from the outset that this special trade status was a short-term opportunity and that the Cambodian garment industry would eventually have to compete with countries like China and Vietnam, both of which have lower labor costs, more agile customs bureaucracies, and a sophisticated entrepreneurial culture. Consequently, the Cambodian government

---

* The General Agreement on Tariffs and Trade (GATT) is a treaty framework that was established in 1947 and lasted until 1994 when it was replaced by the World Trade Organization. <http://en.wikipedia.org/wiki/World_Trade_Organization> [accessed 15 July 2011].

almost immediately undertook negotiations with the United States, which imports more than 70 percent of its textiles, to ensure preferential market access upon the termination of the MFA. The resulting agreement, known as the Trade and Investment Framework Agreement (TIFA), was signed in 1999 and renewed in 2006. It includes a number of features to promote exports from the U.S. to Cambodia, as well as other free trade measures promoted by the U.S. Congress, particularly those to ensure fair treatment for Cambodian workers.

As part of this process, Cambodia also joined the World Trade Organization (WTO) in 2004 and invited the International Labor Organization (ILO) to monitor conditions in Cambodian textiles factories that export apparel to the United States. This review has reported the Cambodia factories to be free of abuses such as child and forced labor and open to the organization of labor unions. Cambodia is the only country in the world where ILO independently monitors and reports on working conditions in clothing factories according to national and international standards, a strategy that has enabled Cambodia to successfully transition from an export country overly dependent upon trade subsidies, to one able to compete based on a fair-labor model. More than 60 percent of buyers in an ILO-commissioned survey, stated that compliance with labor standards was of equal or greater importance than price, quality and lead-time to market.[64] This production model is very attractive to high-profile brands wary of consumer boycotts organized by human rights and labor organizations in developing countries.*

The decision to pursue a "fair-labor" niche in its textile sector has paid multiple dividends to corporate investors and Cambodian society. Based on this strategy, growth in Cambodia's textile sector experienced about 20 percent annual growth between 2004 and 2005; it then fell to less than 3 percent in 2007 and actually shrank in 2008 due to the economic slowdown. In 2007, the garment sector provided jobs to about 350,000 Cambodians in more than 300 factories and provided the country with 80 percent of its foreign exchange earnings. It is estimated that for every position provided by the garment sector, an additional two jobs are created in a supporting service enterprise, mainly

---

*According to *USA Today*, Hong Kong-owned Archid Garment Factory produced 360,000 shirts each year for Gap, Victoria's Secret, Express and Foot Locker <http://www.usatoday.com/money/world/2005-04-04-cambodia-usat_x.htm> [accessed 15 July 2011].

food preparation and transportation. Cambodia has a working population of about seven million people[65] and the textile sector directly and indirectly provides employment to about 15 percent of the working population. Even though wages are low, ranging from $70 to $140 per month, they are above the national average.[66] The overwhelming majority of the individuals employed by the garment industry are young women who have few skills to offer in a limited job market. These young women are more likely than young males to share part of their income with their younger siblings and parents, thus ensuring that a portion of the wealth created by the export sector is channeled through the most underprivileged sector of Cambodian society.

There is another aspect of the Cambodian industrialization process and the idiosyncrasy of its labor model that has brought additional positive benefits to the country. Although the textile industry is based in Phnom Penh, many of the textile mills are situated in rural areas along one of the major transportation arteries of the country. This dispersion of manufacturing facilities is probably related to the scarcity of adequate real estate in the capital, which is situated on the floodplain of the Mekong River where more than 80 percent of the available landscape is under water for about four months per year. The land adjacent to the highways that connect Phnom Penh to Sihanoukville or the regional capitals of Kampong Chhnang and Takeo are dotted with factories and construction sites where soil from nearby quarries has been used to create a raised platform. The need to use landfills to create real estate increases the cost of establishing a large factory and provides additional incentive to locate factories in a rural setting less subject to inundation. The dispersion of manufacturing facilities not only eases urban sprawl, but also promotes rural development by providing disposable income to rural families via their unmarried daughters.

Despite these successes, Cambodia needs to improve its competitiveness to maintain its relatively positive position as a textile exporter to the United States, especially in light of the global economic downturn. Failure to do so could lead to the loss of hundreds of thousands of manufacturing jobs, and the inability to provide jobs for a workforce that is estimated to grow by about 300,000 young people each year. Not only will Cambodia need to continue improving worker conditions to stay ahead of other countries that will copy its successful labor model, but it will also have to improve the productivity of individual workers, as well as reduce corruption and improve administrative efficiencies. One of the most important steps, however, that Cambodia can take to improve its competitive stance in the textile sector is to resolve its long-standing energy deficit and reduce the cost of energy.

As previously noted, Cambodia has some of the highest energy costs in the world; this will change over the medium term as hydroelectric facilities come on-line and the natural gas in the Gulf of Thailand is exploited for the production of electrical energy. Increased biofuel production could also provide an alternative source for transportation fuels. However, many of these alternatives require a fairly long start-up period and the Cambodian textile industry will continue to face high energy costs over the short term. To meet this need for affordable energy over the short term, the government recently decided to approve the construction of a 200 MW coal-fired power plant by the Power Synergy Corporation (PSC) near Sihanoukville between 2009 and 2011 at an estimated cost of about $390 million.[*] However, this facility would be dwarfed by another coal-fired power plant proposed for the Koh Kong province that would generate 1,800 MW of electricity and cost approximately $5 billion. This massive facility, if approved, would be built and operated by a consortium of Thai and Chinese companies, and the energy would be purchased by the Electricity Generating Authority of Thailand. The joint venture, which apparently will be called Koh Kong Power & Light (KKPL), contracted the services of an unnamed broker to negotiate a concession agreement with the Cambodian government in 2008.[†]

These two coal-fired power plants illustrate the trade-offs that the Cambodian government faces in its industrial development. Most environmentalists would oppose coal-fired plants based on multiple criteria, such as local air pollution, GHG emissions and the impact of surface

---

[*] A power purchase agreement was signed on 27 March 2008 by the MIME and Power Synergy Corporation, a joint venture between Malaysian and Cambodian firms as part of an 30-year investment to build and operate a 100 MW coal-fired electric power plant and a 230 kV transmission line to Phnom Penh; the base tariff for the electricity payments will be $0.07212 per Kw-h. Announcement to the Bursa Malaysia <http://www.leaderuniversal.com/mediaCentre/bursaAnnouncements_details.asp?intID=533> [accessed 15 July 2011].

[†] The Italthai Board of Directors approved $20 million in fees in 2008 to be paid via a third party broker negotiating the terms of a concession agreement and regulatory approval for development of the facility. See Thai Development Public Company Limited And Subsidiaries Notes To Interim Financial Statements 30 June 2008 and 2007 (Unaudited But Reviewed) <http://www.itd.co.th/files/financial-q/en/june_08e3.pdf> [accessed 15 July 2011].

mining of coal. Government authorities must also view the benefits that an electric generating facility would bring to the national economy, including direct benefits to domestic consumers, employment possibilities for Cambodian nationals, and tax revenues generated by the enterprise, as well as secondary benefits to the national economy from easing the energy deficit, such as an increase in economic growth and job creation. They should also evaluate, however, the opportunity costs that would be incurred by approving these facilities; in other words, what economic opportunity is being forgone by building these coal-fired plants at these specific localities at this point in time? Finally, an enlightened government might also consider the environmental and socioeconomic costs, which would include increased pollution and acid rain, GHG emissions, the dislocation of local residents, and the loss of biodiversity in both terrestrial and marine habitats.

In the case of both these proposed coal-burning plants, the energy source, coal, would be imported from Australia or Indonesia, so coal mining and its transport would create no economic benefits for Cambodians and would, instead, add to Cambodia's negative balance of payment. The PSC facility would produce energy for domestic consumption, while the KKPL plant would export its power to Thailand, where opposition to new coal plants is a major factor driving the decision to locate this plant on Cambodian soil.[67] Since the owners and operators are largely Thai companies, most executive staff and middle management will also likely be Thai nationals. The PSC plant at Sihanoukville will be located in an existing industrial park with port facilities, adjacent to an existing fuel oil-powered electric plant that generates electricity for the national grid. In contrast, the KKPL facility would be located at an undeveloped site on the coast approximately 50 kilometers south of Koh Kong, which would require new port facilities and extensive conversion of natural habitat, as well as the construction of a high tension line to connect the facility to the Thai national grid. Other energy options — including the hydropower facilities planned or under construction and the (apparently) abundant natural gas reserves situated off shore — are available, but the national electrical authority of Cambodia probably wishes to diversify its power generation capacity given the well-known risk of drought-induced power shortages in countries that are overly dependent on hydropower and the long delay in the development of the off-shore gas fields. Viewed from a utilitarian standpoint, coal-fired power plants are a cost-effective and rapid solution for Cambodia's very real energy shortage.

From the environmental and opportunity cost standpoint, investments in fossil fuel generation are increasingly coming under question around the world because of their environmental impacts. Clean-coal technologies can reduce the effects of acid rain, but there is no published indication that either of the facilities being proposed in Cambodia would incorporate those technologies. In addition, both facilities will be located on the Cambodian Coast, which offers one of the most important opportunities for future economic growth in Cambodia. A large industrial port facility and power station will have a significant impact on coastal scenic beauty, which is a fundamental component of any tourist enterprise. Because the PCS facility would be located within an existing industrial zone, its impact on tourism would be less than the KKPL facility, which is slated for construction on a near-pristine stretch of the Cambodian Coast within Botum Sakor National Park. The enlargement of the PCS facility would contribute to the orderly and planned development of one of Cambodia's nascent industrial centers, while the KKPL would displace local fishing communities and limit the potential development of this beautiful section of the coast for tourism. Reports of the potential investment have essentially disappeared from the media following the border dispute that erupted between the two nations in 2009. As the proposed KKPL plant is still at a preliminary stage of development, detailed feasibility and environmental impact studies will hopefully allow the Cambodian authorities to make a fully informed decision that is in the best interest of the Cambodian people.

The process of industrialization is often considered to be antithetical to the environmental conservation, largely because it is viewed as a source of contamination and often occurs simultaneously with urbanization. However, urbanization can also be beneficial to conservation, as the migration of people from rural to urban environments reduces the competition for natural resources on densely populated rural areas and the migration towards the forest frontier. Urbanization and industrialization have their own environmental challenges, but they also have technological solutions — if the government and private sector work together to develop the necessary policies and invest in the requisite infrastructure, and if citizens assume their responsibilities and obey laws concerning urban planning and waste disposal. For Cambodia, industrialization has played an especially important role by providing employment to rural inhabitants, thus lessening the pressure on urban communities. This model of rural industrialization has the potential to improve the lives of its citizens and keep Cambodia competitive in the global market place.

Climate Change

It is now widely accepted that the planet is being irreversibly altered by the emissions of greenhouse gases (GHG) that are changing its biogeochemical cycles, and in turn, causing climate change. The most obvious and easiest understood aspect of this phenomenon is global warming. Although temperatures will continue to fluctuate from year to year, the next few decades will see a steady increase in temperatures with both local and global impacts. One of the most worrisome manifestations of global warming will be the loss of the planet's temperate and tropical glaciers, and — possibly — the Greenlandic Ice Cap. This melting, combined with temperature-related changes in the volume of ocean surface water, may cause sea levels to rise by as much as one meter in the next 50 years.[68] This change will profoundly affect the Mekong Delta, which starts at Phnom Penh at less than five meters above sea level, with perhaps one-third of the delta less than two meters above sea level. Rising sea levels will cause salt water to intrude into coastal aquifers, increase the intensity of seasonal flooding, and expose a greater area to risk from periodic storm surges.[69] Although the physical impacts of these changes will be felt mainly in Vietnam, the economic impact will be felt well beyond its borders, as it will affect the productivity of what is one of the world's greatest rice producing regions.

Air temperatures will also increase in mainland Southeast Asia between 2 to 6°C over the next century.[70] A change of this magnitude will lead to seasonal heat waves with the potential to take human lives and will negatively affect the biological processes in the region's natural and agricultural ecosystems, threatening the survival of individual species, as well as impacting the productivity of almost all crops. The rising temperature will also affect crop growth and yield by changing water-use efficiencies. Irrigation and dry land cropping are both constrained by the availability of water; warmer temperatures will lead to greater evaporation and cause plants to transpire at higher rates, thus removing more water from the system.

Given the impact of rising temperatures on water in the natural systems, the most important regional aspect of climate change, however, may very well be how it will change what many consider the most reliable climatic phenomenon on the planet — the Asian

Monsoon. The monsoon can be visualized as the flow of energy (and water) from the Indian Ocean and the South China Sea (or in the case of Cambodia, the Gulf of Thailand) to the Asian continent. It is inherently stable because it is a function of geography and the earth's orientation as it orbits the sun (see Chapter 2) — neither of which are going to change under any climate scenario. But global warming and increased sea surface temperatures will cause greater evaporation over the ocean and lead to greater energy inputs into the monsoonal system. A warmer planet will have earlier summers across all latitudes, which will create a more rapid and intense build up of temperatures over the Asian landmass that drives the onset of the monsoon. Consequently, most climatologists predict an increase in the intensity and duration of the Asian Monsoon as a consequence of global warming,[71] which in turn will lead to a net increase in precipitation across Southern Asia, Southeast Asia and East Asia.[72]

At first glance, this increase in precipitation and a longer wet season could be "good news," especially where agricultural production is limited by scarce water resources. However, another common prediction is for an increase in "extreme events," as the increased energy from warmer sea surface temperatures causes an increase in convection,* which will increase the intensity and severity of tropical storms. Fortunately, Cambodia's geography protects it from the most massive and destructive types of tropical storms in Asia, known as typhoons, which originate as tropical depressions only over relatively large expanses of open water (e.g., Eastern Pacific, South China Sea and Indian Ocean).† Nonetheless, Cambodia will still experience something similar in smaller proportion, as storms originating in the Gulf of Thailand will also increase in intensity and frequency. Storm surges in coastal areas and heavy rainfall in the streams and rivers of the Cardamom Mountains will increase flooding that can lead to loss of life, destroy

* Convection is defined by meteorologists as the process by which heat is vertically transferred between layers of the atmosphere. It is responsible for the formation of clouds and precipitation. <http://en.wikipedia.org/wiki/Convection_(meteorology)> [accessed 15 July 2011].

† Cambodia and Thailand are essentially protected by the archipelagos of Malaysia and Indonesia, which prevent tropical depressions from accumulating enough energy to form typhoons, and by Vietnam and Myanmar, which absorb the full impact of the typhoons that originate in the South China Sea (Vietnam) and the Indian Ocean (Myanmar). This characteristic makes Cambodian beaches particularly attractive to tourist investors.

agricultural production, and seriously degrade strategic infrastructure. Moreover, the impact of this wetter monsoon would be magnified by the increased stress on the river systems that feed into the Mekong River: hydrological simulations predict that an increase of 10 percent in precipitation over the Indochina Peninsula would lead to an increase of almost 20 percent in the annual flows of the Mekong River;[73] consequently, future severe storms would occur in a river basin that is already stressed more than usual by higher water levels during the wet season, setting the stage for "natural" disasters.

The term extreme event also refers to the opposite end of the precipitation spectrum: the prospect of lower-than-average rainfalls. Historically, drought in Southeast Asia is associated with the ENSO phenomena (*El Ñino – Southern Oscillation*), during which a series of linked events in the Pacific Ocean leads to severe drought in Indonesia and Malaysia; however, it also is associated with a wider but less intense reduction in rainfalls all across southern Asia (see Chapter 2). Droughts linked to the ENSO cause a prolonged and more intense dry season (October to June); during an "average" ENSO event, the reduced precipitation essentially coincides with the dry season and is often not markedly observed. However, large or mega-ENSO events can last up to two years and severely impact the socioeconomic well-being of the affected countries.[74] The ENSO phenomenon has been extensively studied over the past two decades, but its future role in mediating global climates remains uncertain. Different climate models predict different outcomes, including a weakening of the ENSO phenomenon coupled with a strengthened monsoon, and a contrasting scenario where increasing sea surface temperatures eventually override the (cooling) influence of the ENSO phenomena in the eastern Pacific and redistribute equatorial rainfall patterns worldwide, leading to a weakening of the Asian Monsoon.[75]

These scenarios are further complicated by a recently discovered phenomenon known as the *Indian Ocean Dipole* (IOD), which is similar in many respects to the ENSO, because it is an oscillation of meteorological states in the eastern and western Indian Ocean. Essentially, the monsoon winds that flow across Sumatra from the Indian Ocean in a northeastward direction promote the upwelling of cool waters from the Sunda trench, which cool surface waters and inhibit convection, leading to drought in western Indonesia. Although the ocean currents and wind jets that drive the IOD are independent of the ENSO, the two can interact to either increase or counteract the *El Ñino*-induced droughts that occur throughout the Australasia.[76]

Recent studies of the paleoclimatic history of these two ocean-basin systems have revealed unexpected connections between protracted droughts in Indonesia during the mid-Holocene (30,000 years ago), when the region experienced relatively wet rainfall conditions similar to *La Niña* today. One very important discovery from this high resolution study of a very long and ancient climatic history revealed that the monsoon is not as stable as previously assumed and that it can experience abrupt changes over relatively short geological time scales. If the Asian Monsoon intensifies, as expected due to global warming, it could interact with the IOD to create more frequent droughts in western Indonesia that could be magnified periodically by more widespread ENSO-driven droughts.

What does this mean for Southeast Asia and Cambodia? Perhaps more important than any one potential regional climate scenario is the realization that past climates have been radically different than today, and there is no reason to expect that future climates will look like the one we are currently experiencing.[77]

CHAPTER 6

The Value of Nature

The term "ecosystem services" has been coined to describe goods and services that natural or semi-natural habitats provide to human society. Many of these are essential for human well-being and survival — most obviously clean air and water, but also fertile soils, natural fisheries and the carbon stored in both terrestrial and marine ecosystems. The Millennial Ecosystem Assessment recognizes four distinct types of ecosystem services (Table 6.1). When ecosystem services are missing or significantly degraded, society must pay a cost to restore them or to mitigate the impacts of their absence, or it simply suffers the consequences of the degraded services, whether that be reduced human health and well-being, or lost economic opportunities. For example, water treatment plants can mitigate the impact of water pollution, while chemical fertilizers or extended fallows can restore the productivity of degraded soils; however, severe air pollution is less easy to mitigate and usually leads to respiratory ailments that shorten human life spans.

The goods and services that ecosystems provide are a perfect example of a public asset held by the state as an endowment for future generations. Their value is difficult to estimate because they are largely intangible; even though they are real, we may not see them and so usually take them for granted. Because ecosystem services are in the public domain, they are perceived as being free and, in some instances, a basic human right, such as air and water. As the planet has become transformed from a mosaic of largely natural ecosystems to one overwhelmingly dominated by human-mediated ecosystems, the value of nature's intangible services has become an issue of increasing public importance. While anthropogenic landscapes also provide ecosystem services, they rarely replace all of the benefits provided by natural

Table 6.1. Types of Ecosystem Services identified by the Millennium Ecosystem Assessment and examples that pertain to the Greater Cardamom Landscapes in both natural and anthropogenic ecosystems.

Category	*Type*	*Example*
Provisioning	food & fiber	timber, fish, crops
	water	irrigation & drinking water
	natural products	medicinal plants & resins
	energy	fuel wood & hydropower
Regulating	climate regulation	carbon sequestration
	decomposition	decay of forest & crop residues
	biological services	pollination predator – prey interactions
	watershed services	flood control & sedimentation
Supporting	soil formation	crop productivity
	ecosystem structure	food webs in aquatic ecosystems
	primary production	timber, fish, crops

habitats, and in some cases, they significantly degrade the natural assets that are the foundation of a sustainable economy. Consequently, economists have begun devising methods to estimate the value of ecosystem services and to find ways to incorporate the cost of maintaining these services into local, national or global economies.

This policy concept has been termed "payment for ecosystem services," and, in its most simple form, requires those who use the natural resource to pay the producers of that resource, whether the resource is clean and reliable water supplies, or the sequestration of CO_2 into some sort of long-term carbon sink. Unfortunately, such simple solutions are not very common, so the challenge is to devise methods where subsidies, regulations, and market mechanisms can compensate the guardians of natural ecosystems and ecosystem services that are essential for maintaining a productive landscape, an economically competitive nation or a climatically stable planet. Since ecosystem services are by definition a public good, the role of the state is essential in designing and implementing the compensation systems.

Given the rapid and economically-motivated changes that are facing the Cambodian nation, as described in Chapter 5, understanding the true value of Cambodia's natural resources is key to resolving the

Cardamom Conundrum. Without this knowledge, truly informed decisions cannot be made and there is the risk that Cambodia, like many developed and developing nations before it, will overlook the true value of nature and compromise the long-term economic value of its forest, fish, wildlife and water resources, in exchange for short-term economic gains based on traditional development choices.

Water and Watershed Management

Consider the rivers of the Cardamom Mountains, which are a strategically important national asset for Cambodia. Much emphasis has been placed on their potential for hydropower, with hundreds of millions of dollars of investments planned or underway. However, these highland watersheds also provide drinking water to hundreds of thousands of people living in approximately 3,500 villages and towns in the region.* In many communities, surface water is also diverted to irrigate crops; thus, it contributes to the subsistence of tens of thousands of small farmers, whose surplus production in turn supplies the food needs of urban populations. This interconnected network illustrates the chain of dependency on Cambodia's rivers and highlights the need for an integrated watershed management plan that protects these river systems from degradation caused by deforestation and erosion on mountain landscapes. Such a conservation plan would not only recognize the economic value of Cambodia's water and forest resources, but would safeguard the nation's strategic investments in hydropower and in the agricultural enterprises situated in the downstream sectors of these watersheds, both of which depend explicitly on the ecosystem services provided by rivers and intact watersheds.

One crude method to monetize the value of these ecosystem services would be to assign an economic value to the water from the Cardamom Highlands that is used by household, industrial, and agricultural consumers. Some conservation economists argue that water consumers should pay a fee to conserve the natural ecosystems that mediate their water flows; however, although it might be theoretically conceivable to formulate such a fee, the practical and political aspects of collecting it are formidable — if not insurmountable. Any fee

* Although the dependency on streams and rivers as a direct source of water is decreasing thanks to endeavors to provide secure water to the population, the physical link between shallow aquifers and surface water reinforces the need for an integrated watershed management strategy.

would meet with fierce resistance from land tenants with historical use-rights to a public resource that they probably view as "their water," particularly since many residents in the Cardamom Mountains are poor and such a fee would strain their already inadequate financial resources. Large-scale landholders with economic concessions in the agro-industrial or mining sector could be obligated to pay for their water use, as they are new users with no historical use-rights and because their concessions are specifically for land or minerals, rather than for water. Even that sort of rational use and payment would need to be clarified, however, by hydrological and economic analyses to ensure that granting water use-rights to new users would not affect communities with historical use-rights.

The general public might also be resistant to pay for watershed management based on their understanding of the Cambodian climate. Namely, since the rains over the Indochina Peninsula are linked to the South Asian Monsoon, mean annual precipitation should be unaffected by changes to land cover. Even if the Cardamom Mountains were completely deforested and converted to agriculture, rain would still fall, rivers would still flow, and inhabitants would still enjoy the benefits of the region's water resources. While this simple view of the monsoon has validity (the total volume of water transported through the watershed would remain the same), it overlooks the role of natural ecosystems in regulating and distributing water flow: intact forest ecosystems capture, store and regulate water flows; the leaves and branches of trees intercept rain, absorbing its impact, while understory plants and the detritus on the forest floor soak up water and slow runoff. Forest ecosystems add massive amounts of organic material to the soil that act like sponges, as water percolates through the soil to underground aquifers. If the forest were cleared, rapid runoff would create greater variations in river water flows: higher flows would occur after rain storms, followed by longer periods of low water.

This does not mean that deforestation will lead to severe floods or prolonged drought, as many environmental activists frequently contend. These unfortunate catastrophes are related to extreme weather events that overwhelm the capacity of even fully intact ecosystems to regulate water flows. They may be aggravated by poor watershed management, but they are not usually the result of poor management.[1] Rather, the impact of deforestation would be felt during average years when the increase in high water flows during the rainy season would aggravate low water flows during the dry season. Chronic low water flows will lower farm productivity and impact the ability of hydropower facilities to produce low-cost energy (see below). The economic impact of deforestation would not necessarily be catastrophic and punctual (e.g., floods and extreme drought) but endemic and persistent (e.g., lost opportunity and reduced productivity). Although catastrophic impacts capture people's attention, the low-grade degradation of a natural resource over time is far more damaging to a society and its economy.

Poor watershed management caused by deforestation also brings about soil erosion. In spite of the dense canopy cover of humid tropical forests, more than 80 percent of the soil surface is exposed to the air. The multiple layers of forest canopy — trees, lianas, epiphytes, saplings, and shrubs — capture almost 90 percent of the photosynthetically available light before it reaches the forest floor;[*] thus, the limited light environment allows only a sparse cover of herbs to become established, which sets the stage for severe erosion if the forest canopy is cut down and burned. To make matters worse, forests are often cleared near the end of the rainy season, so that downed trees can dry out over a period of months. The cleared land is then put to the torch just before the onset of the next rainy season, which typically arrives with heavy rains. On flat topography soil erosion can often be effectively controlled after land clearing, but on hilly terrain, such as the Cardamom Mountains, erosion control is much more difficult — if not impossible — and forest clearing immediately causes tons of sand, clay and silt to be washed into streams, greatly degrading the water quality. Although the impact is fortuitous rather than intentional, forest clearing for agriculture actually removes the key resource for agriculture — soil — by maximizing soil erosion. Not only does this compromise agricultural production, but the effects on the economy are also felt further downstream. For example, increased sedimentation on the margins of the *Tonle Sap* has been linked to the intensification of land-use on the surrounding landscapes and may negatively impact fish populations and the livelihoods of tens of thousands of people.[2]

The degree of erosion after land clearing also depends on the agricultural production system that is adopted. Annual cropping has the worst impact, because the soil surface is plowed or otherwise disturbed each year, sending massive amounts of sediment downstream.

[*] Light is composed of different wavelengths on the electromagnetic spectrum; chloroplasts in plant leaves preferentially absorb those wavelengths that drive photosynthesis, allowing other components (such as green light) to filter through the leaves.

Minimum tillage methods can reduce both wind and water erosion, but they cannot eliminate soil erosion on hilly landscapes. Perennial pasture grasses are relatively effective at slowing erosion, but overgrazing can reduce their effectiveness. Plantation forestry can likewise provide many of the ecosystem services provided by a natural forest, but eucalyptus and pine plantations produce natural herbicides that impede the growth of understory species, again increasing soil erosion. Slash-and-burn agriculture, followed by an extended fallow period, allowing for a second-growth forest, is relatively effective in controlling erosion; nonetheless, the periodic clearing of secondary forest to plant crops leads to erosion and increases the sediment load of river systems.

The value of watershed management, when considered as an ecosystem service, is also important to the short- or long-term potential of the hydropower facilities under construction in the Cardamom and Elephant Mountains. Variation in water flows will directly affect the ability of these multimillion dollar facilities to maximize energy generation. Dams essentially function as energy storage devices, converting the kinetic energy of water flowing downhill into potential energy in a reservoir.* In areas like the Cardamoms, where precipitation is highly seasonal, reservoirs must be able to store sufficient water to generate electricity at approximately the same rate during both wet and dry seasons. Forest cover reduces high water pulses during the rainy season, so that more water can be captured during the transition to the dry season and released to generate electricity throughout the year. Severe drought can wreak havoc on hydropower systems: As water levels drop, power generation must be scaled back. In some cases, failure to sufficiently do so may lead to a complete shutdown of a facility. Australia, New Zealand, China, Chile, and Panama all had recent examples of drought-induced energy crises between 2006 and 2008. The most illustrative example, however, comes from Brazil, where hydropower provides about 85 percent of the nation's electrical power needs and where a prolonged drought in 2000–01 reduced electricity supplies to critical levels, forcing the government to impose energy rationing on industries and reducing economic growth to near zero.[3] While that is a dramatic example, even less severe water shortages can affect the

* Recent comments by government authorities, including the Prime Minister, referring to the Cardamoms as the "battery" of Cambodia are fully accurate given the potential to transform the kinetic energy of its waterways to energy and harvest it as electricity.

profitability of a hydropower facility, limiting its ability to meet contracts, and forcing power utilities to purchase more expensive energy generated from fossil fuels. From this perspective, watershed management becomes a matter of risk management, where the aim is to keep water flows as stable as possible.

Deforestation and intensive land-use in mountainous terrain also causes increased sediment loads in reservoirs that shorten the useful lifespan of a hydropower facility. As a reservoir slowly fills with sediment, its capacity to store water is reduced, and electricity generation is increasingly curtailed during low water periods, eventually leading to the same economic limitations as in a severe drought. Rather than being a rare event triggered by severe weather, however, sedimentation buildup becomes a chronic problem that limits a hydropower facility's ability to generate energy during critical periods of the year. Eventually, the facility will be shut down when the reservoir can no longer store enough water to power the turbines, cover its operating costs, and return a profit.[4]

The dams under construction in the Cardamoms have relatively high seasonal water flows and low sediment loads. The high seasonality puts them at risk from drought, and forest conservation is the most cost-effective method for regulating these water flows. Fortunately, the current low sediment loads bode well for these strategic investments, which should enjoy long periods of functional operations. However, if the Cardamoms Mountains are deforested, this optimistic prognosis would change immediately and the expected functional life-time would be decreased at the same time as the risk from seasonal low waters or severe drought would be increased.

One way to estimate the economic value that the rivers and forested watersheds of the Cardamom Mountains provide would be to calculate the lost revenues that low water levels might cost the nation's hydropower facilities if deforestation led to the loss of those services. For example, what would it cost Cambodian society to replace hydropower with more expensive fossil fuel energy? Or what would be the cost to the economy if the generating capacity had to be reduced without sufficient time to arrange for alternative energy sources? If drought affects the region with the same periodicity of *El Niño* events, then the Cardamom hydropower facilities will face low water levels about once every seven years. Assuming that poor watershed management leads to a 10 percent reduction in energy generation, the total installed capacity of the 11 hydropower projects identified for fast-track development (estimated to generate about 4,000 GW-h*, see

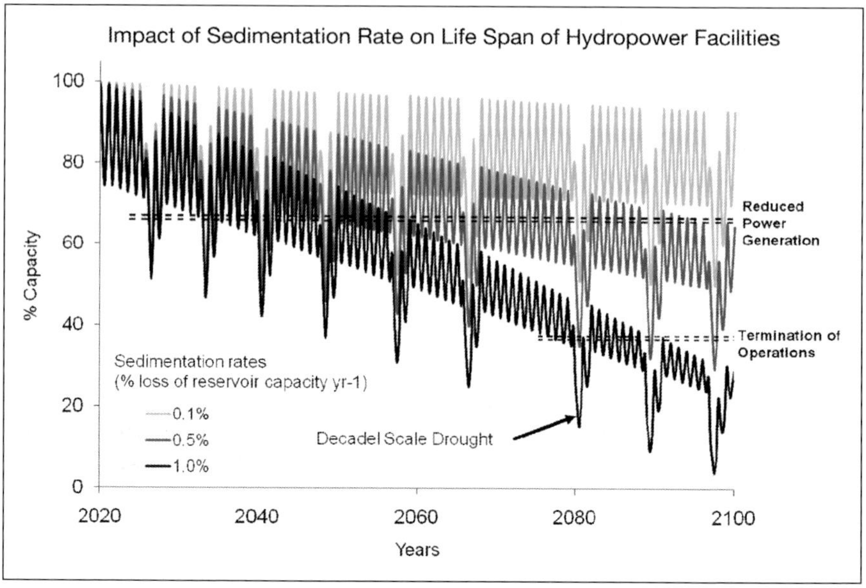

Figure 6.1. Schematic representation of the annual and periodic variation in water levels for a typical hydropower facility. Deforestation and intensive land-use would lead to a higher rate of sedimentation and decrease the life span of the facility.

Table 5.1) would be reduced by about 400 GW-h. The estimated cost of a kW-h generated by these dams was recently calculated at between \$0.05 and \$0.10 per kW-h; thus, the estimated cost of a periodic drought leading to a 10 percent reduction in power generation would be between \$40 and \$80 million (Figure 6.1).

Another method for estimating the ecosystem service value of forest and watershed conservation would be to calculate the lost revenues from a premature closing of a hydropower facility due to sedimentation. The annual rate of sedimentation is described as the percent of the total storage capacity lost each year due to sedimentation of a reservoir, which can vary depending on climate, geology, land-use, and watershed management. In a recent survey, the lowest reported rate was for the United Kingdom with only 0.1 percent sedimentation per year for all of that nation's reservoirs. This implies that the average

* GW-h = Mega Watt-hours, or 1 W-h × 10^9, kW-h = 1 W-h × 10^3.

reservoir in the British Isles can expect to last hundreds of years.[5] The country with the highest reported annual rate of sedimentation was China with an unhealthy 2.5 percent, which theoretically implies that mean lifetime of a reservoir in China is a mere 20 years.[6] Fortunately for the Chinese, these data are an overestimate because they come from the loess landscapes of Northern China that are particularly susceptible to erosion; the rates for the rest of China are much lower and are probably near the global average of about 1 percent, which translates into a reservoir lifetime of something slightly less than a century.[7] The useful life of a reservoir ends long before it is completely full of sediment, averaging between 50 and 75 years — unless extraordinary (and expensive) means are taken to keep it free of sediment or to remove the sediment deposits.[8] Revenues from the suite of 11 dams planned or under construction in the Cardamoms are estimated at between $200 and 400 million annually (e.g., 4,000 GW-h × $US 0.05–0.10/kW-h); if these facilities were decommissioned after 65 instead of 75 years, the total lost revenues would be between $2 and $4 billion dollars.*

As mentioned in Chapter 5, the hydropower facilities under construction in the Cardamoms are part of a build, operate and transfer (BOT) concession system. Concessions will last 35 years, a good portion of the dam's expected lifespan, during which time revenues will go to the investor; however, most of the lost revenues will be experienced by the Cambodian state after the termination of the concession (Figure 6.2). A small annual investment in watershed management that reduces sediment loads would yield tens of millions of dollars in future revenues, both to the operator over the near turn, as well as for the state over the long term, even when future revenues are discounted for inflation. Financing a payment for this ecosystem service should be borne by the beneficiary of that service — in this case, the operators of the hydropower facilities and, through them, the consumers of hydropower electricity, including urban households, industrial users and tourist resorts, as well as foreign users who are expected to purchase excess production in future years. A modest rate of $0.002/kW-h would yield about $600,000 annually from a 50 MW facility and pose a rate increase of only 1–2 percent above

* It is extremely difficult to predict the useful lifetime of a hydropower facility. Because its demise and decommissioning would not be abrupt but gradual, these hypothetical values might vary as much as 50 percent.

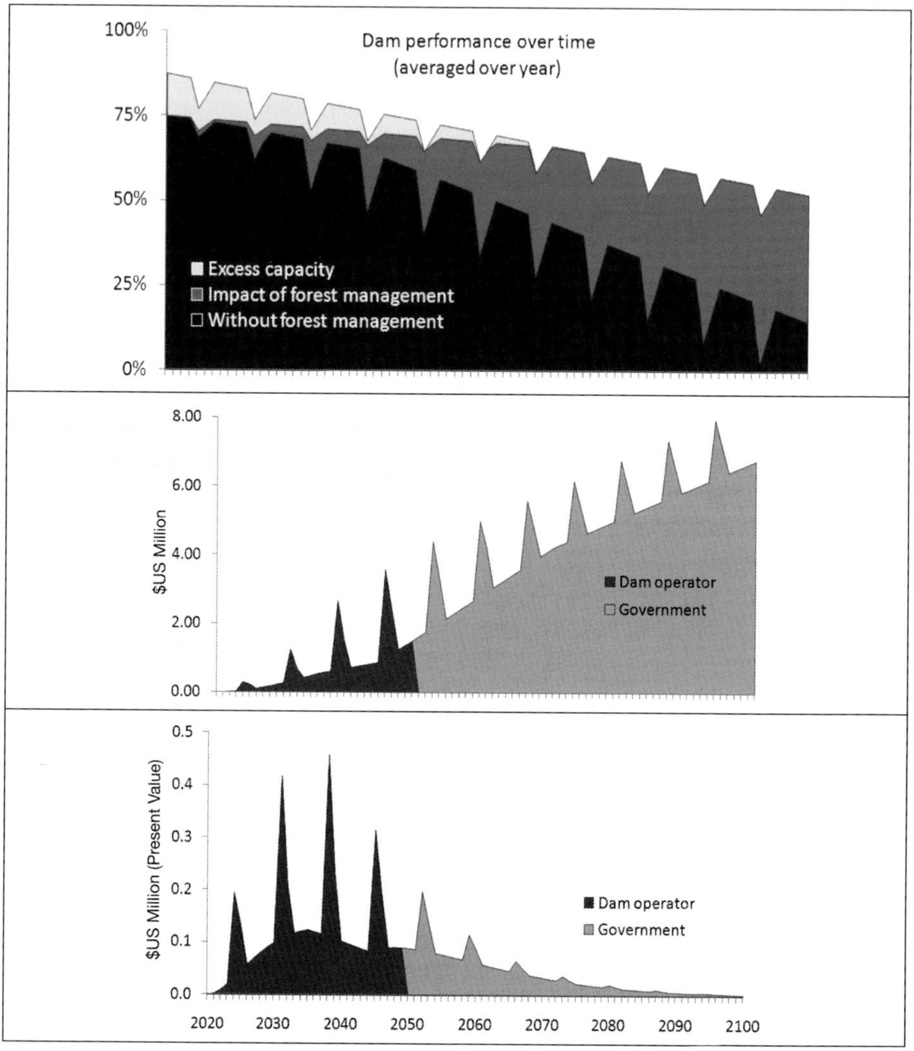

Figure 6.2. The impact of sedimentation on energy output and revenues for a 50 MW hydropower facility generating ~300 GW-h annually when calculated at the nominal rate of $US0.06 per kW-h. Increased sedimentation lowers energy generation over time (simplified based on Figure 6.1). This leads to a significant loss of revenues over time when expressed in constant dollars. When discounted for inflation at an assumed rate of 5 percent per year, the lost revenues can be expressed as "net present value" or NPV. The darker area shows the distribution of revenue losses under a "Build, Operate and Transfer" (BOT) scheme, demonstrating that the national electrical authority faces greater long-term losses, but when discounted for inflation, the short-term operator has significant motivation for supporting sound watershed management.

current electricity tariffs in Cambodia that vary between $0.12 and 0.08/kW-h. This would represent a classic case of a Payment for Ecosystem Services (PES) scheme and such modest rate increases would not likely be noticed by the public and, if they were, could be justified as a reasonable investment that would yield significant returns in the future.

Terrestrial Carbon — Sequestration and Storage

As the world looks for solutions to climate change, renewed attention has focused on the capacity of terrestrial ecosystems to sequester and store carbon, which is a constituent in many green house gases (GHG). Carbon, in the form of CO_2, is not only released by the combustion of fossil fuels but also when forests are cleared for agriculture and other land uses. In fact, tropical deforestation represents between 12 and 17 percent of global GHG emissions.[9] And recent studies have shown that forest conservation and reforestation are much more cost-effective strategies to abate GHG emissions than the massive investments that would be required to replace the industrial infrastructure and energy matrix that currently underpins our global economy. Although our modern world will eventually need to make this technological transition to renewable energy, sequestering and storing carbon in our natural ecosystems can play an important interim role in mitigating the impacts of climate change. This is a critical ecosystem service that tropical forests can provide, and one that — as it is becoming identified and valued in a burgeoning international carbon market — could provide financial incentive for developing countries like Cambodia to conserve their forests.

Carbon exists in terrestrial ecosystems in a variety of "pools," the principal ones being "above ground biomass," which is overwhelmingly composed of the stems and tree trunks, and "below ground biomass," which is made up mostly of roots.* The other major carbon pool is soil organic matter, a mixture of carbon-based molecules that are formed by the partial decay of biomass. The term "biomass" refers to the materials produced by living organisms. It is about half-composed of carbon, with the rest being various amounts of hydrogen, oxygen,

* The IPCC has identified five carbon pools to be eligible for inclusion in projects that conserve or enhance terrestrial carbon: above and below ground biomass, litter, deadwood, and soil organic carbon; deadwood and litter are sometimes referred to as "necromass" by ecosystem ecologists.

nitrogen and other minerals required to maintain the organism's physiological function. Biomass is created when plants absorb CO_2 and fix the carbon onto sugar molecules via photosynthesis. This process simultaneously releases O_2 into the atmosphere.* The most abundant biomass molecule is cellulose, a polysaccharide that is the principal constituent of plant tissues, including wood, leaves, and roots.† The relative size of the different carbon pools varies among ecosystems; in most forest ecosystems, above-ground biomass is the largest, while below-ground biomass and soil organic matter are more important in grasslands. In some wetland ecosystems, soil organic matter accumulates to form peat, which can store massive amounts of carbon.

Carbon flows from one pool to another via ecological and physiological processes: it is eventually returned to the atmosphere via cellular respiration as CO_2. Theoretically, a mature "primary" ecosystem should be at equilibrium, so that photosynthesis approximately equals respiration in plants and soil microorganisms. However, the flow of carbon is in constant flux from one pool to another and the flow rate determines whether there is a net sequestration or emission of carbon. Many, if not most, ecosystems are not at equilibrium due to disturbances from both natural events, such as hurricanes and floods, but more commonly from human activities; examples of the latter abound and are very relevant to Cambodia, including deforestation, logging, fuel wood collection, grazing and fire. In some parts of the world, urbanization and inexpensive foodstuffs provided by global food supply chains have led to the abandonment of farmland that has been replaced by young forests that are very effective carbon sinks. Climate change itself is also causing a shift in the equilibrium of ecosystems as it alters biophysical cycles. For example, higher temperatures increase the rate of soil respiration and CO_2 emission, while increased concentrations of atmospheric CO_2 can accelerate the rate of photosynthesis. These changes may be shifting the arctic tundra to become a net emitter of CO_2, and causing intact tropical forests, such as the Amazon, to function as a net carbon sink.[10]

The dynamic flux of carbon pools illustrates the risks that can exacerbate global warming, as well as the opportunities that are available to mitigate it. Although ongoing land-use change and deforestation

* The release of O_2 is why the popular press often refers to forest ecosystems as the "lungs" of the Earth and why financial instruments linked to forest carbon are often mistakenly referred to as "oxygen bonds."

† Animal tissue is also biomass, but all biomass originates from plant photosynthesis and 99 percent of biomass in terrestrial ecosystem are found in plant tissues.

can cause significant GHG emission, initiatives that support forest conservation can slow or reverse that trend. Terrestrial ecosystems, including agricultural systems, can be scientifically managed to reduce emissions and enhance sequestration, both in above-ground and below-ground carbon pools. There is also growing recognition of this opportunity in the international community and an awareness that international treaties must incorporate policies and economic incentives to maximize long-term stores of carbon in natural ecosystems and to add new carbon stocks in managed landscapes.[11] The possibility of paying countries for providing this important ecosystem service is being developed within the context of the United Nations Framework Convention on Climate Change (UNFCCC) and will provide nations such as Cambodia with an alternative framework for estimating the value of their forest resources and with new sources of revenues to conserve and manage those resources.

Reduced Emissions from Deforestation and Forest Degradation (REDD)

The global community has realized that reducing emissions from deforestation and forest degradation (REDD) is a cost-effective way to reduce GHG emissions over the medium term.[12] The REDD mechanism — whose rules and regulations are currently being negotiated under the auspices of the UNFCCC — could provide many tropical countries with an alternative forest development option that does not depend on logging or forest conversion. Potential revenues have been estimated to be worth from $2 to $28 billion per year over the next 30 years, depending on the international value of forest-based carbon credits and the potential demand from developed countries who would purchase these credits to offset their current emissions.* Since tropical deforestation is almost entirely limited to developing countries, there is an understandable desire that these revenues be used to reduce poverty and promote sustainable development in these nations.

In general, countries tend to pass through a series of development phases that can be tracked by mapping the change in forest cover

---

* Based on an estimated 10 percent reduction in deforestation rates over 30 years (2005–2030) @ $2–5 t^{-1} CO_2 ($1.2 B). A 50 percent reduction in deforestation rates would cost $10–21 t^{-1} CO_2 ($50 B). From: Kindermann, G. *et al. PNAS*, 105, 10302 (2008) <http://www.pnas.org_cgi_doi_10.1073_pnas.0710616105> [accessed 15 July 2011].

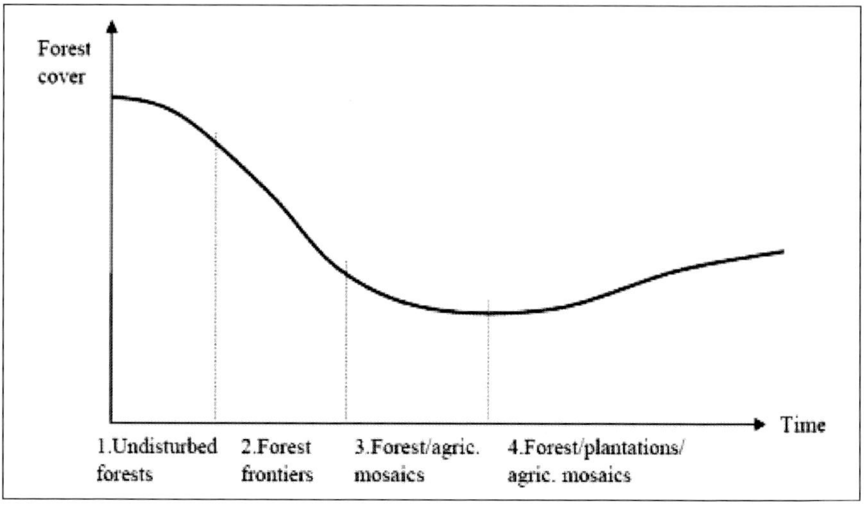

Figure 6.3. A schematic representation of how forest cover changes over time depending on a country's development trajectory: From a state of high forest cover and low deforestation rate (1), deforestation rates increase as countries convert forest to agriculture or plantation forestry (2), as forests resources are consumed, deforestation rates fall and eventually stabilize near zero (3), many countries eventually reverse the loss of forest resources either through forest plantations or reforestation (4). From Angelsen (2007).

(Figure 6.3). In the middle of the twentieth century, most tropical developing countries had relatively high forest cover, but as their economies grew, they cleared their forest to create agricultural economies or to exploit timber resources to fund investments in infrastructure and industry (e.g., Brazil and Indonesia). During the early phases of this development, the rate of deforestation increases sharply, but as forest resources dwindled, deforestation rates likewise slow and eventually approach zero; slowing deforestation is partly the result of legal and physical constraints that prevent further deforestation, but it is also the inevitable consequence of the reduction in the amount of forest available for clearing (e.g., India and Thailand). In some cases, as countries become more prosperous, forest cover increases due to deliberate reforestation efforts (e.g., Costa Rica and China) or to demographic phenomena that cause farmland to be abandoned and reforested via natural processes (e.g., El Salvador).[13] Only very few countries have managed to maintain high forest cover well into the twenty-first century and are still characterized by relatively low deforestation rates (e.g., Guyana and Liberia).[14]

As the name implies, the goal of the proposed REDD mechanism is to reduce emissions from deforestation and forest degradation and discussions originally focused on avoiding deforestation, which is technically relatively easy to identify and monitor. However, unsustainable forms of forest exploitation also lead to forest degradation, which is responsible for large, if largely unquantified, CO_2 emissions; consequently, REDD discussions quickly evolved to include the need to monitor forest degradation as well. Forest degradation has been defined as any type of activity that reduces the carbon stocks in a forest ecosystem, including logging, wildfire, and fuel-wood collection. These activities are responsible for massive amounts of carbon emission. It is estimated that up to 50 percent of above-ground biomass may be lost through selective logging and even more if logging is followed by wildfire, as is often the case. In some situations, progressive degradation essentially leads to deforestation and 100 percent loss of above-ground biomass.

As discussions evolved to involve more countries, it became increasingly apparent that the REDD mechanism would need to include all tropical forest countries, regardless of where they may be situated on their development trajectory. For example, a country with high forest cover and low deforestation rates, but with a nascent and expanding agricultural sector would find it much more difficult to reduce its deforestation rates than a nation that is nearing the point where most of its forest cover has already been converted. Consequently, a modified mechanism, now known as REDD+ has been devised, so that it addresses the needs of all of these countries and provides each with an effective mechanism to conserve forest cover and maximize the carbon stored in managed forest landscapes and includes compensation mechanisms and will scheme to support the diversity of countries and their different development needs.

As a first step, each nation that wishes to participate in the REDD+ mechanism needs to know the dimensions of its carbon assets and the rate at which it is consuming them. Countries can then develop strategies for reducing their emissions or increasing their rates of sequestration. In addition, the Conference of the Parties (COP) to the UNFCCC* has established as a precedent that emission reductions certified under a REDD+ mechanism must be evaluated in the context of national changes in forest cover. This stipulation discounts the impact of "leakage," a term that describes a displacement of emissions

---

*The COP is the main negotiating forum for the climate change treaty; COP meetings are held every two years; COP-3 yielded the Kyoto protocol, while COP-15 approved the resolution on REDD.

rather than a genuine reduction. For example, the creation of a protected area may successfully curtail forest conversion within that area, but if the underlying drivers of land-use change have not been addressed, deforestation rates may increase in surrounding landscapes and cancel out any reductions achieved through the creation of the protected area. Identifying leakage is problematic, however, because land-use is highly dynamic. For example, the increased clearing next to the protected area may have been the result of a new highway and economic growth that would have occurred regardless of the creation of the protected area, so in that instance it would not be considered leakage.

As stated above, developing countries must go through a number of steps to prepare for the future carbon trading system; one of the first is to document forest cover and estimate the rate of land-use change. To establish a baseline, historical change can be documented using archival satellite imagery that covers multiple years (preferably decades), while new imagery can monitor ongoing change. Field-based inventories that measure carbon stocks in vegetation and soil must be undertaken and integrated with maps from the satellite images to produce national or regional estimates of carbon pools.[*15] The next step is more difficult, as individual countries must devise policies that effectively reduce deforestation at the national level. This is an issue of national sovereignty, and individual nations will choose different approaches to achieve this goal — or not. It is highly probable that some approaches will fail, while a number of countries may choose to opt out of the system and forgo compensation as they pursue a traditional development options. Land-use zoning studies are often proposed as a way to guide future investment and reduce unwise land-use, but these must be combined with financial incentives to encourage the desired land-use. The creation of protected areas and forest reserves will also undoubtedly play a role in almost all national strategies, and improved revenues can improve the management of these areas. To be effective, all of these options will require land-use zoning, while protected areas and other regulatory approaches will require effective and transparent governance — a condition that almost never exists on agricultural frontiers.

---

[*] The IPCC has outlined three methodological approaches (Tier I, II, III) to provide countries with objective methods for conducting their national forest carbon inventories even when primary data is lacking or scarce. These three tiers provide a pathway to improving the robustness of national inventories over time, and to improving the accuracy of forest carbon estimates, thereby enhancing their marketability.

Key to the success of any strategy will be the ability to engage stakeholders on agricultural frontiers whose economic activities are directly linked to forest clearing and forest degradation: REDD+ revenues must be used to motivate them to change their activities. Since much land clearing is driven by a demand for agricultural commodities, some of the REDD+ revenues might be dedicated to improving the productivity of existing farmland. This could be done, for example, by providing technological assistance through official development assistance (ODA) in collaboration with government ministries. Investments in land titling programs would improve farmers' access to credit, which is often a constraint to improving farm productivity, particularly for small landholders. Similarly, improving the regulation of banks and other lending institutions or modifying their lending policies could curtail loans to large land holders whose business model is largely based on land speculation.

Cambodia is well poised to benefit from the future market for forest-based carbon credits, thanks to its large, relatively intact, forest estate. Because of its unfortunate history, it did not pursue the development — and deforestation — trajectory that many other countries elected in the second half of the twentieth century. The large tracts of forests that cover the Greater Cardamom Region provide the foundation for a new investment opportunity. Cambodia has changed radically over the past few years as foreign and domestic investment has grown exponentially; this presents a different kind of challenge as this kind of rapid growth is generally characterized by increases in land-use change. Without incentives from REDD+, or some similar mechanism, it is highly likely that Cambodia will follow a traditional development scenario where the extent of deforestation largely depends on the rate of economic growth and, perhaps, the ability of the government to conserve the forest within protected areas. If neighboring countries are any guide, Cambodia will be extremely fortunate to have as much as 30 percent of intact forest after 40 years of growth and development.

REDD+ is not based strictly on forest conservation; rather, it is contingent on a reduction of emissions that occurs by slowing deforestation or limiting forest degradation. In this context, emission reductions will be certified by comparing GHG emissions to some sort of reference value, which simple logic assumes would be a historical baseline. This logical approach does not necessarily work for Cambodia, however, because historic deforestation rates are going to be lower when compared to current and potential future rates. If the potential financial compensation is based on the difference between the past and present (or future), then it will be extremely difficult to generate revenues based on historical deforestation rates — much less satisfy a REDD+ compliance mechanism that demands verifiable emissions reductions.

There are no reliable statistics on Cambodia's land-use change from the 1980s and 1990s, but annual deforestation averaged about 200,000 hectares in the 1990s, and slowed to about 135,000 hectares per year between 2002 and 2006.* If current market forces and socioeconomic conditions in Cambodia continue, deforestation will soon exceed 200,000 hectares per year or, more likely, reach even greater values given the current rates of growth in the region and the nascent boom in biofuels and the proliferation of economic land concessions in rural Cambodia (Figure 6.4).

These numbers tell only part of the story, because they reflect only gross land-use change without taking into account the carbon stocks stored on different types of land cover. As stated previously, REDD+ is not based on forest conservation, *per se*, but on a reduction of emissions by slowing deforestation — and forest degradation. To be eligible for compensation, Cambodia would have to decrease not only land-use change, but net GHG emissions. For example, early stages of land-use change in Cambodia during the 1990s occurred primarily on low-biomass landscapes, particularly abandoned farmland and adjacent native woodland (Table 6.2). Deforestation has progressively moved to high-biomass rain forest, so overall GHG emissions today are higher than they were in the 1990s, even though the overall rate of deforestation has decreased (Figure 6.4). Given the current economic scenario and Cambodia's growth trend, the demand for land will most likely increase rather than decrease over the next 10 years, with a continued focus on high biomass landscapes, which will make it all the more difficult for emissions to meet historical baselines, even if the country is successful in reducing its overall rate of deforestation. The country would have to make even greater reductions in the deforestation rate to receive any significant compensation. Given the choice of receiving

* Studies by WWF (1995) and the FA (2002 and 2006) all used different criteria to define forest types; Cambodia's forests represent a gradient from open grassland and woodland to deciduous and evergreen forests. Each map has slightly different boundaries between strata, particularly those delineating dry forest from woodland. Most land-use change occurs at the dry end of the gradient causing statistics to vary greatly; in addition, changes due to seasonality in deciduous forest have been mistaken as deforestation and vice versa, introducing uncertainty into the deforestation estimates.

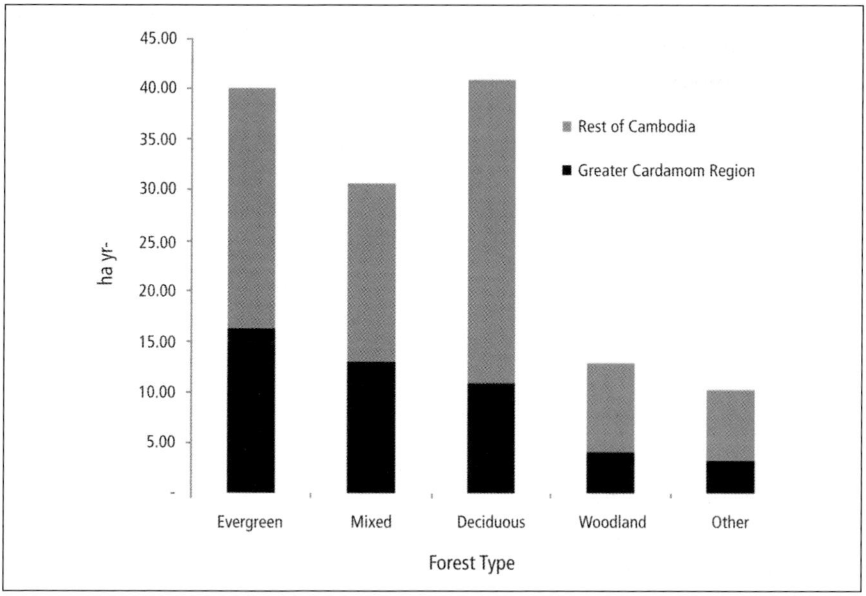

Figure 6.4. The proportion of deforestation in Cambodia that occurs within the Greater Cardamom Region was approximately 37 percent of all deforestation which occurred in Cambodia between 2002 and 2006.

inadequate compensation for avoiding deforestation and facing discontent in rural areas, the government might understandably choose to forgo participating in the REDD+ mechanism and concentrate its efforts in developing as much agricultural land as possible.

Because the pitfalls of historical baselines are not unique to Cambodia, the term "baseline" has been replaced by the term "reference case scenario." A reference case could be a historic baseline, but it could also refer to a future scenario that objectively reflects the economic and social reality of individual countries. If the objective is to reduce future emissions, then REDD+ must encourage as many countries as possible to participate and be sufficiently flexible to accommodate different social and economic settings. In the case of Cambodia, a potential reference case scenario could be developed by relating land-use change over the past decade with economic growth and infrastructure investments, in order to demonstrate the potential emission level of a "utilitarian scenario" where land-use change is maximized for agricultural production. REDD+ credits could be certified when the national level of GHG emissions from deforestation and forest degradation falls below this projected and hypothetical level.

Table 6.2. Rate of land cover change in the Greater Cardamom Region between 1995 and 2006 according to land cover type.

Land Cover Types	*Land Cover (ha)*			*LCC 1995–2006*		*LCC 1995–2002*		*LCC 2002–2006*	
	WWF-1995	*FA-2002*	*FA-2006*	*ha yr-1*	*% rate*	*ha yr-1*	*% rate*	*ha yr-1*	*% rate*
Evergreen Forest	1,679,134	1,596,690	1,532,190	14,694	0.88%	11,778	0.70%	16,125	1.0%
Mixed Forest	347,916	321,976	270,246	7,767	2.23%	3,706	1.07%	12,933	4.0%
Deciduous Forest	919,817	831,417	788,067	13,175	1.43%	12,629	1.37%	10,838	1.3%
Woodland*	595,117	346,267	330,597	26,452	4.44%	35,550	5.97%	3,918	1.1%
Other Forest	195,259	154,128	141,266	5,399	2.77%	5,876	3.01%	3,216	2.1%
Flooded Forest Mosaic**	288,660	218,823	256,580	3,208	1.11%	9,977	3.46%	(9,439)	–4.3%
Mangrove***	70,785	44,447	42,215	2,857	4.04%	3,763	5.32%	558	1.3%
Rubber Plantation	372	83	83	29	7.77%	41	11.10%	–	0.0%
All Upland Forest & Woodlands	3,737,243	3,250,479	3,062,366	67,488	1.81%	69,538	1.86%	47,028	1.4%
Nonforest	1,253,100	1,837,944	1,990,750	73,765		83,549		38,201	

Notes: *Changes in Woodland in first temporal period (1995–2002) may be overestimated because FA may have classified open woodlands as nonforest.

**Changes in Flooded Forest Mosaic are not reliable because the three different maps used different stratification criteria to define the categories.

***Changes in Mangrove during first temporal period is probably overestimated due to differences in scale and precision of the WWF map.

Table 6.3. Mean annual carbon emissions from Cambodian upland forests and potential value as carbon credits.

		Above Ground Biomass (2006)			*Annual Emissions*		
	Land Cover (ha) 2006	*Carbon stock ($t\ ha^{-1}$)**	*Carbon ($10^6\ t$)*	*CO_2 eq. ($10^6\ t$)*	*CO_2 emissions ($10^6\ t$)*	*Millions of $US***	*$US ha^{-1}*
Evergreen Forest	1,532,190	75	115	421	4.4	22.1	1,373
Mixed Forest	270,246	65	20	74	3.1	15.4	1,190
Deciduous Forest	788,067	40	59	216	1.6	7.9	732
Woodland	330,597	20	25	91	0.3	1.4	366
Other Forest	141,266	20	11	39	0.2	1.2	366
All Upland Forest & Woodlands	3,062,366		230	841	10	48	805

Notes: *Carbon stocks assume a 30 percent loss in biomass from previous logging and fuel wood collecting.
**Estimated at $5/ t CO_2.

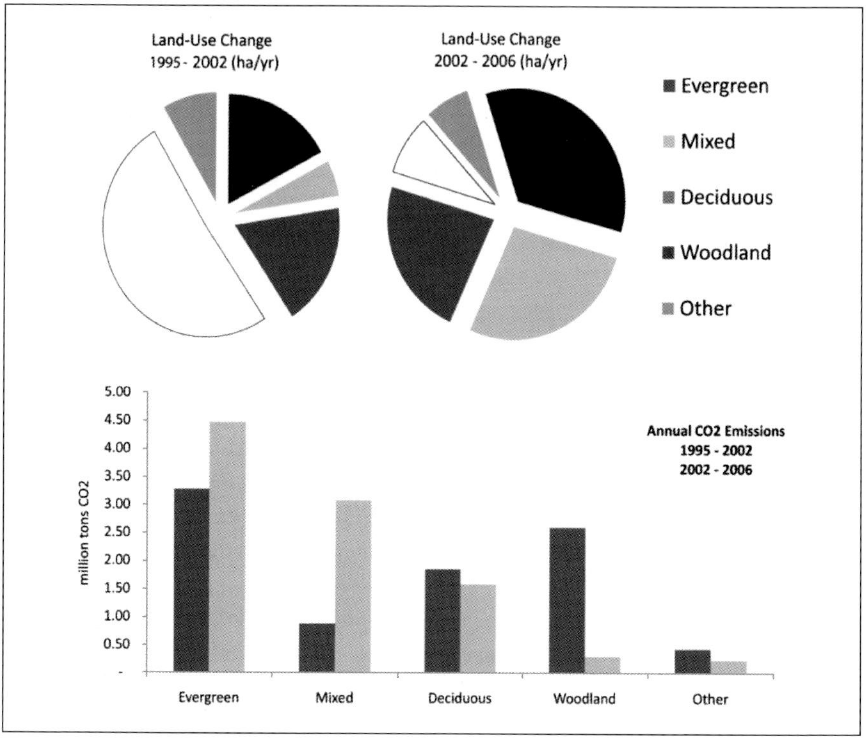

Figure 6.5. Within the Greater Cardamom Region, deforestation has shifted from low biomass vegetation types to higher biomass vegetation types, although overall levels of land-use change have slowed, carbon emissions have remained approximately the same.

REDD+ and the Cardamom Mountains

The Cardamom region will be a key component in any national REDD+ strategy. First and foremost, the region represents Cambodia's largest carbon asset, with an estimated 230 million metric tons of carbon (Table 6.3). Second, the area has experienced significant recent deforestation (see Figure 5.6) and forest degradation from unsustainable logging practices over the past two decades. These historical facts create the conditions necessary for calculating relatively high emissions based on a historical baseline reference case scenario. Moreover, these phenomena are ongoing and can be expected to continue and even increase in the near term. Deforestation in the Greater Cardamom Region is responsible for approximately 37 percent of all deforestation in Cambodia (Figure 6.5), so the national REDD+ effort must succeed in the

Cardamoms if it is to succeed in the country as a whole. At the same time, the region is essentially a microcosm of Cambodian society, so whatever works here will probably work elsewhere.

To estimate the potential dimension of REDD+ revenues from the Greater Cardamom Region, we compared five future scenarios for the region based on distinct development pathways. These scenarios are assumption-based models that project different levels of deforestation and degradation for different types of land-use and landscapes. Each scenario encompasses the entire Greater Cardamom Region, which was stratified into 13 units based on topography, soils, roads, demography, forest type, land management category and proximity to the agricultural frontier. The first four basic scenarios are differentiated largely by deforestation rates, but also by how land-use was distributed among the 13 landscape units. A fifth scenario shows the impact of a logging ban and the potential revenues from allowing degraded forests to recover carbon stocks. Emission reductions were estimated by comparing the model outputs for specific time periods; differences in forest cover were converted into CO_2 emissions based on default levels of carbon stocks and standard conversion factors (Figure 6.6). Like the scenarios presented in the first chapter of this book, these five scenarios are purely hypothetical; nonetheless, since they are based on a defined set of conditions and a specific geographical area, they provide an objective estimate of the potential revenues a REDD+ initiative might generate in the Greater Cardamom Region.

Historical Development Scenario (HD). This reference case scenario assumes that future land-use change will mirror past trends at the national and regional level. It assumes that there is a fundamental demand for new farmland and that this demand will remain constant over time. Forest conversion occurs first in the more accessible areas on the eastern and northern plain adjacent to existing agricultural landscapes; as this resource is diminished, land-use change moves into remote areas. The absolute rate of change is based on the historical mean rate of 46,650 ha yr^{-1} or 1.45 percent of forest cover annually; however, since the area of remnant forest decreases over time, the relative rate will increase slightly (Figure 6.6, top). By 2050, only about 36 percent of the original forested cover will remain intact. Forest cover on the eastern and northern plains will have been reduced from a total of 55 percent to between 10 and 12 percent, while the two wildlife reserves will have lost more than 55 percent of their original forest cover; even the remote regions of the Central Cardamom will lose as much as 25 percent of their forest habitat (Appendix 6.1).

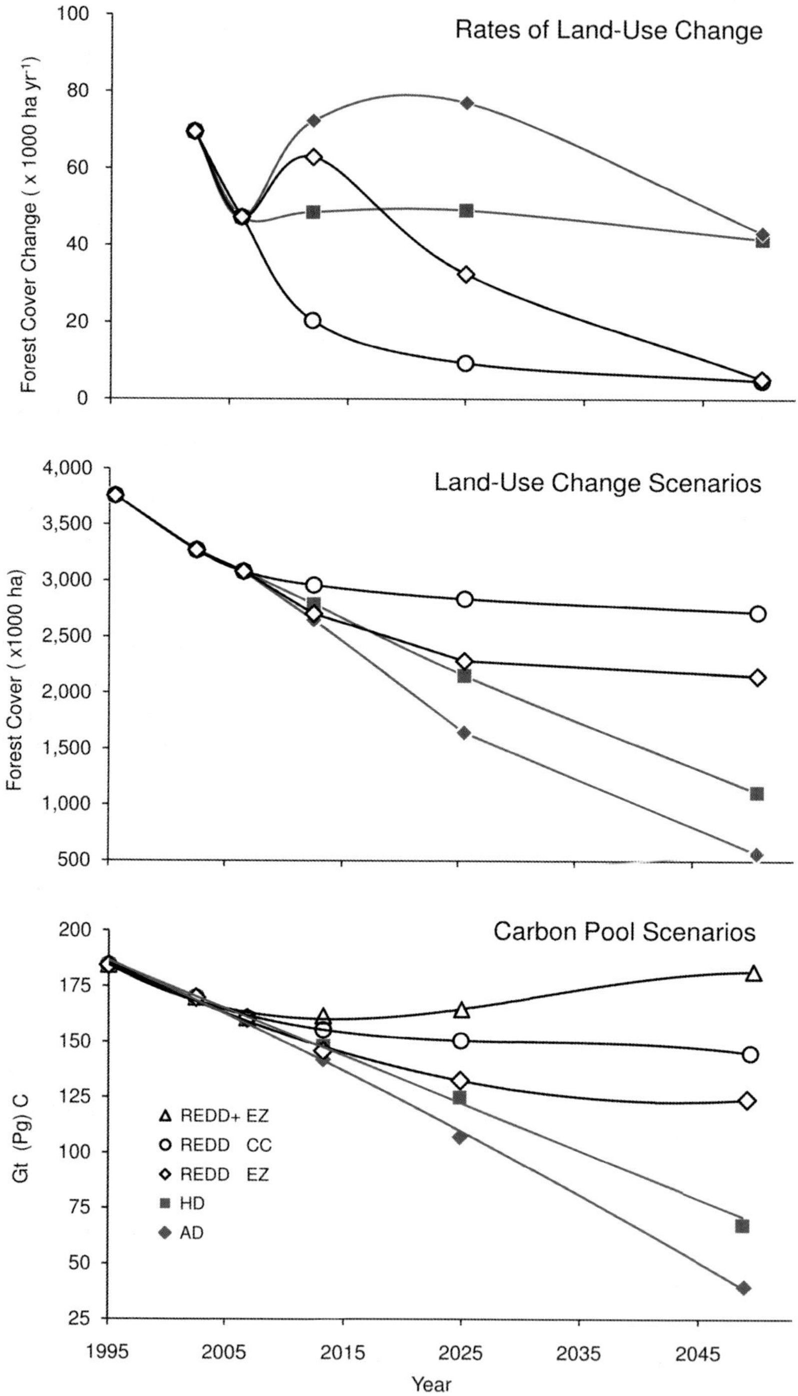

Rates of Land-Use Change
Forest Cover Change (x 1000 ha yr-1)
100
80
60
40
20
0
Land-Use Change Scenarios
Forest Cover (x1000 ha)
4,000
3,500
3,000
2,500
2,000
1,500
1,000
500
Carbon Pool Scenarios
Gt (Pg) C
200
175
150
125
100
75
50
25
REDD+ EZ
REDD CC
REDD EZ
HD
AD
1995
2005
2015
2025
2035
2045
Year

Accelerated Development Scenario (AD). This model is based on the assumption that land-use change is linked to economic growth, which remains high in Cambodia over the near- and mid-term; this causes deforestation rates to increase significantly over the next decade and remain high well into the future. As in the previous scenario, land with arable soils and those closer to previously settled areas are converted first, but the increasing demand for land causes deforestation to increase rapidly on landscapes that are less apt for intensive production systems. The relative rate of land-use change increases and remains high (2.6–2.9 percent); the absolute rate of deforestation reaches a maximum value of 77,300 ha yr-1 in 2025, but then declines as less remnant forest is available for conversion (Figure 6.6, top). By 2050, only about 18 percent of the original forested cover will remain intact (Figure 6.6, middle); large areas will have been almost completely deforested, including the northern and eastern foothills and the coastal plain. The forest estate, including most forested landscapes outside of protected areas, has lost almost 90 percent of its forest cover, and even the more remote areas of the Central Cardamom Mountains lose nearly 40 percent of their forest cover (Appendix 6.1).

Reduced Emissions Scenario Based on Conservation Criteria (REDD CC). This scenario assumes that the Cambodian state decides to aggressively pursue an emissions reduction strategy based on forest conservation. The entire region is provided with economic incentives to slow deforestation and conserve native habitat, including on landscapes that are relatively suitable for agriculture and that are under severe pressure by settlers and land speculators. The rate of land-use change is reduced progressively, dropping from 1.45 percent to less than 0.5 percent in the first 20-year period (Figure 6.6, top) and more than

Figure 6.6 (p. 190). The reductions in carbon emissions can be calculated based on a projection of future development trajectories scenarios. Deforestation rates might remain the same under a historical development scenario (HD) or increase under an accelerated development scenario (AD), while an ambitions REDD program based on strict conservation criteria would reduce rates (*REDD CC*), while economic zoning would allow the conversion of low biomass landscapes while protecting high biomass forest (*REDD EZ*) (top). The total forest cover varies over time depending on which development trajectory predominates in the region (middle). The total carbon stored in the region is a function of forest cover and the biomass of different forest types; however, if forests are not degraded from logging and fuel wood extraction, they would recover biomass lost in previous decades and yield additional carbon abatement benefits (REDD+ *EZ*) (bottom).

88 percent of the original forest cover would still be intact in 2050 (Figure 6.6, middle). Efforts to consolidate the protected area system are successful, leading to the conservation of the extensive dry forests located in the Phnom Aural Wildlife Reserve, while an integrated land management initiative ensures that approximately 50 percent of the forest is retained on the foothills surrounding the Cardamom Plateau and along the Cambodian Coast (Appendix 6.1).

Reduced Emissions Scenario Based on Economic Zoning (REDD EZ). In this scenario, the Cambodian state chooses to regulate land-use so that rain forest habitats are protected as a priority, while landscapes considered apt for agriculture and those where social conditions would make it difficult to stop land-use change are zoned for conversion and settlement. The relative and absolute rates of change actually increase over the short term approaching 63,000 ha yr-1 (2 percent) by 2015, but then fall dramatically as the landscapes zoned for agriculture are cleared and the conservation strategies implemented in protected areas and forest reserves are consolidated (Figure 6.6, top). In spite of initial high deforestation rates, emission reductions are achieved over the mid-term due to the government's commitment to protect high-biomass habitats, while tolerating the conversion of low biomass landscapes. By the year 2050, approximately 70 percent of the original forest cover would still be intact, including more than 80 percent of the two wildlife reserves adjacent to the agricultural expansion zones and the coastal plain (Table 6.4).

The dimensions of potential REDD+ revenues can be estimated by making pair-wise comparisons between one of the two high deforestation scenarios (HD or AD) as a reference case and one of the low deforestation models (*REDD CC* and *REDD EZ*). Essentially, the area of each forest type is multiplied by a carbon stock value (t C ha^{-1}) to estimate the carbon stored in that habitat and these values are summed over the entire landscape to estimate the total amount of carbon stored in the terrestrial ecosystems of the region (Figure 6.6, bottom). Emission reductions are obtained by subtracting the carbon stores from the two contrasting scenarios and the converting that value into tons of CO_2, which is the basic unit of international carbon markets (Figure 6.7).

The market value of a carbon credit varies greatly depending on its source and the type of market. Compliance markets serve nations that have mandated CO_2 emission reduction targets and, typically, have stringent requirements as to what type of carbon credits are allowed. Values in the compliance market are typically greater than

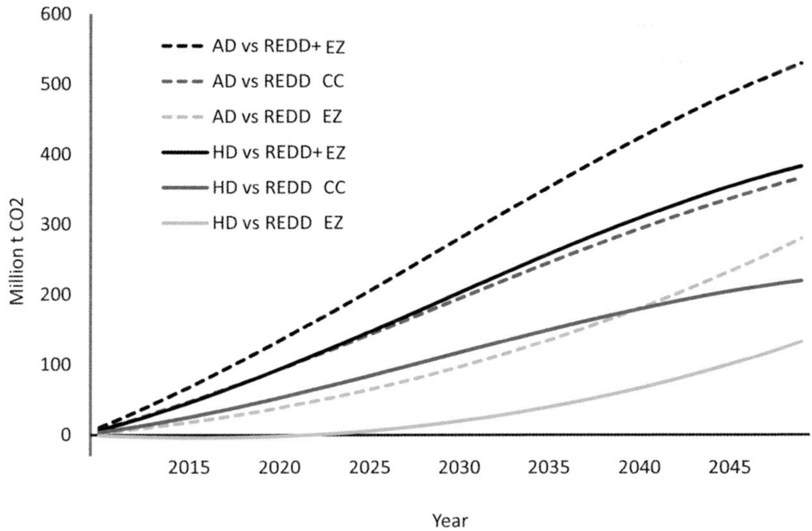

Figure 6.7. A comparison of the cumulative emission reductions from six different scenario comparisons show that carbon abatement benefits from a no logging regime are approximately equivalent to the reductions gained from avoided deforestation, showing that a balanced approach that allows for conversion of arable land suitable for agriculture combined with efforts to sustainably manage forests can yield a positive outcome (REDD+ EZ).

those found in voluntary markets. The largest compliance market is the Emission Trading System of the European Union, where companies and nations trade a variety of emission reduction units*, but the European system does not (yet) allow for the trade of emission reductions credits derived from forest conservation (e.g., REDD). Voluntary markets, on the other hand, serve countries and economies that have yet to adopt mandatory emission reduction targets. Prices for CO_2 emission reductions in voluntary markets have ranged between \$5 and \$8 per ton CO_2, with between 65 and 125 million tons of offsets traded annually over the past three years.[16] Valuations in compliance

* An emission reduction unit that refers to a tradable credit is typically expressed as a ton of CO_2 but may also include other greenhouse gases, as well as sequestered carbon, all of which are expressed in equivalents amount of CO_2 in terms of GHG impact.

Table 6.4. Comparison of carbon stocks and emission reductions between two different reference case scenarios and three future development scenarios show the range of potential revenues from a potential REDD+ mechanism. The potential revenues values reflect the total annual revenues over 40 years of a hypothetical REDD agreement, but the future revenues have been discounted for inflation and reflect their net present value (NPV) in 2010.

	Reference Case Scenario	
	Historical Development	*Accelerated Development*
(A) Future Development Scenario	*$US*	*$US*
REDD – Economic Zoning	1,368,184,113	2,794,949,789
REDD – Conservation Criteria	2,193,169,246	3,661,169,246
REDD+ (Economic Zoning & No Logging)	3,826,235,570	5,294,235,570
(B) Future Development Scenario	*$US (NPV)*	*$US (NPV)*
REDD – Economic Zoning	109,133,831	435,747,942
REDD – Conservation Criteria	499,787,098	858,678,259
REDD+ (Economic Zoning & No Logging)	877,149,339	1,236,040,500

markets were generally twice this amount with volumes almost an order of magnitude greater: Approximately 6 billion tons (Gt) of CO_2 carbon credits worth an estimated $US120 billion was traded in 2009 in compliance markets — up from only 235 tons in 2005.[17] Forest based carbon credits, which largely consist of emission reductions from REDD+ pilot projects, are almost entirely traded in voluntary markets and, as a consequence, their values lag behind the value of emission reductions from the industrial or transportation sectors. Future growth in the compliance market is contingent on the parties to the UNFCCC reaching an agreement and fulfilling the commitment to incorporate a REDD+ mechanism. Even if those efforts fail, however, other solutions will come to the forefront, and the potential for forest-based carbon offsets is considered to be good.

Based on these prices and the above calculations, potential revenues from REDD initiatives in the Greater Cardamom Region would range between $1.5 and $3.6 billion over 40 years based on avoided deforestation and a nominal price of $5 per ton CO_2 (Table 6.4A).

The lower value is based on a comparison between the two conservative scenarios: the historical development scenario (HD) and the economic zoning scenario where areas appropriate for agriculture would be cleared (*REDD EZ*) and generate annual revenues that would vary between $10 and $80 million. The higher estimate is based on comparing the accelerated development (AD) scenario with an aggressive strategy to conserve all forest landscapes (*REDD CC*), a strategy that would generate annual revenues between $70 and $110 million. One option might be to discount the future economic value of REDD revenues and endow a trust fund using REDD+ revenues discounted for inflation over the first decade as part of a concerted effort to implement a national REDD+ strategy (Table 6.4B).

Choosing to pursue either an aggressive conservation policy (*REDD CC*) or the compromise pathway that tolerates some forest conversion (*REDD EZ*) are hypothetical choices and could play a role in a marketing strategy for attracting potential investors in a Cambodian REDD initiative. The real emission reductions and the corresponding payments (including from a trust fund) would be based on performance, which would be monitored and verified over the life of the initiative.

The real REDD policies and outcomes will emerge from a political process that involves multiple stakeholders. There are obvious political benefits to allowing current patterns of land-use and land-use change to continue, particularly on the low-biomass landscapes of the agricultural plains, whereas implementing conservation measures on the sparsely populated high-biomass and biodiversity-rich habitats of the Cardamom and Elephant Mountains would probably not meet strong resistance from the population. Nonetheless, all natural habitats provide important ecosystem services, so measures to conserve dry forest and woodland vegetation should not be ignored purely because they are low carbon, as they would also benefit Cambodian society, particularly rural residents who depend on forest resources.

Both REDD scenarios discussed in the previous paragraphs are based entirely on reduced emissions from avoided deforestation and neither considers the potential of avoiding forest degradation as a way to enhance carbon stores. Doing so could significantly increase the value of Cambodia's forests as a financial asset in the context of REDD+. In a primary forest, carbon stocks are assumed to be at equilibrium (photosynthesis = respiration), and reduced emissions from avoided deforestation are based on that maximum equilibrium value. In an exploited or degraded forest, however, carbon stocks have been reduced because timber extraction and the associated collateral damage both reduce carbons stocks. The above-ground biomass of a mature, unexploited rain forest should store between 125 to 275 tons of carbon per hectare, semi-intensive logging can reduce this to less than 75 tons of carbon per hectare. Once logging stops, however, trees will regenerate and increase in both density and girth; a degraded forest will increase its biomass between 2–5 tons per hectare each year until its carbon stocks are approximately equal to those of an undisturbed forest — assuming the forest is not subject to another harvest cycle. The definition of a sustainably managed forest, thus, is one where the timber and waste is approximately equal to the annual growth rate of the forest and where biomass remains constant over time, even if the forest is harvested periodically. If the forest is set aside as a protected area, its carbon stocks would also recover and the reduced emissions from avoided deforestation would likewise increase in value, as the following scenario illustrates (Figure 6.6, bottom).

Recovering Carbon Stocks in Previously Exploited Forest (*REDD+ EZ*). In this scenario, the Cambodian state chooses to pursue the economic zoning option described above where forest habitats on arable soils suitable are cleared for agriculture and the high-biomass, humid forest landscapes of the Cardamom and Elephant Mountains are conserved. This scenario, however, also adopts a policy to reduce forest degradation and foregoes the exploitation of its forest estate for 40 years. In spite of initial high deforestation rates, total carbon stores are immediately offset from recuperating forests, which continue to generate carbon stocks that are almost double those projected in scenarios based only on avoided deforestation (Figure 6.6, bottom). The largest gain is in humid forest landscapes that have been most severely affected by unsustainable logging practices (e.g. the Forest Estate). Dry forest would also increase their carbon stores over time and generate REDD+ revenues, but only if measures were taken to suppress wildfire and promote the use of charcoal cook stoves. This scenario, which includes both avoided forest degradation and forest regeneration, could provide Cambodian society with revenues ranging between \$3.2 and \$5.3 billion (Table 6.4) and annual income flows between \$100 and \$150 million. To make that claim, however, the state would have to conduct relatively precise forest inventories that document the increase in forest biomass over time. An additional long-term economic benefit would be the opportunity for Cambodia to recover its high value timber resources, which could then be exploited via sustainable forest management (SFM) after the 40-year REDD+ period expires.

The Importance of Monitoring

Carbon related revenue generated via REDD+ will require a comprehensive documentation system on the part of individual nations. The carbon stored in natural and planted forests must be accurately assessed, and its flows into and out of those ecosystems monitored. This will require a multiphase program that improves the knowledge base of Cambodia's forest and non-forest resources, which can be updated to provide accurate estimates of its carbon stocks and, at the same time, provide the information need to improve the management of its natural resources. Cambodia's forest authority has already started this process by documenting forest cover and forest cover change.[18] Their cartographic studies must now be coupled with efforts to document the actual carbon stocks in protected, managed and unmanaged forests. In addition, the forest cover map needs to be stratified according to differences in biomass, and each stratum will need to be sampled by plot-based forest inventories. Plots must be monitored periodically to document increases in biomass, particularly in forests that are recovering from a previous round of forest exploitation. Methodologies must be developed to quantify carbon emission related to forest exploitation and how those emissions are reduced by sustainable management practices. Although this monitoring system will have to cover the entire nation, it must also provide services at the local level, as action to reduce carbon emissions or to increase carbon sequestration will occur on public landscapes across the nation. If revenue flows to these localities are to be based on performance and compliance, recipients will need a reference baseline upon which to base payments and to demonstrate that they have preserved or enhanced those carbon stocks. This local monitoring will need to be integrated into the national system, so as to manage the distribution of benefits via a transparent and fair system.

The Economic Value of Biodiversity Conservation

There are many ways to assess the economic value of biodiversity. Perhaps the easiest and most robust measurement is the market value of items sold directly to human populations as food. In Cambodia this includes both the inland and marine fisheries, which are important components of the domestic economy. Terrestrial biodiversity also has market value as fiber and building materials, or as a source of energy; plant products play an important role in traditional medicine, and there is an increasingly large global market for "natural" products for cosmetics and other uses. Unfortunately, this same market demand can lead to overexploitation when it comes to wildlife, as we see with the local and global wildlife trade that has decimated mammal populations throughout the Greater Cardamom Region, which is fundamentally driven by the very high economic value of animals and animal parts.

It is important to recognize that biodiversity also has other, non-market value that warrants consideration as it is weighed against traditional sources of revenue such as extraction and development. For example, the scenic view of a wooded ridge can be essential in the business plan of a resort hotel, while the myriad forms of life found on a coral reef attract visitors and allow coastal communities to diversify and add value to a tourist package based largely on beach recreation. Similarly, the world's large community of avid bird watchers is an important upscale tourist market, while young backpackers in search of a gibbon colony or elephant wallow can bring revenues to remote forest communities. Beyond the strictly economic potential that we can measure directly by markets and exploit by economic enterprises, biodiversity also has other inherent value, as exemplified by the Buddhist tradition of respecting all living things, a Christian ethic of conserving a divine gift, or a humanistic philosophy of preserving a natural heritage. Recognizing the different ways of valuing biodiversity is an essential first step in convincing society of the need and the wisdom in conserving biodiversity.

Fisheries

The most important biological resource in the Greater Cardamom Region is its fisheries, whose remarkable productivity is a direct result of the area's marine, coastal and freshwater biodiversity. The lower Mekong Basin is home to an estimated 500 species of fish, while the coastal waters support more than 470 species.[19,20] Although commercial exploitation is concentrated on a dozen or so large fish in both ecosystems, the entire fish fauna is part of interconnected food web that supports the productivity of both the commercial and subsistence fisheries. The economic scope of these fisheries spans the breadth of the economy, ranging from small-scale subsistence fishermen who rely on rudimentary fish tackle, to mid-scale family enterprises with more sophisticated fishing methods to harvest a greater diversity of fish and aquatic habitats. At the top of the economic and technological ladder are the large-scale commercial fishing companies that have access to

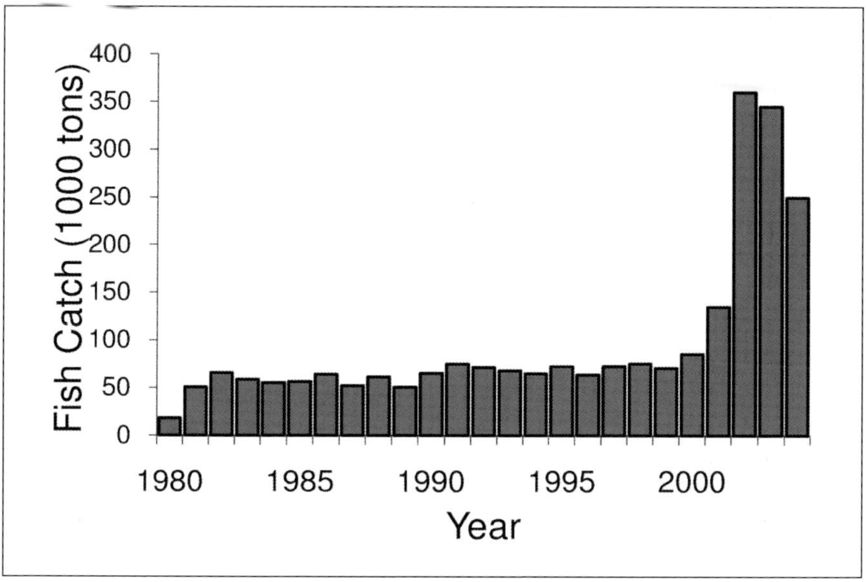

Figure 6.8. The perception of the value of the annual fish harvest from the Tonle Sap and Mekong River systems was dramatically altered when the Fisheries Administration began estimating the harvest of small-scale subsistence fishermen.

the fishing lots auctioned by the national government. Finally, there is a state-owned company that has a monopoly on the export of fish and fish products.

The remarkable productivity of Cambodia's inland fisheries depends on water flows into and out of the Tonle Sap, which occur due to the unique topography of the lower Mekong River Basin and the pronounced seasonality of the Asian Monsoon. The dimensions of this inland fishery are unusually large in proportion to the national economy. The Mekong River Commission estimates that between 289,000–431,000 tons of fish are harvested each year, of which two thirds originate directly from the Tonle Sap. In 2002, Cambodia's Fisheries Administration expanded the scope of its annual census to include not only the commercial fish harvest but also the catch from rice fields and small-scale subsistence fisherman. This led to an increase of almost 100 percent in total reported fish harvest, demonstrating the true dimension of this resource within Cambodia's national economy and its importance to hundreds of thousands of rural Cambodians (Figure 6.8).

The financial contribution of the freshwater fishery has been estimated at $500 million, corresponding to 12 percent of Cambodia's GDP. Fisheries are labor-intensive and have an outsized importance within the national economy because it provides so many jobs to individuals. An estimated 812,000 individuals are directly employed as fishermen, while secondary employment from transportation, processing and commercialization generates an additional 2 million jobs.[21] In a country of 14 million inhabitants, that means that just about every single family is involved in the fisheries economy, either directly or indirectly. How many of these jobs are located within the Greater Cardamom Region is difficult to know with any level of precision, but since the Tonle Sap forms the northern border of the region and the marine fisheries are a major economic activity on the coast, it is safe to assume that fisheries are second only to rice farming as a producer of income and jobs in the region.

Marine fisheries are similarly important to the economy of the Cardamom region. The Fisheries Administration estimates that around 50,000 tons of seafood are harvested each year in the waters off the Cambodian Coast; this include more than 14,000 tons of fish and 18,000 tons of shrimp, with smaller amounts of crabs, shell fish and squid.[22] However, these figures seriously underestimate the real total catch in Cambodia's coastal waters because they only include data from commercial fishing fleets with motors larger than 33 hp. Thus, they miss the catch of thousands of fishermen who use rudimentary fishing techniques. Moreover, as much commercial activity in offshore fisheries is the domain of non-Cambodian trawlers, their catch is offloaded in their home ports and thus not registered (nor taxed) by Cambodian authorities. By way of comparison, Thailand harvests about 1,000 tons of fish and seafood per kilometer of coastline, while in Vietnam this value is estimated to be between 800 and 400 tons per kilometer.[23] Cambodia reports less than 115 tons per kilometer, a clear indication that harvests are being underreported by at least a factor of three, if not more.

No clearer example can be found of the economic value of Cambodia's biodiversity than in its fisheries. Overexploitation or a disruption in the hydrology of the Mekong River would have massive impacts on the national economy. Reduced fish yields would negatively impact millions of Cambodians, including those who depend directly or indirectly on this resource for their livelihoods (Figure 6.9). The general population would have to replace this protein source with another one, which would probably cost more to produce (or import), provide fewer jobs, and be of lower quality as a nutritional staple.

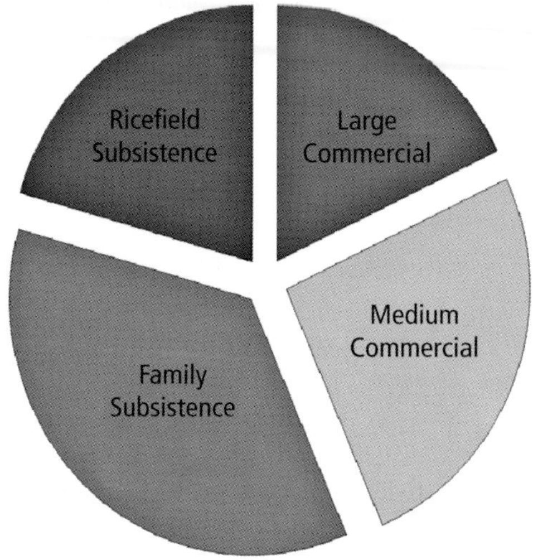

Figure 6.9. The importance of subsistence fishing in providing protein to the Cambodian population cannot be overemphasized, as is the role that inland freshwater commercial fisheries have in providing both jobs and food resources to the nation.

Forests

Cambodia's forests are extraordinarily rich ecosystems that contain many species that add value to the nation's economy and to the livelihood of its people. The true economic value of Cambodia's forest is difficult to estimate precisely, because the commerce surrounding forest products, including the trade in timber, has been underreported or unreported. Some products, such as medicinal plants or bamboo, are traded almost entirely on the local or domestic scale. Nevertheless, the economic value of these forests — ranging from the international timber market to the use of non-timber products for local construction, or medicinal herbs and resins collected by local communities — is critical to Cambodia's economy.

The largest and most obvious economic valuation of Cambodia's forests comes from its worth on the global timber market, although this is difficult to assess accurately due to the clandestine nature of Cambodia's past timber trade. Harvests between 1986 and 2000 were reported at more than 5 million cubic meters according to the Forest Administration.[24] Given the current market price for round wood in Southeast Asian markets, this amount would be worth more than $1 billion dollars today.[25] This figure is a massive underestimate, however, because the volume of timber harvested was considerably underreported. Moreover, if the timber had been processed domestically for construction lumber or finished products such as furniture, doors and windows, its contribution to the national economy would be many times greater. The potential value of a restored and sustainable timber industry can also be estimated according to forest area, which in Greater Cardamom Mountains is reported to be approximately 1.5 million hectares. Tropical forests in Cambodia produce about 4 tons of biomass per year;[26] if only 25 percent of this growth were captured as timber, the total annual production in wood would approach $200 and $400 million per year, based on current prices for high quality sawn wood in Malaysia and Indonesia.[27] Although the forests of the Cardamom's were overexploited during the 1990s, significant timber resources within the area remain and are recuperating their value over time. Conserving the value of that type of biodiversity would have clear positive outcomes for the country.

The other major forest product with economic value in Cambodia is fuel wood for cooking and charcoal production; Cambodia is one of the ten most fuel-wood-dependent countries on the planet, with about 90 percent of its total energy consumption coming from this energy source. The impact of this activity on land-use change and forest degradation was discussed in Chapter 5; the economic value of that consumption can be estimated by multiplying the reported annual fuel wood consumption of 9.4 million m^3 by $0.03/kg, which is the current price for fuel wood in most Cambodian provincial cities.[28] This yields an annual value of $282 million, which is a relatively modest sum considering it provides energy to 90 percent of the nation's population. Considering that the Greater Cardamom region incorporates about one third of the national area, its contribution to the nation's fuel consumption would be about $90 million annually, although it is probably greater because much of the firewood for Phnom Penh comes from the dry forest landscapes on the eastern flanks of the Cardamom Mountains and the Phnom Aural Wildlife Refuge.

Cambodia's forest is also the source of other non-timber products (Figure 6.10), the two most important commercial species being rattan

Figure 6.10 (p. 203). The Khmer depend on forest resources as a source of food, fiber and fuel: including a multitude of house artifacts (top left), energy for both domestic and industrial use such as brick making (top right), and food especially fresh fish, but also bamboo shoots (left middle and bottom); artisans add value to hard wood for statuary and furniture (right).

and bamboo. It is difficult to estimate the direct economic activity of these two plant products, because they are largely traded in informal markets by locals who buy and sell finished or semi-processed artisanal products. It is precisely this link to local markets that makes them attractive as a development option, because the economic benefits flow to people situated close to the resource. The international trade in bamboo and rattan — furniture, mats, screens and baskets — was reported to be more than $2.1 billion in 2007,[29] of which Cambodia exported a mere 376 tons with a reported value of $500,000.[30] This number massively underestimates the true commercial value because it is based only on world commerce and ignores the even greater volume of goods traded within domestic economies. By some estimates, the true level of global trade is over $10 billion annually[31] and the potential for growth is many times that amount, as the demand for "natural products" continues to expand. Values have not changed much since the last few years of the previous decade.[32] Nonetheless, few would doubt the importance of bamboo to Cambodia's domestic economy, where it is used for scaffolding on a construction site in Phnom Penh or for a broad array of household products sold by vendors in urban markets.

Resins and gums are other non-timber products originating from several distinct dipterocarp tree species, which when taken together rival rattan and bamboo in importance within the domestic economy. The most important, known as Chor Tuk resin, is extracted from the bark of the endangered forest canopy tree *Dipterocarpus alatus*. It is used by Cambodia's fishermen to waterproof their boats and is exported as a feedstock for paint and varnish manufacturers in Vietnam and Thailand. The resin is collected by creating a hole in the bark of the tree and searing it with fire to stimulate the flow of resin, which is then collected over three or four days; resin is harvested once a year and trees can produce resin for up to 60 years.[33] Approximately 20,000 tons of resin is collected annually across Cambodia at a reported price of $172 per ton, amounting to approximately $3 million in revenue that is shared along a supply chain including tree tappers, wholesalers, transporters, and domestic retailers. The value of the *dipterocarp oleo* resin export market is unknown, but if the annual production was commercialized at the reported value paid for exported resin, it would total approximately $6.5 million.[34]

Other resins include Chor Chong (*Shorea vulgaris* – Dipterocarpaceae), which is used as fuel and lacquer from the Kroeul tree (*Melanorrhoea laccifera* – Anacardiaceae); Gamboge gum, used as a pigment for water colour paints (*Garcinia hamburyi* – Clusiaceae) and laka pigments which provides a strong red die that is actually derived from insects that have infected the bark of any one of a number of small trees or shrubs. Historically, one of the most valuable natural resources from the Cardamom Mountains was Agarwood, a resin extracted from the *chan krassna* tree (*Aquilaria crassna* – Thymelaceae). Also known as "*oud*," this essential oil is almost as highly prized as sandalwood, both of which have been traded in the Orient and Middle East for at least 3,000 years. Agarwood resin is produced as a defence mechanism by trees in response to a fungal infection (*Phaeoacremonium parasiotica* – Ascomycota); poor quality oils are traded internationally for about $20 per kilogram, but highly refined and distilled oils can cost as much as $7,000 per kilogram.[35] Prior to the civil war, the traditional communities of the Cardamom Mountains harvested about 50 tons of Agarwood annually; at today's prices this would equal about $8.5 million.[36] Resins continue to represent an important source of cash income for forest communities to supplement what is in most other respects a subsistence livelihood.[37]

Some nontimber forest products are so valuable that they challenge the security forces who oversee the forest resources of the Cardamom Mountains and adjacent protected areas. A case in point is the recent effort to control the harvesting of sassafras oil from the Mreah Prew Phnom tree (*Cinnamomum parthenoxylon* – Lauraceae); sassafras is used to fabricate "ecstasy," a hallucinogenic drug in the methamphetamine group of pharmaceuticals. Ecstasy first became popular in the late 1980s as a "designer drug" among young people in the United States and Western Europe and, over the past decade, has become the most popular illicit drug in Australia. Its immediate precursor is safrol, an essential oil that is found in the resins of numerous species in the laurel family;* the concentration of safrole in Mreah Prew Phnom wood is unusually high, making it very profitable to exploit that species for the production of this illicit drug. In Australia, ecstasy retails between $40 per 150 mg dose, and the 33 tons of safrol incinerated by the Cambodian police in February 2009 (Figure 6.11) had an estimated street value of approximately $7 billion dollars.[38] The revenue that would have flowed to Cambodia from this illicit activity would have been a miniscule part of that estimated value, however,

---

*This includes the North America sassafras tree that is the ingredient of root beer; it is also a component of camphor which is used both as a cooking ingredient and as an embalming fluid.

Figure 6.11. The wildlife trade takes a heavy toll on the flora and fauna of the Greater Cardamom Region. Pangolins (top left) and turtles (middle right) are in much demand as culinary delicacies in neighboring countries, while the trade in crocodile, otter and cat skins is commonplace. The bark of the Mreah Prew tree is used to extract a precursor for the illicit drug ecstasy at illegal laboratories in the Cardamom Mountains (bottom).

because international drug cartels siphon off most of the proceeds so the price paid to remote drug labs is a fraction of retail price.

Although the exploitation of illicit drugs cannot be condoned — much less used — as part of a sustainable forest management strategy, it demonstrates the pharmaceutical potential of the species in Southeast Asia's rich tropical forests. Beyond the potential to discover new pharmaceuticals, traditional remedies based on medicinal plants can also make real contributions to the national economy. Between 600 and 800 species of plants in Cambodia are believed to have medicinal value, many of which form the backbone of Khmer traditional medicine,[39, 40] whose practitioners are the primary health care providers to the majority of Cambodian citizens. Unlike the market for illicit drugs and the trade in mammals, the demand for medicinal plants rarely endangers native species populations, partly because their consumption is not driven by a global market, but also because the most popular medicinal plants are either cultivated in home gardens or grow in disturbed habitats near villages.

The value of traditional medicine to the national economy is difficult to estimate; there are few reliable statistics as to the fees charged by practitioners or the level of efficacy of their treatments which are based on a combination of magic combined with mixture of plant and animal derived concotions. That at least some of these remedies are effective was recently confirmed when a group of anti-malarial drugs prescribed by traditional healers was identified to be effective against various strains of malaria, including the *Plasmodium falaciparium* strain that is resistant to many western medicines.[41] This result demonstrates the value of traditional knowledge in the search for potential pharmaceuticals; the centuries of use essentially represents a long-term screening for effectiveness and potential side effects due to toxicity. Even if many of these plant remedies are less effective than western remedies, they nonetheless constitute part of the local economy by generating income for the families of the medical practitioner and by providing both tangible and intangible benefits to the patients. Medicinal plants are just one example of the multiple plant resources upon which rural communities depend for their livelihoods. Native or semi-wild plants also provide food for families, fodder for animals and construction materials for homes.

Wildlife

Wildlife has long had an economic value on the market. Traditional Asian medicine has a long history of using animal tissues both as prophylactic and curative treatments. Practitioners of Khmer traditional medicine, known as *Kru Khmer*, likewise use animal parts for their remedies, though practitioners have sought out less expensive plant substitutes as the availability of animal parts has decreased.[42] Demand remains strong, however, among affluent Chinese and Vietnamese expat communities and middlemen who export animals and animal parts to their home countries primarily to meet the demand of traditional medicine practitioners and high-end urban restaurants that serve exotic meats. The prices for animal parts in Cambodia range from a few dollars per animal, such as an otter or loris, to hundreds of dollars for a single animal organ, such as a bear gall bladder.[43,44] Unlike the relatively low per-unit market value of most biodiversity products, such as bamboo, rattan, or even fish, the per-unit market value of animal tissue for medicinal purposes or as a culinary delicacy is relatively high (Figure 6.11). Moreover, the animals in high demand are the most endangered vertebrate species — elephants, tigers and bears, as well as dugongs, sun bears, and pangolins — and are threatened by the failure to control export markets. Indeed the illegal trafficking in wildlife is similar in many ways to the trade in illicit drugs.[45]

Unlike with other forest resources, the purpose of knowing the economic valuation of wildlife is not intended to improve the sustainable management of the resource (as, for example, with timber) but rather to slow down, or stop, a commercial process that, ultimately, has negative economic consequences. Efforts to improve the management of timber or nontimber resources can bring positive economic benefits to communities, but the trade in wildlife is usually an unsustainable exploitation of a shared resource by individuals that does not benefit the larger community — a phenomenon that is referred to as the "tragedy of the commons."[46] The wildlife trade in Cambodia is an example of this phenomenon, but with two major factors that make it even less economically attractive for the nation, and, consequently, easier to combat: (1) the individuals who are consuming the wildlife resource are not local residents, and (2) only relatively few people in Cambodia are actually wildlife hunters, so the economic benefit to Cambodia is not large and, theoretically, should be amenable to regulation and law enforcement action to terminate the most abusive behavior.

The illegal trafficking in wildlife is a different and more profound problem when compared to subsistence hunting, which has only a negligible impact on populations of highly endangered vertebrate species. Most of these animals have already been eliminated in more accessible areas due to overhunting and are now found only in more remote areas of the forest that require specialized knowledge in tracking and trapping. Illegal trafficking in wildlife, although driven by urban and export market, is undertaken usually by specific individuals in forest communities. Most illegal hunters are not typical members of forest communities, where the overwhelming majority of families are dedicated to subsistence farming. Instead commercial hunters often have low social status and lack access to land, so market hunting is one of the few options for supporting their families. Successful efforts to eradicate the wildlife trade are usually linked to initiatives to help these families improve their livelihoods by facilitating their access to other forest resources or by hiring them as forest rangers or park guards. The distinction between market and subsistence hunters, and the dominance of foreign demand in driving the wildlife trade, creates an opportunity to portray conservation initiatives as a way to preserve valuable local resource from outside exploitation. Many of the most threatened species are also charismatic and have significant cultural value within Khmer society, as demonstrated by recent efforts by Buddhist organizations to promote wildlife conservation.

The wildlife trade also has an economic cost beyond the direct and indirect impacts caused by the loss of biodiversity — the increased risk of disease transmission. Two recent examples have highlighted the link between wildlife trade and infectious diseases: the HIV virus that causes AIDS and the coronavirus responsible for "severe acute respiratory syndrome" (SARS).[47] In both cases, these lethal pathogens originated as a much less virulent virus restricted to wild populations of primates (HIV) or felines (SARS); viral particles were able to cross the physiological and immunological barriers that normally exist between species — either through the mixing of blood via open wounds or bites (HIV) or by inhaling virus particles distributed via feces (SARS). The cost of a new emerging disease is massive, both in the potential loss of human life and in the precautionary measures that must be taken to prevent the spread of the pathogen. In 2003 the SARS epidemic devastated the tourist industry in Hong Kong, resulting in losses that surpassed $3.6 billion in revenue and 27,000 jobs. Tourists are exceptionally sensitive to disease risk and the potential of an emerging disease from Cambodia's wildlife trade could devastate its vibrant and strategically important tourist industry.

Landscapes

Natural beauty is an enormously valuable asset, and Cambodia is an amazingly beautiful country. Whether seen from the air or from the ground, the country is a mosaic of stunning anthropogenic and natural landscapes. These scenic landscapes add value to the life of the average citizen, as well as forming the linchpin to a strategically important tourist industry. The appreciation of the economic value of these assets is evident, ranging from the housewife who maintains a flower garden to the hotel owner who strategically places his dining hall to over look a coastal panorama. Investing in and preserving natural beauty is an inherent part of Khmer culture, and Cambodians create wealth by consuming services, such as gardeners or landscape architects, that improve their living spaces by making them look more like natural places. Moreover, as the economic opportunities from tourism have become more apparent, so has the realization that the conservation and enhancement of natural beauty should be an integral part of any tourist investment — including public investments such as roads and energy generation that support the tourist sector, which in 2007 enjoyed revenues exceeding $1.7 billion.[48]

There is a tendency to subdivide the tourist industry into sectors or "niches," to better market tourist products. This strategy has been used by entrepreneurs who wish to use tourism as a model for sustainable development and channel some revenues to support the environment and benefit rural communities. Thus, tourist packages labeled as ecotourism, adventure tourism and cultural tourism have all been incorporated into Cambodia's increasingly diversified tourist industry. However, it is not just these specialized tourist niches that profit from the natural beauty of the country; all tourists expect to visit areas that are aesthetically pleasing. Consider, for example, the popularity of the Ta Prohm Temple near Angkor Wat, where massive fig and silk cotton trees tower over an ancient Khmer temple, while their roots and lower branches envelop the exquisite carvings of Buddhist mythology. Even the most common Cambodian landscape, the rice plain, offers lessons on the aesthetic value of landscape: one of the most popular travel guides makes a favorable comparison between Cambodian and Vietnamese by noting that the very intensive farming practices of Vietnam have produced a monotonous landscape, while in Cambodia the abundance of sugar palms and tree-lined hamlets is much more pleasing to the eye.

What constitutes an aesthetically pleasing landscape? In most all cases it is the combination of certain topographic features, such as mountains, hills and watercourses, while panoramic vistas can be particularly striking. One common element in almost all scenic landscapes is the presence of vegetation, even on landscapes that have been almost completely transformed by human activities. Natural areas such as the Cardamom Mountains and the Cambodian Coast are exceptionally beautiful — and particularly valuable — and should be developed according to a strategic plan that preserves this beauty. Agricultural landscapes with intact riparian corridors, tree covered hillsides, ponds, and watercourses are widely recognized as being more attractive than those without these features. Most people would agree these are better places to live and raise a family, where children can climb trees to harvest edible fruits and swim in ponds to cool off.

Cultural Values

The Buddhist tradition includes among its fundamental philosophical tenets the interconnectedness of all life forms and the benefit of living in harmony with nature. In Cambodia, Buddhist temples are at the center of village life, and pagodas have a long tradition as a place of teaching as well as a forum where communities seek consensus on important issues. There are approximately 4,000 pagodas in Cambodia, mostly located in rural areas. About 50,000 monks and 10,000 students live in these communities, and each pagoda attends to the spiritual needs of between 200–500 families. The Buddhist tradition of respecting nature is more than just an intangible concept linked to spiritual well-being and cultural identity; in the case of Cambodia, the Buddhist pagodas also contribute to the material well-being of their adherents. The grounds of most pagodas contain a small pond that stores water throughout the year and is connected to the neighboring rice fields via a system of canals. These small quasi-natural habitats provide a "keystone" ecosystem service, functioning as fish sanctuaries where fish can repopulate the rice paddies after the onset of the monsoon and survive the dry season by feeding on the seasonal abundance of algae and invertebrates. The intake from rice field fisheries is estimated at between 50,000 to 100,000 tons annually; only a fraction of this would be harvested without the existence and contribution of pagoda ponds.[49] Many pagodas also incorporate a forest remnant within their grounds, particularly those associated with ancient Khmer temples located on the small hills known as *phnoms* that are scattered across the alluvial plain of the Mekong and Tonle Sap (Figure 6.12). Virtually all of these isolated hills are covered with some type of forest vegetation, and even though it is usually severely degraded, it provides the local population fuel wood, construction

Figure 6.12. Pagoda forest, such as the former royal capital at Oudong in Kandal Province are common throughout the country and provide a unique opportunity to increase forest resources that will directly benefit communities.

material and medicinal plants. Assisting Buddhist monks improve the management of these forest remnants which bring significant benefits to most communities.

The spiritual and cultural benefits of pagodas to the Khmer people defy any objective economic valuation; however, the economic value can be estimated. Assuming, for example, that each family enjoys about $100 worth of tangible benefits from the pagoda each year, then the contribution to the national economy would be equivalent to approximately $200 million. Beyond this, the Buddhist tradition commitment to sustainable development has also taken on a new, more modern, dimension over the last decade, as Buddhist monks and teachers have become advocates for conservation and sustainable development, educating the Cambodian people about the negative aspects of the wildlife trade, illegal logging, and uncontrolled deforestation, while promoting community forestry, watershed management and other issues.[50]

Leveraging Nature for Economic Growth

This chapter has provided an overview of the many ways Cambodia's natural resources — its mountains, water, forests, fisheries, landscapes — generate wealth for the nation, a wealth that is often undervalued and underappreciated when development opportunities are evaluated. The new rational for estimating the economic value of ecosystem services is the notion that development and nature conservation are not at odds. Rather, economic value is derived from and depends on natural resources. Traditional development depends on nature too — land, water, air, timber. The difference is not whether societies benefit economically from natural resources, but rather on where and how it is valued and over what time scale that value is monetized. The financial return over the short term is an important and rational way to judge opportunities, but a longer-term perspective is needed to ensure that a nation's natural and human-mediated ecosystems continue to provide the goods and services upon which short-term opportunities eventually depend. Over the next few years, Cambodian society will make decisions that will determine its development trajectory and will probably incorporate elements from all three of the potential future scenarios described in the first chapter. Individuals, institutions and communities will make these decisions and, hopefully, they will understand that although some trade-offs between conservation and development are unavoidable, many more are not, and that some actually provide synergies that can lead to positive outcomes. These positive outcomes are exactly what the nation needs if it is to discover the solution to the Cardamom Conundrum.

Appendix 6.1. Deforestation rates (percent annual change) for four reference scenario models to estimate potential land-use change and emission reductions in the Greater Cardamom Region. Landscape units were categorized according to topography and soils, proximity to the current agricultural frontier, population density, roads networks, and land management types (see Figure 6.6). Final Cover is the projected total forest cover in 2050.

		Historical Development Scenario				*Accelerated Development Scenario*			
		Deforestation rate (%)				*Deforestation rate (%)*			
	Historical Baseline	*2006–2012*	*2000–2025*	*2006–2012*	*Final Cover (% total)*	*2006–2012*	*2000–2025*	*2026–2050*	*Final Cover (% total)*
Thai Border	4.4	4.4	3.3	2.1	10.3	6.6	4.4	3.3	2.2
Agric North	2.8	3.4	4.1	2.4	10.4	4.9	5.9	3.0	2.9
Agric East	2.6	3.3	3.3	2.6	11.6	4.5	5.4	2.2	7.2
Sihanouk Coast	2.8	3.5	4.4	2.2	10.6	4.5	4.7	3.0	5.1
Koh Kong Coast	2.8	3.7	4.6	2.3	9.3	4.5	4.7	2.4	7.9
Phnom Aural	0.2	0.2	0.7	2.2	38.9	0.6	1.8	3.7	5.6
Phnom Sankos	0.2	0.3	0.9	1.8	46.5	0.5	1.5	3.0	19.2
SCPF	0.7	0.8	1.0	2.1	36.8	1.5	4.5	3.4	5.6
Coastal Islands	0.6	0.7	0.9	1.8	43.6	1.1	1.4	2.7	23.2
N Forest Estate	0.4	0.5	1.2	2.2	35.3	1.3	3.3	3.3	8.6
S Forest Estate	0.2	0.2	0.9	2.6	30.6	0.8	2.7	3.5	8.2
Elephant Mts.	0.2	0.2	0.7	1.4	56.9	0.4	0.7	2.2	38.8
CCPF	0.2	0.2	0.3	0.9	72.1	0.4	0.5	1.2	61.5
Total	1.45%	1.6%	1.8%	1.9%	36.1%	2.3%	2.9%	2.6%	18.4%

Appendix 6.1. Continued.

		REDD Conservation Criteria Scenario				*REDD Economic Zoning Scenarios*			
		Deforestation rate (%)				*Deforestation rate (%)*			
	Historical Baseline	*2006–2012*	*2000–2025*	*2006–2012*	*Final Cover (% total)*	*2006–2012*	*2000–2025*	*2026–2050*	*Final Cover (% total)*
Thai Border	4.4	2.2	1.1	0.6	32.8	4.4	2.2	1.1	19.4
Agric North	2.8	1.4	0.7	0.4	52.2	5.7	3.8	1.9	12.3
Agric East	2.6	1.3	0.6	0.3	55.8	5.1	3.4	1.7	15.7
Sihanouk Coast	2.8	1.4	0.7	0.4	52.4	4.2	2.1	1.1	27.6
Koh Kong Coast	2.8	1.4	0.7	0.7	47.2	4.2	2.1	1.1	27.4
Phnom Aural	0.2	0.1	0.1	0.0	95.9	0.6	1.2	0.4	87.2
Phnom Sankos	0.2	0.1	0.1	0.0	95.0	0.7	1.5	0.5	84.2
SCPF	0.7	0.4	0.2	0.1	85.5	0.1	0.0	0.0	89.8
Coastal Islands	0.6	0.3	0.2	0.1	87.8	0.3	0.2	0.1	91.3
N Forest Estate	0.4	0.2	0.1	0.1	91.3	0.1	0.0	0.0	94.3
S Forest Estate	0.2	0.1	0.0	0.0	96.2	0.0	0.0	0.0	97.4
Elephant Mts.	0.2	0.1	0.1	0.0	95.3	0.0	0.0	0.0	96.9
CCPF	0.2	0.1	0.0	0.0	96.0	0.0	0.0	0.0	97.6
Total	1.4%	0.7%	0.3%	0.2%	88.2%	2.0%	1.2%	0.2%	69.8%

CHAPTER 7

The Cardamom Solution: Envisioning a Green Economy

People across the globe have come to understand that global warming is a real phenomenon and not some remote hypothetical possibility. They see how extreme climate events can affect people's lives, and they understand that future climate change represents a threat to our collective well-being. Global society is demanding change, which is now taking shape in international policy forums and global markets. Nations are seeking ways to limit their dependence on fossil fuels and are investing in a broad range of low carbon technologies or emission abatement strategies. This response is not limited to rich nations with strong environmental lobbies and technologically sophisticated enterprises, but is also clearly evident in emerging economies where enlightened leaders see economic opportunities for future growth. For example, China leads in the manufacturing of solar panels and wind turbines; Brazil is the acknowledged giant of the biofuel industry, while South Korea is seeking to develop super efficient electrical technology. Costa Rica, Guyana and Indonesia have embraced forest conservation as an integral part of their national development strategies. These countries are radically different in their demographic and economic profiles, but each has recognized the need to prepare for the future and to embrace a low carbon development strategy that is forward looking and innovative. Cambodia is well-poised to evaluate its options and can benefit from embracing a low-carbon development strategy.

Cambodia is at an important stage in its history as it transitions from a nation with a large rural population and limited economic development, into a small but dynamic nation fully integrated into a regional economy characterized by growth and innovation. It has a unique opportunity to adopt policies that will propel it into the future as a leader in the development of a green economic model. Cambodia is already starting from a low carbon past, in part because it does not have an existing infrastructure legacy that is overly dependent on fossil fuels. Consequently, it has the option to invest in low-carbon options from the outset; its current commitment to hydropower is a good first step and provides a solid foundation for a low-carbon economy. If these investments are diversified over the mid-term to include other renewable energies, Cambodia can avoid the traditional carbon-intensive energy paradigm. What has been seen by traditional planners as a liability — the lack of fossil fuel generation capacity — can be turned into an asset, since the investment in renewable energy can benefit from subsidies via the Clean Development Mechanism (CDM) of the Kyoto Protocol, or other forms of climate finance being developed by the United Nations.[1] Similarly, energy and transport systems can be targeted for technology transfer, so that investments in basic infrastructure can skip the high carbon models typical of developed countries that have led to the current global crisis. A low carbon development strategy would also make Cambodian products more competitive in global markets; consumer countries are in the process of introducing Life Cycle Analysis (LCA) methodologies into regulatory regimes to measure and monitor the carbon footprint of consumer goods.* Further evidence of market pressure is evidenced by large retailers who now promote LCA labeling to facilitate consumer choice. A move towards a low-carbon economy would promote diversification and make the national economy less susceptible to the fluctuations of global markets. At the same time it would expand opportunities for domestic growth and job creation across the economic spectrum, from rural farmers to urban professionals.

Cambodia is also at an advantageous position at this moment in history regarding the global awareness of climate change. Because of its relatively large forest estate and documented recent high deforestation

---

* Also known as Life Cycle Assessment, the LCA attempts to measure and quantify a product's full environmental costs, from raw materials to final disposal. It considers resource consumptions, energy and waste and is increasingly being used to estimate the "carbon footprint" of different products.

rates, it will be able to negotiate significant revenues from the international REDD+ system for avoiding deforestation and forest degradation (see discussion in Chapter 6). Once the major outlines of the REDD+ mechanism are determined, other emission abatement strategies — particularly efforts to increase carbon sequestration on agricultural landscapes — will also generate interest and be accompanied by renewed efforts to improve the sustainability and productivity of agriculture in the developing world. Cambodia has a large agricultural sector that is operating well below its productive capacity, so if revenues from REDD+ can be used to invest in technologies that boost agricultural productivity on available land, then demand for new farmland also can be eased. If forest conservation and agricultural development is pursued strategically using modern spatial planning methodologies, then Cambodia can integrate scenic beauty as a priority in its national development policy and enhance the value of its tourist ventures as they expand and diversify across the nation.

To be successful, this alternative strategy must generate greater wealth for Cambodian citizens when compared to traditional options — especially over the short term, because that is the time frame in which most economic decisions are made. Just as importantly, this wealth must be equitably distributed to the individuals and communities — both rich and poor — who will be asked to forgo traditional opportunities (i.e., the conversion of forest landscapes for agriculture) that are neither green nor low-carbon. Key to Cambodia's success will be the ability to leverage the revenue flows from REDD+ and other carbon markets with other financial resources. Although REDD+ represent an important opportunity for slowing GHG emissions, they still represent modest revenue flows relative to the investments that Cambodia needs to escape poverty. The term "leverage" has many definitions, and in financial markets is mostly used most to describe the practice of borrowing money to magnify the return on investment capital; however, economists also use the term to describe the strategy of deploying limited resources in combination with other funds, so as to maximize the impact of an investment.

Key to both concepts is the word "investment." Unfortunately, there is a tendency for REDD+ revenues to be viewed as compensation mechanism that distributes money among stakeholders, particularly forest communities, who will be asked to forgo forest conversion and the development opportunities linked to it. This approach tends to treat REDD+ revenues as rent, rather as investment capital, which is an inherently limiting perspective. A more innovative strategy would be to treat REDD+ revenues as investment capital, which can be deployed to influence economic decisions and promote sustainable production strategies that provide a long-term competitive advantage to those that participate in efforts to conserve forest. REDD+ resources are already being leveraged with different types of official development assistance, but because the mechanism is not yet in force, it has yet to generate significant revenue flows from the private sector. The CDM/AR has also opened a window for the private sector to access carbon markets, but those revenues are extremely limited and have not been successful in fostering climate-friendly investment in land use.

Cambodia's low-carbon future can be constructed on four main investment strategies: forest conservation, agricultural intensification, renewable energy and mass transit. These pillars are mutually linked and potentially synergistic. For example, revenues from carbon markets linked to forest conservation could be used to improve agricultural extension services, access to fertilizers and irrigation technology, helping to promote an expanded and more diversified agricultural sector. This could include second-generation biofuels derived from tree plantations, which would also function as a carbon sink and increasing carbon storage on productive landscapes. Forest conservation and land-use planning could design a watershed management strategy that maximizes hydroelectric power generation in the Cardamom Mountains, which might be supplemented by wind turbines stationed on the ridge tops of the Elephant Mountains. Urban mass transport systems can be designed to operate on electrical energy, while heavy transport and rural machinery can use biodiesel instead of petroleum-based liquid fuels. As the electrical energy grid expands across the countryside and cultivated biomass become more abundant, families can lessen their reliance on wood and charcoal for cooking fuel that is currently extracted from forest landscapes. This will increase the biomass sequestered and stored in Cambodia's forests, which can be used to attract more REDD+ revenues.

Cambodia's demonstrated ability to attract direct foreign investment has placed it in a good position to pursue this strategy. The low-carbon development model based on revenue flows from the developed world will be successful, however, only if investors have confidence that their capital is secure. Cambodian society has demonstrated that it understands the importance of meeting business and financial commitments, as well as the economic benefits that accrue from an atmosphere of trust and integrity. The example of business leaders and public officials who embraced high workplace standards in the textile industry is illustrative; not only did they respond to international concern, but successfully defended market share in a highly competitive industry. That type of political leadership and broad societal commitment will be required to compete successfully in international carbon markets, where investors will be seeking the greatest potential return and the lowest risk of project failure. The Greater Cardamom Region is an ideal laboratory for implementing and testing this new development paradigm of a low-carbon economy. Its landscapes and populations are typical of the country as a whole, and a development strategy for this region can function as a template for the rest of the country.

Forest Resources

Cambodia has a unique opportunity to simultaneously exploit and conserve a renewable natural resource that is appreciating in value due to climate change. Decisions about how to invest in and profit from the nation's forest resources are, in fact, quite timely for the Cambodian government, because it is in the process of implementing a new management strategy that will change the nature of the public-private partnership. Having decided to terminate the old concessionaire-based system (see Chapter 5), the government has made forest management now clearly the responsibility of the Forest Administration (FA). In the old system, public forests were allocated to private companies, who paid royalties (or not) and assumed responsibility for managing the forest, as well as harvesting timber and converting it to lumber, doors, windows and other consumer products. The Cambodian state now will assume the responsibilities as the steward of the forest, while allowing private companies to pursue opportunities in the transformative and commercial phases of the forest product industry. Private firms will still play a role in the forest as service providers harvesting trees, but much larger opportunities exist for those entrepreneurs who turn timber into wood products. In the future, government authorities will act as natural resource managers and regulators by allocating harvest quotas based on scientific measures of forest productivity and ecology.

This reform of Cambodia's forest sector has been occurring in parallel with efforts to promote government decentralization and delegate some authority for monitoring and managing forest resources to local authorities, a process that is occurring in tandem with other efforts to formalize the rights of rural communities to forest resources. There remains more than one million hectares of suspended forest concessions that have yet to be allocated to a specific management or land tenure regime. It is not unreasonable to presume that if this natural forest area is not seen as profitable to the nation, it will be converted to farmland or plantation forestry. In addition, there are nearly 2 million hectares of forest within Cambodia's 15 protected areas, many of which are multiple-use reserves where the rights of communities must be respected and where settlement is an expanding phenomenon due to lax enforcement and contradictory government policies. Carbon markets can provide revenues to the state and provide society with the economic incentives needed to conserve both state and community forests, protect biodiversity and improve the provision of ecosystem services over the long term.

Sustainable Forest Management

The sustainable exploitation of forests resources at a rate and in a manner that maintains their biodiversity, productivity and regeneration capacity over time has been promoted over the past two decades as a development paradigm that is compatible with forest conservation.* Proponents of sustainable forest management (SFM) argue that if the state allows a productive forest to be exploited for timber and other resources, it will provide economic justification to maintain forest cover, especially in areas such as the Cardamom Mountains that are not suitable for intensive agriculture. The SFM model has many different manifestations, but in tropical forest ecosystems there are two key criteria: (1) the volume of timber harvested should not be greater than the forest's productive capacity, and (2) the harvest should not be so intensive as to compromise the ecological integrity and biodiversity of the natural forest ecosystem.

In practical terms, this means that natural processes of plant reproduction and tree recruitment are relied upon to maintain both ecosystem integrity and the productive capacity of the forest as a timber resource. This is a radically different approach when compared to reforestation programs that tend to dominate the public imagination. The common citizen believes that reforestation is equivalent to

---

* The full definition of sustainable forest management by the FAO is the stewardship and use of forests and forest lands in a way, and at a rate, that maintains their biodiversity, productivity, regeneration capacity, vitality and their potential to fulfill, now and in the future, relevant ecological, economic and social functions, at local, national, and global levels, and that does not cause damage to other ecosystems. <http://www.fao.org/docrep/003/x6896e/x6896e0e.htm> [accessed 15 July 2011].

planting tree seedlings or saplings into the forest ecosystem following logging. This perception comes from temperate ecosystems where relatively uniform forests are clear cut and replaced with tree plantations; in contrast, tropical forestry operations remove only between 5 and 60 percent of the standing trees. Reforestation programs based on planting trees into a natural tropical forest almost always fail, which occurs due to the inherent complexity of these ecosystems. For starters, the shock of transplantation weakens the young plant, which must compete with dozens of adjacent plants that have germinated and rooted *in situ*. To be successful, a tree-planting program would need to transplant tens of millions of saplings and conduct weeding interventions over several years to provide the transplanted saplings with a competitive edge over their neighbors. Instead, sustainably managed natural forests rely on mature healthy trees that are not logged during periodic harvest cycles; if left standing, these "seed trees" will produce tens (or hundreds) of thousands of seeds each year over several decades and provide a more economical and ecologically viable option for maintaining populations of desirable species. Because it relies on natural processes, the SFM model thus ensures the conservation of biodiversity and the provision of important ecosystem services for watershed management, as well as high quality timber that characterizes native species and maximizes the economic value of natural forest habitats.

The SFM system also incorporates straightforward business management principals exemplified by "reduced impact logging" (RIL), which depends on the careful design of logging roads and planning of tree felling. These common-sense practices lower operating costs and minimize collateral damage from sloppy tree felling practices, thus preserving the value of the standing forest.[2] Like all sustainable production models, the SFM system goes beyond ecological and economic principals to include social criteria that protect forest communities and other stakeholders with legitimate rights to forest resources, which might include subsistence hunting or the collection of non timber forest products, as well as the provision of water or water-related services that flow from forest landscapes.

Theoretically, SFM could be an integral part of a low-carbon economy in a nation with rich forest resources like Cambodia. The challenge will be to ensure that SFM operations are actually sustainable and do not degrade the forest's timber resources over time. The concept of sustainable forestry developed as a response to calls for reform in the forest products sector, where the most common practice was for concessionaires to intensively log a natural forest and strip it of all economically valuable timber, after which it was either abandoned or converted to a tree plantation or an industrial farm. This led to the development of SFM certification systems where timber companies voluntarily seek to comply with environmental and social standards.* Unfortunately, certification has had only limited success in improving forest management in many parts of the world. Economic considerations predominate, and even conscientious concessionaires tend to adopt a pseudo-sustainable model where they employ the precepts of sustainable forestry, such as RIL, but implement a management plan based on 20- to 30-year logging cycles.[3] This harvest cycle is absurdly short, considering that most of the trees removed from forests are between 100 and 200 years old. Experienced foresters are fully aware of this management paradox, but argue that they will sequentially change their target species in subsequent harvest cycles. In fact, it is more likely that short harvest cycles will lead to a progressive degradation of the forest's economic potential and, once stripped of its economically valuable specie, the degraded forest will be converted to a tree farm. This is essentially a slow motion version of the business-as-usual practices that have characterized the forest product sector over the last 50 years in tropical developing counties.

A truly "sustainable" SFM model would be based on logging cycles of 60 to 100 years,† which would also dramatically increase the carbon stored in forest ecosystems and make them even more valuable for climate change mitigation (see Chapter 6). Other proactive management practices can also lead to higher overall carbon stocks. For example, the eradication of low biomass vines increases tree growth, while the selective killing of fast-growing species encourages slow-growing hard wood trees with greater market value. These "silvicultural" treatments increase the commercial value of a forest but are only marginally cost-effective and require a decade or more to generate a return on the investment in increased timber revenues. Carbon subsidies from a

---

* One common standard is the Forest Stewardship Council (FSC), "an independent, non-governmental, not-for-profit organization established to promote the responsible management of the world's forests" <http://www.fsc.org/> [accessed 15 July 2011].

† The optimal option in timber management would be to manage timber species populations individually by culling trees according to age, growth rates, reproductive cycles, and, most importantly, population structure. The use of generic harvest cycles has been adopted as an easily monitored system appropriate for developing countries with limited technical capacity and poor governance.

REDD+ mechanism could support these activities and provide an economic incentive to forest stewards to adopt and perfect the SFM model as the economically most attractive land-use option for public forests.

The Forest Estate (REDD+SFM)

Although the peripheral areas surrounding the forest landscapes of the Cardamom Mountains would best be managed as community forests, the large central core of forest might best be managed by the FA for the benefit of the nation as a whole. Population density is relatively low in this area and the need to compensate the region's inhabitants for lost economic opportunities is commensurably small. The needs of local communities could be attended to by offering residents meaningful employment with the FA, at hydroelectric facilities, or through the tourist industry. The administrative and regulatory framework prescribed by the forest law of 2001 provides ample leeway for the FA to assume the managerial role in a REDD+ initiative, which would incorporates the SFM concepts and maximize benefits for the state. Beyond revenues from sustainable timber and REDD+, an ecologically intact forest in the Cardamom Mountains would contribute many other benefits to the regional and national economy, including the long-term viability of the strategic hydropower facilities and an expanded tourist industry linked to coastal development.

Currently, approximately 30 percent of this core forest estate is located within forest concessions, even though these are not being actively exploited as a consequence of the nationwide moratorium on logging.* Concessionaires were perceived as using unsustainable logging practices to the detriment of the Cambodian state, so the government declared a moratorium and rescinded approximately half the area licensed to private companies. The remaining concessions, although not formally terminated, have been obligated to suspend activities pending the review of their management plans (see Chapter 5). These formerly exploited production forests could form the basis of a REDD+SFM initiative.

Since the most valuable timber species have already been harvested from these forests, their economic worth has been greatly reduced, increasing the probability they would be converted to agriculture

* The moratorium on logging within concessions does not apply to bamboo concessions, such as the one in Battambang Province.

or, more probably, to commercial tree plantations if companies were allowed to pursue traditional development options. A REDD+SFM initiative would provide another revenue-generating option for the state and allow forests to recover valuable hardwood stocks. The REDD+ revenues would be partly based on avoided deforestation, but also on avoiding further degradation from unsustainable logging practices by adopting a SFM model. Over the lifetime of a REDD+SFM project, these landscapes would become net carbon sinks and generate additional revenues via carbon sequestration (see Chapter 6). The capacity for a previously logged or degraded forest to recover lost carbon is essentially the (+) in the REDD+ policy mechanism; however, for Cambodia to claim this additional carbon sequestration as marketable carbon credits, it requires documenting current biomass stocks, in order to establish a scientifically sound baseline for future compensation.

Protected Forests (REDD+Conservation)

A REDD+ project for Cambodia's forest estate also should incorporate the protected forests of both the Central and Southern Cardamoms, particularly as risks from illegal logging increase from renewed legal logging in adjacent areas and from improved access as a result of the construction of roads to the region's hydropower plants. Although the potential emission reductions from these protected forests are not as large as those from the production forests, they are still significant and warrant inclusion in a region-wide strategy to reduce emissions from deforestation and forest degradation (Figure 6.7). At the end of a 40-year REDD initiative, the Cambodian state could choose to review the status of the "protected forests" and determine whether all or part should be open to exploitation via a sustainable harvesting regime. In several decades, the economic value of tropical hardwoods will be such that the investment in forest conservation might very well pay dividends equal or superior to the revenues earned over the medium term from carbon markets. Finally, the future value of biodiversity may be realizable in international markets, or Cambodian society may decide that the conservation of its forest heritage is important, just as the preservation of its cultural legacy of Angkor Wat is in the national interest, and choose to preserve the entire area as a nature reserve.

Community Forests

One of the main criteria that have emerged from international discussions about REDD+ is the need for "benefit sharing mechanisms" to ensure

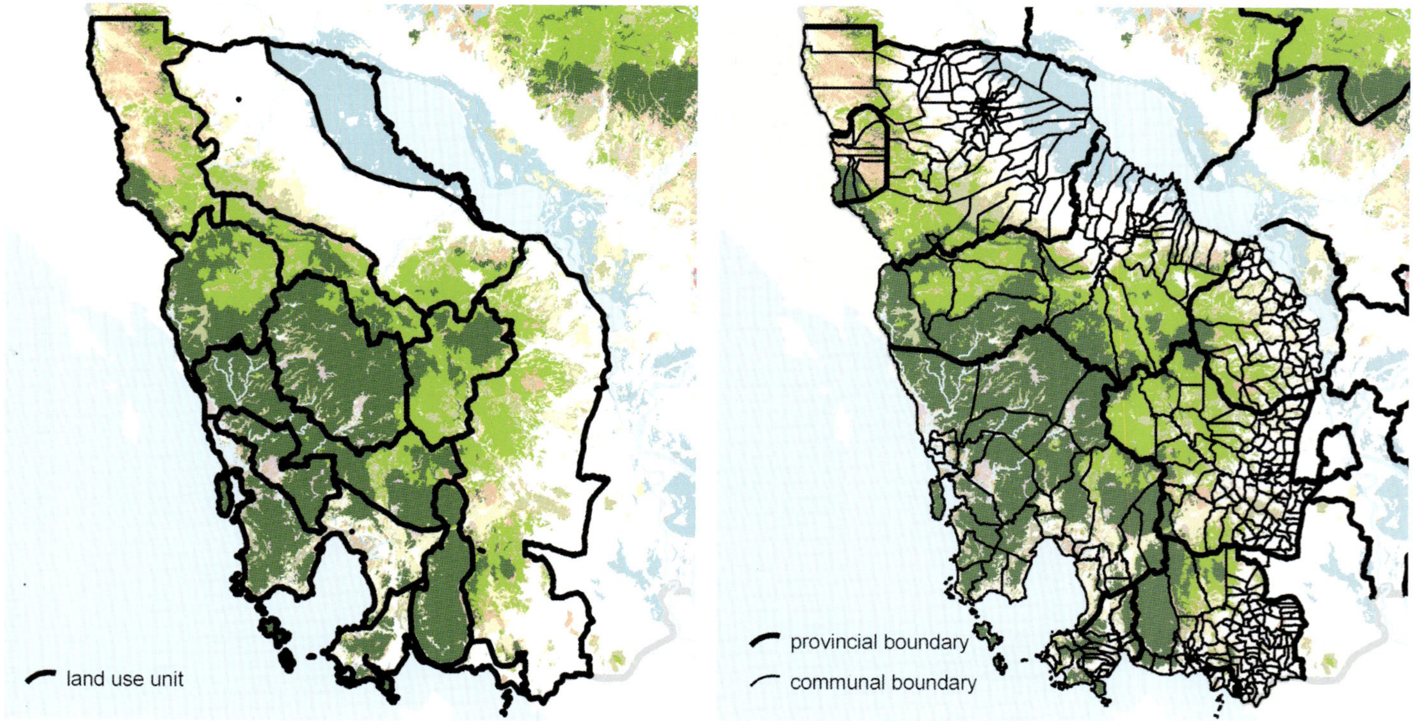

Figure 7.1. The Greater Cardamom Region can be classified into separate land-use entity based on political, physical, biological, social and economic criteria. (Left) The stratification used to estimate future potential deforestation and forest degradation from which potential REDD+ revenues were calculated. (Right) The potential for community forestry initiatives as part of a REDD+ revenue sharing mechanisms is highlighted by identifying the 140 communes with more than 1,000 hectares of remnant forest cover.

that revenues reach the communities who depend on forest resources and would suffer if their access to these resources were restricted or denied. Cambodia's existing community forest program, established in the Forest Law of 2001, is one way to direct REDD+ revenues to the local level. A network of community forests established along the margins of the forest estate in the Greater Cardamom Region could function as a buffer zone to protect the core area from overexploitation. As part of an integrated strategy, REDD+ revenues could be used to benefit local communities on the front lines of forest use and land-use conversion, provide them with an economic incentive to manage their forests and assist them in developing enterprises that sustainably exploit those resources.

Currently, there are only 66 community forests in the Greater Cardamom Region, although, somewhat surprisingly, only 20 of these 66 community forests are situated within one of the 140 communes with sufficient forest cover to potentially qualify as a potential REDD+ target. The remaining 46 communes are located in areas where most of the land has already been converted to intensive agriculture. A national REDD+ initiative provides an opportunity to radically expand this network of community forests; for example, 140 communes in the region have more than 1,000 hectares of forest cover, while an additional 30 smaller communes have less than 1,000 hectares of forest cover, but with at least 25 percent of the total area covered by forest* (Figure 7.1). REDD+ offers a means to radically expand the national network of community forests; if each of these communes allocated 70 percent of its remnant forest cover to a REDD+ initiative, that would amount to more than 700,000 hectares of forest situated on the forest frontier with a population in excess of one million inhabitants. The introduction of SFM practices would benefit all these communities that rely on forests for fuel and construction material, while investments in processing facilities and distribution centers could increase the trade in forest products and materially improve their standard of living. Thus, community forestry would contribute to growth of the national economy and provide a politically sustainable strategy for protecting forest over the long-term.

Two forest products with great potential in community forests are bamboo and rattan, both of which have robust international and

---

*There are a total of 465 communes and 53 districts in nine provinces of the Greater Cardamom Region.

domestic markets (see Chapter 6). These non timber forest products are used for a wide range of consumer goods, which themselves require a value added production chain that employs semi-skilled and skilled artisan labor, both of which are readily found in rural Cambodia. As wild populations of bamboo and rattan decline across the world, the demand for these natural fibers from cultivated or semi-wild plantations will expand and a greatly expanded system of community forests in Cambodia could supply this growing demand — by conserving existing populations and by promoting the growth and reproduction of these plants in natural or quasi-natural forest ecosystems.

In the case of bamboo, Cambodia's already existing wild populations could play an important part in an expanded production model, as evidenced by the ongoing harvest of bamboo forests in Battambang Province. Mountain communities have a long cultural tradition of harvesting bamboo, which can recover quite quickly from periodic harvests because of its clonal habits — assuming the forest is not cleared and converted to intensive agriculture. Bamboo production can be more than a sustainable forest management strategy, however, and can be the foundation of an agro-forestry production model in communities with limited forest cover or where bamboo is scarce. Bamboo has been cultivated for centuries all across Asia and most of the production from China and Thailand, the world's largest exporters of bamboo products, originate from cultivated or intensively managed native populations. There are five genera of cultivated bamboos in Southeast Asia: *Bambusa* (37 species), *Dendrocalamus* (29 species), *Gigantochloa* (24 species), *Thyrsostachys* (2 species)[*], and *Schizostachyum* (35 species), and there are many species that are native to Cambodia. The goal should not be to import genetic material, but to establish bamboo plantations using rhizomes[†] gathered from local wild populations.

Rattans, on the other hand, are not easy to cultivate, so their production is more likely to depend on semi-wild populations maintained in managed forests. Rattan populations have survived centuries of harvesting throughout Southeast Asia, but are now in decline due to deforestation in the two major exporting countries, Malaysia and

[*] This genus is endemic to the Indoburma Hotspot.

[†] A rhizome is a botanical term for a specialized stem that typically grows horizontally either underground or superficially on the soil surface; in bamboos there are two types of rhizomes: (1) *Leptomorph* rhizomes are elongate and slender and typical of species with dispersed stems, and (2) Pachymorph rhizomes are short and fat and typical of cultivated bamboos that form dense clusters.

Indonesia. Thailand, the other major producer of rattan products, now obtains most of its raw material from Laos.[4] Rattans, like all lianas, are structural parasites on trees and take advantage of forest clearings to "climb" trees. This ecological strategy is very successful in disturbed forests where logging creates numerous forest gaps; the elongate horizontal rhizomes of rattans are ideal for spreading along forest floor clearings, which allow vertical stems to grow up into the canopy. However, these same traits impede the formation of self-propagating dense colonies that are characteristic of cultivated bamboo and which makes that plant such a successful and easy to cultivate tree crop. Like the overwhelming majority of native tropical forest species, rattans require a forest matrix to survive, and the best option for producing commercial quantities of rattan may be via the intensive management of semi-wild populations in highly disturbed forests. This production model could easily be introduced and fostered in the community forests of the Cardamoms and other areas of Cambodia.

The concept of community forests in Cambodia has focused largely on conserving upland forests and providing rural populations with fuel wood, construction materials and other forest products; but the flooded forests of the Tonle Sap could also be a good potential site, and source of REDD+ income, for community management. The southern shore of the Tonle Sap, which falls within the Greater Cardamom Region, has between 150,000 to 300,000 hectares of flooded forests and woodlands, although the precise area of flooded forest is poorly documented due to conflicting definitions of forests and woodland.* Although the carbon stocks of these flooded forests are unknown, there is reason to believe they are significant. Swamp forests tend to have less above-ground biomass than upland forests, but the amount of detritus and soil organic matter is usually two to three times greater than in upland forests. Mature upland forest will typically store between 15 to 30 percent of the carbon stored in the surface vegetation, while the soil organic matter in a swamp forest is typically greater than above-ground biomass.† The reduction of emissions by avoiding the degradation of the flooded forests of the

* The distribution and conservation of vegetation types on the Tonle Sap floodplain is complicated by the poorly documented phenomena of clearing for agriculture and the role that fallow fields have in the land-use dynamic and fish reproduction in Tonle Sap fisheries.

† Peat forests, which do not occur in Cambodia, can have as much as 10,000 tons per hectare of soil carbon.

Tonle Sap could yield REDD+ revenues that might be used to subsidize their conservation. Moreover, since flooded forests are vital to maintaining the fisheries resource of the Tonle Sap and Mekong River, their conservation would contribute to the food security of the nation and improve the management of an important natural resource.

Reforestation, Afforestation and Agroforestry

International discussions on abating CO_2 emissions have focused renewed attention on countries that have largely exhausted their native forest resources. In part, this can be attributed to the need to be inclusive and build political consensus among the signatories to the UNFCC; however, it is also an objective realization that there are significant climate benefits in helping these countries increase forest cover. Reforestation refers to planting trees or restoring forest habitat in areas that have been deforested historically. It is also often part of a deliberate program to maintain a productive forest landscape after intensive logging, such as the case of temperate forest ecosystems in developed countries where clear-cutting is a common form of forest management. However, reforestation also refers to the practice of allowing recently deforested or abandoned agricultural land to revegetate and form a secondary forest via natural processes.

Afforestation refers to initiatives to plant trees in landscapes that have never been forested, or which were deforested in past centuries or millennia. In this case, areas with extremely low levels of biomass are transformed into high biomass landscapes with the potential to store large quantities of carbon over decades. Agroforestry refers to mixed production systems where the cultivation of trees is deliberately combined with farming and grazing on individual land holdings. It provides many benefits for improved sustainability, by increasing soil nutrients, controlling erosion, and improving the conditions for animal husbandry, but more important is the opportunity to diversify the production systems and reduce risk for small farmers. The prospects for agroforestry in Cambodia are manifest and have the potential to improve the livelihood of hundreds of thousands of rural poor.

In Southeast Asia, reforestation or agroforestry also sometimes describes plantation forestry or plantation agriculture, both of which require a great deal of organizational capacity and the economies of scale, as well as large capital investments in processing facilities. Plantation forestry has a long tradition in Cambodia based originally on the colonial model of producing rubber for export and industry. Rubber plantations are being renovated and expanded throughout the country, growing from 85,000 to more than 180,000 hectares in 2011, which exceed the government's plans in 2008 to reach 150,000 hectares by 2015. The demand for rubber and its resurgence in Cambodia is part of broader phenomenon where global markets for commodities, including vegetable oil, biofuels, pulp for the paper industry, and hardwood, have expanded dramatically over the last decade due to increased consumption from China and other emerging economies. However, the plantation sector is no longer dominated by distant colonial masters; it is still a global industry, but now is dominated by multinational companies based in Southeast Asia. Another important trend is the shift from large plantation in equatorial regions of the Malay Peninsula and Sumatra from rubber to oil palm due to the improved profit of that crop. The shift in production has benefited greatly smallholders in those countries, which remain the world's largest rubber exporters, and producers in more seasonal climates where rubber out competes oil palm. Cambodia's rubber producers are benefiting from these changes, including the plantations sector and smallholders; the government is seeking a balanced model where 50 percent of expansion is allocated to individual family farms. The Greater Cardamom Region has appropriate climate and soils and its many small farmers on upland landscapes can beneficially incorporate rubber into integrated agroforestry production schemes.

The Clean Development Mechanism (CDM) of the Kyoto Protocol, which underpins a large part of the current international market in carbon credits, currently provides opportunities for nations to generate revenue from the carbon market through Afforestation and Reforestation (AR) projects. Unfortunately, the CDM/AR has not been successful and after nearly a decade of planning and investment in training and pilot projects, only 22 projects have been successfully registered with the UNFCCC Secretariat.[5] Most analysts attribute this poor performance to an overly complex set of verification and monitoring rules and a project-based approach that has created an overly-bureaucratic administrative process. Hopefully these lessons will inform the development of REDD+, so that both public and private efforts to increase forest cover can be encouraged and subsidized. The plus in the proposed REDD+ mechanism is intended to provide a partial solution to the deficiencies in the CDM/AR mechanism by creating national carbon budgets, which will be used to govern national baselines. Under this proposal, each nation would have a ledger-like system that would credit carbon sequestration from increases in forest cover and reduced emissions from avoided deforestation, while debiting new emissions from land clearing for agriculture, infrastructure or mineral development. The potential to use REDD+ revenues or a reformed CDM/AR to promote reforestation, plantation forestry and agro-forestry would bring multiple benefits to the Cambodian economy, particularly in terms of job creation, poverty alleviation and sustainable production.

Pagoda Forests and Private Property

There are other pieces of Cambodia's land tenure mosaic that can provide additional opportunities to generate revenue from carbon markets. One of these is the pagoda forest, a community asset found on Cambodia's agricultural landscapes, typically in areas composed of cultivated trees. The carbon stored in a pagoda forest could be augmented by enlarging its surface area and increasing biomass stocks by reducing fuel wood collection, which could be promoted by the introduction of more efficient cooking stoves or alternative fuels. Of particular interest are efforts to promote the use of charcoal, which is more energy efficient when compared to firewood and bring additional mitigation benefits when it is produced in modern kilns that limit GHG emissions. A widespread switch to charcoal as a source of cooking fuel would also stimulate a charcoal supply chain that would provide jobs, thus further benefiting rural communities. The amplification of the rural electricity grid could also massively reduce the consumption of firewood, because electric rice cookers are usually the first household appliance Cambodians purchase once electricity becomes available. Pagoda forests are typically confined to the grounds of the village temple, but the concept of a community asset managed by the local temple monks could be expanded to other common areas, such as roadsides, stream banks and the numerous hills or *phnoms* that are scattered across the alluvial plain.

Also, if the goal is to maximize carbon density on all landscapes, then a mechanism that rewards private land holders for conserving forests or planting trees should be included as part of a national land-use strategy. Large and medium-sized private landholdings are becoming increasingly common in Cambodia, even though most are actually long-term concessions of state-owned land. The motivation for acquiring a concession or property is a combination of speculation and production, with the obvious aim of maximizing an investment return. Most analysts assume that private holdings will be converted to agriculture or plantation forestry, as that is the traditional path to generating revenue. A national REDD+ initiative could influence those decisions, because the combination of revenues from more sustainable production system might be more attractive when combined with carbon revenues. At the very least, it could provide additional motivation for private landholders to invest in production systems based on perennial woody species.

Agricultural Diversification and Intensification

Even though REDD+ revenues will provide economic incentive to conserve forests, the pressure to produce food and fiber could easily outweigh those incentives and encourage people to convert natural habitats to farmland, especially if the underlying drivers of deforestation are not addressed. In other words, the future land market might replicate the past, but with land being more expensive and land-use conversion even more lucrative. One way to counteract that trend is to make agriculture on existing lands more profitable — either by improving yields or by decreasing the cost of production — so that investment flows to these lands instead of to the agricultural frontier for new land clearing. The cost of agricultural production in Cambodia is among the lowest in the world, but yields are also well below regional averages; consequently, any improvement in productivity will increase profitability and improve prosperity for Cambodia's rural residents, many of whom are desperately poor. If resources from carbon markets are invested in initiatives that improve agricultural productivity, then Cambodia can more easily obtain its goals of reducing poverty, while simultaneously mitigating the impacts of global warming.

Although most REDD+ resources will be spent on frontier landscapes where the threat of land-use change is greatest, some portion could be allocated to established agricultural landscapes. Since these lands house the settlers who migrate to the forest frontier, improving their productivity would help arrest, or at least slow down, the long-term threat to forests from migration and land-use change. Some REDD+ revenues might also be used to resolve specific problems that are known to impede agricultural development, such as improved extension services, land-use planning and resolution of land-tenure problems, which also influence land-use and exacerbate deforestation on the agricultural frontier.

More than increased production on available land, however, Cambodia is also in a position to incorporate low-carbon practices into its agricultural strategy itself, so that it can simultaneously meet economic goals of reducing poverty and environmental goals of reducing net emissions. Cambodian agriculture has received massive amounts of development assistance over the past two decades, with more than 162 major initiatives valued at over $850 million.[6] This aid has converted Cambodia from a country that could only marginally meet its own food needs to a net exporter of agricultural commodities. In the wake of the global recession of 2008–09, the Cambodian government has renewed its commitment to agriculture, with the goal of diversifying the national economy and making it less subject to the volatility of international markets. These objectives are consistent with a low carbon economy, if the investments in agriculture incorporate energy efficient practices, focus on crops that sequester carbon, and promote practices that increase carbon stocks in soils. In fact, incorporating carbon sequestration into Cambodia's agricultural development strategy would generate significant economic benefits from improved land management practices that would exceed those from the pending REDD+ mechanism.

A low-carbon economic development strategy that seeks to capture revenues from future agricultural carbon markets will require the same institutional capacity for monitoring and certification as implementing REDD+ initiatives or validating CDM/AR projects. Not surprisingly, methods used to estimate carbon stocks or measure soil organic matter in forest ecosystems are similar, if not identical, to those that are needed to establish carbon baselines for agricultural landscapes.

Rice: The Foundation of the Cambodian National Economy

Rice is the most important foodstuff in Cambodia, accounting for more than 80 percent of the caloric intake of the average citizen, while the sale of any surplus rice is probably the most important source of cash income for the subsistence farmer. There are several forms of rice farming in Cambodia, including upland rice, a form of slash-and-burn agriculture common in mountain regions, as well as floating and receding rice, both of which are adapted to the seasonal fluctuations of the country's lakes and rivers. The vast majority of rice, however, is grown in the hundreds of thousands of paddies that cover the rural countryside; more than 3.1 million hectares of the country are dedicated to rice paddies, representing almost 78 percent of Cambodia's cultivated area.[*][7] Cambodia is at the heart of the Southeast Asian rice bowl, and its production has climbed steadily since the early 1990s. From a low in the 1970s, annual production reached a record-breaking 6.8 million tons in 2008 (Figure 7.2). The country attained self-sufficiency in rice production in the mid-1990s and exported about 500,000 tons in 2002; the volume of rice exports has surged over the past three years, with some estimates surpassing 2 million tons in 2008 or about 30 percent of national production.[†][8]

Along with its economic benefits as a crop, rice production has numerous other advantages, not the least of which is its cultural significance within Khmer society and its inherent ecological stability as an agricultural system; rice cultivation in flooded paddies is one of the few agricultural systems that actually improve soil quality. It also provides ecosystem services in the form of flood control, because it captures and holds water that would otherwise flow into the already swollen rivers and streams during the monsoonal wet season and as a supplement to the strategically important natural fisheries (see below under Fisheries and Aquaculture).

Most of Cambodia's increase in rice production has been achieved by intensifying yield on available land — from 1.3 tons per hectare to today's average of 2.3 tons per hectare — but the area under cultivation has also been expanding, both incrementally around the margins of the Tonle Sap alluvial plain and from the conversion of natural wetlands into agro-industrial estates.[‡] The government optimistically projects rice production to reach 15 million tons by 2015, with exports of 3 to 5 million tons. As remarkable as this goal seems, it

---

[*] The MAFF states the area cultivated in rice as 2.2 million hectares; the difference is due to the method of reporting and analysis. The GIS value is based on a spatial analysis from satellite imagery and includes the total area of the rice plain, including roads, fallow fields, shrub islands, villages, etc., while the MAFF report is based on data of fields sown and harvested that was collected by provincial authorities and extension agents and fed into a national database.

[†] Most rice exports leave the country as unmilled (paddy) rice, which Vietnamese middlemen purchase directly from farmers, so export figures must be deduced from production and national consumption estimates. In 2009, the Cambodian government initiated a program to add value to its rice exports by investing in storage facilities and rice mills, so that rice can be exported directly from Cambodia under the supervision of government authorities.

[‡] Kuwait has offered to finance a large irrigation scheme on the *Stung Sen* river in *Kampong Thom* province that would provide water to a 50,000-hectare rice farming operation north of the *Tonle Sap* (Economist, 23 April 2009).

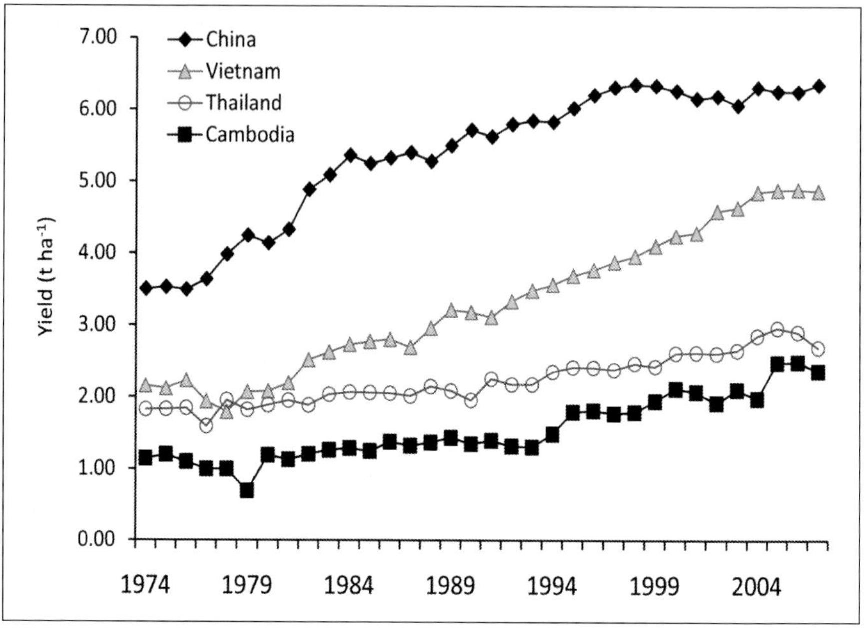

Figure 7.2. From a low in the 1970s, Cambodia's annual rice production reached a record-breaking 6.8 million tons in 2008; productivity lags behind neighboring countries indicating opportunities to improve yields.

is not unrealistic in the context of current and potential rice yields. Cambodia is clearly on a pathway for increased rice production and its rice yields are well below neighboring countries;[9] consequently, there is definitely potential for improvement (Figure 7.2).[10] Moreover, this goal can be achieved as part of a low-carbon development strategy where increased production originates on already cultivated land.

Agronomists refer to the difference between potential and actual crop yield as the "gap," which is usually attributable to one of five interrelated constraints:

(1) Biophysical: the inherent limits of natural resources such as soil, water, pests and weeds, with water being the most important;
(2) Crop management: tillage practices, nutrient and water management, and post harvest practices to avoid spoilage;
(3) Socio-economic: knowledge, traditions, financial capital and labor;
(4) Institutional: government policy, markets, land tenure, and access to credit;
(5) Technology transfer: presence and competence of research and extension services.[11]

Although Cambodia faces several challenges to meet the government's ambitious goals for increased rice production, the most straightforward way to close the "gap" would be to increase the area under irrigation, which is currently only about 15 percent of the total area under cultivation. The government has adopted several strategies to increase irrigation, one of which is based on a feasibility study conducted in 2007 that evaluated the possibility of using the proposed dams in the northern watersheds of the Cardamom Mountains in conjunction with an irrigation network on the alluvial plain in Pursat and Battambang Provinces. As pointed out in the previous chapter, forest conservation and sound watershed management practices on landscapes located upstream of the dams would benefit the proposed hydropower facilities, as well as improve the functioning of the potential irrigation network located downstream of the dams.

Even with improved water-use efficiency and increased storage capacity, the ability to improve rice yields via irrigation may be limited in Cambodia by its inherent geography. Vietnam rice growers achieve high yields largely due to the abundant surface water that naturally floods the extraordinarily flat topography of the Mekong Delta. In contrast, the much lower yields of Thailand reflect its reliance on rain-fed paddies throughout the northeastern part of the country.* With a similar reliance on seasonal rains, Cambodia's intentions to develop large-scale irrigation projects may be limited by topographic constraints, which might restrict water distribution systems to lands immediately adjacent to the main stem of the Mekong River. Any large-scale removal of water from the Mekong system carries significant environmental and social risk, because diversion of water flows would alter the hydrology of the Tonle Sap and, potentially, reduce the productivity of that strategically important natural fishery. Consequently, any such effort should be undertaken only following appropriate and exhaustive environmental evaluation.[12] Other efforts to increase rice production on available land may also have environmental impacts that likewise require careful consideration. For example, the intensive use of pesticides can lower fish yields (see below) and be toxic to farm

---

* Thailand is one of the largest rice exporters in the world, and its low yields are partly due to the reluctance of farmers to switch to high-yield varieties over the traditional cultivars (Basmati rice) preferred by domestic and export markets. Most Basmati rice comes from the Khorat Plateau, which is also a region constrained by a lack of water and irrigation technology in the dry season.

families; a more positive approach are integrated pest management technologies* that can increase both rice production and fish harvests and thereby improve profitability without detriment to human health.

There is an alternative irrigation system that may provide a more promising and profitable avenue for improving and diversifying the Cambodian rice farm economy: drip irrigation. In Cambodia, this system would access water from the aquifer situated underneath the Tonle Sap, which is a large and unexploited water resource that is also inherently stable because it is replenished annually during the flood season via filtration. This water source would complement the existing production strategy for the approximately 3 million farm families that depend exclusively on rain-fed rice paddies and provide them with new alternative cash crops with both domestic and export markets. Drip irrigation is both an energy efficient and scalable technology that can be adopted by both large and small farmers. It was perfected in the semi-arid climate of Israel, but it is also well suited to the long dry season of Southeast Asia and the limited water flow capacity of the Tonle Sap aquifer.† Field trials in Cambodia have showed that the technology could be successfully implemented in traditional farming systems.[13] Future development should focus on identifying and testing pumps and energy sources that could provide the water flows required by an average family farm. Implementing a dry season irrigation model would require significant capital outlay by farmers, most of whom lack access to credit, so this production model would require some sort of remedial action on the part of the government and its development partners. Not insignificantly, the widespread adoption of drip irrigation would create an entirely novel service industry that would create additional jobs and economic opportunity in the private sector.

Upland Agriculture

Modifications to rice farming practices will focus on land-use intensification, but changes in upland farming will be the consequence of an

* Integrated pest management (IPM) focuses on long-term prevention of pests using biological control, habitat manipulation, modification of cultural practices, and use of resistant varieties. Pesticides are used only after monitoring indicates they are needed, and treatments are made with the goal of removing only the target organism. <http://www.calrecycle.ca.gov/Organics/Glossary/> [accessed 15 July 2011].

† The sediments below the Tonle Sap have a relatively high clay content, so pore space and water holding capacity is limited.

expansion of the area under cultivation. Consequently, upland agriculture will become an increasingly important part of the agricultural economy over the next decade; most of its expansion occurring on landscapes currently covered by dry forest and scrubland. Approximately one million hectares, or 20 percent of the Greater Cardamom Region, is covered by this type of native vegetation, almost 90 percent of which is found on flat or gently rolling land with arable soils. The conversion of these dry forests to agriculture is not only inevitable, but also appropriate in a country with an expanding population characterized by pronounced levels of rural poverty. However, the production model that is implemented on these landscapes will affect both Cambodia's ability to eradicate poverty and its ability to develop a low carbon economy, which is vital in establishing and maintaining a competitive position in global export markets. Improvements to rice cultivation must be part of an integrated strategy to create a low carbon economy, but upland agriculture must also adopt technologies that minimize the emission of green house gases. The landscape mosaic that emerges from the conversion of these agricultural frontiers must be sustainable over the long term and, at the same time, function as an ecological buffer zone between the intensively farmed rice plains and the humid forests of the Cardamom Mountains.

The crops that will eventually dominate Cambodia's upland agriculture will depend on local and global markets, as well as on the agricultural development policies adopted by the Cambodian state. Although agriculture has been generously supported by international assistance over the past two decades, most of this assistance has focused on the rice plains where the overwhelming majority of farmers live. Technical assistance for upland agriculture has been based on "tried and true" approaches to improving food security, such as providing extension support for crops such as maize, soybean, sweet potato, peanut, sesame and pulse crops. Measures promoted to increase yields have included the introduction of improved crop varieties, rhizobium inoculation for legumes, and improved fertilizer management, as well as crop rotation, reduced tillage, and integrated pest management. There is little to criticize in past efforts and "more of the same" should be part of any new strategy moving forward, but Cambodia will also need to take into consideration changes in global markets, including the growing demand for sustainable agricultural commodities.

One of the most notable developments over the past few years has been the increasing interest in cassava, due to its consistently high yields under intensive cultivation and the income it generates via export markets. Cassava yields between 9 to 10 tons per hectare on Cambodian farms, compared to 2 to 3 tons per hectare for rice and maize. Moreover, the potential to double these yields is within the technological reach of many small farmers, as evidenced by the success of farmers in Thailand who routinely surpass 18 tons per hectare.[14] The remarkable productivity of this crop is due to its ability to grow in relatively poor soils, drought tolerance, and the lack of any serious pests.* Although it is mainly grown for its large starchy roots, the foliage is also palatable to livestock and provides high quality forage when processed as silage.

The example of Thailand is especially encouraging where cassava cultivation has given rise to a sophisticated industry and a value-added chain that originates on small farms. Since cassava harvesting is manually intensive, small farmers can compete successfully with larger operations. Roots are chipped and dried on the farm and consolidated by merchants, who transport and resell them to factories for further processing into starch pellets or, increasingly, as ethanol for biofuel (see below). About half of the Thai production is destined for domestic consumption, while chip and starch exports have increased from 7 to 18 million tons over the last decade. Investment in cassava processing facilities has been concentrated in the economically depressed regions of northern Thailand, bringing growth to a region that lags behind the rest of the country in terms of development.[15]

The growing interest in cassava as a cash crop in Cambodia is reflected by the increase in the area under cultivation, from less than 15,000 hectares producing 150,000 tons in 2001[16] to more than 40,000 hectares producing 500,000 tons by 2005.[17] The appeal of cassava has also been recognized by the incipient agroindustrial sector. For example, one company reported cultivating more than 27,000 hectares of cassava in 2008 and claimed to have established alliances with other concessionaires for an additional 50,000 hectares.[18] The official list of economic land concessions approved by the national government reveals that 13 concessionaires intend to cultivate cassava;[19] half of these are located within the Greater Cardamom Region.

As profitable as cassava might be, it would be unwise for Cambodia to develop a system of upland agriculture based on any single

---

* Cassava is native to America where pests are controlled by a wide variety of parasites and insect predators that co-evolved with cassava; in Africa the cassava green mite (CGM) has caused major crop losses before it was controlled by varietal resistance and use of predatory mites.

species. Although relatively pest-free at the onset of a cultivation boom, any crop grown as a monoculture will suffer from some type of pathogen infestation over the medium-term — often with catastrophic consequences for the farmers who have elected to grow it. Similarly, a production system overly reliant on exports will lead inevitably to boom and bust cycles driven by global markets beyond the control of any government. A more sustainable model for upland Cambodia would be a diversified production system with a mix of both perennial species and annual crops with strong national and regional markets. Perennial species would bring extra benefits by not requiring annual soil tillage, thereby lowering operating costs and reducing soil erosion. Perennial woody species (trees and shrubs) would play a special role in a low carbon economy because they store significant amounts of carbon in their roots and stems.

Many fruit-producing trees are excellent perennial crops for Cambodia's upland farms, and their fruits enjoy a privileged position in the Southeast Asian cuisine, putting them in high demand in Phnom Penh, as well as in regional urban centers such as Singapore, Bangkok and Ho Chi Minh City. Some of the most marketable species are durian (*Duria graveolens*), mangosteen (*Garcinia mangostana*), milk fruit (*Manilkara zapot*a), jack fruit (*Artocarpus heterophylla*), dragon fruit (*Hylocerus undatus*), rambutans (*Nepheleum lappaceum*), longan (*Dimocarpus longan*), custard apple (*Annona reticulata*), star apple (*Carambola*), lychee (*Litchi chinensis*) and jambu (*Zyzigium jambos*), as well as mangoes, bananas and citrus (oranges, mandarins, grapefruit, etc.). Similarly, a broad range of fresh vegetables, tubers, pulses and beans, and starchy fruits such as squash are well suited to garden plots. Although rains are more than sufficient in the rainy season, drip irrigation systems would revolutionize agricultural systems during the dry season. The key to making this type of intensive cultivation feasible is largely logistical, because it requires an integrated transportation infrastructure and refrigeration technology to ensure that perishable commodities can reach urban markets within a few days, at most. Less susceptible to spoilage are spices such as black pepper, capsicum, ginger, coriander, cumin, and cardamom, all of which play a central role in South Asian cuisines and, even better, have relatively high unit values.

Rediscovering Khmer Silk

A common goal of agricultural development is to promote the transformation of raw commodities into manufactured goods with greater value. The more transformational stages that can be conducted within the country, the more economic benefit will be extracted from the production chain. Silk and silk products are classic examples of such a value added production chain and Cambodia has a centuries-old tradition of sericulture, as the cultivation of silk is known. Khmer women have a long tradition of silk weaving and hand-made silk is a staple item purchased by many tourists. The total annual consumption of silk in Cambodia is estimated at about 400 tons and provides income to tens of thousands of rural women.[20] Cambodia could reap greater economic benefits from this industry, however, by fostering the production of raw silk within its own country. Most current production is based on imported silk yarn from China and Vietnam, because Cambodia has not rebuilt its silk-producing capacity following its implosion during the long civil war. The potential economic benefits from a restored silk farm sector could create tens of thousands of jobs and provide as much as $900 per hectare per year to rural families. Perhaps just as importantly, the resurrection of silk farming as a rural trade is an opportunity to recover a lost cultural tradition with deep roots in Khmer culture.

Sericulture originated in China approximately three thousand years ago, and its trade has played a major role in the history of humankind. Silk has played an important role in the development of trade among nations and when Marco Polo traveled to China it was on a long established trade route known as the Silk Road. Even though other fibers now dominate the textile industry, the unique qualities of silk still make it the preferred fiber in products of exceptional value. Raw silk production requires both plant and animal husbandry, as leaves are harvested from mulberry bushes (*Morus spp.*) and fed to silkworms, which are the larvae of the domesticated moth (*Bombyx mori*). As the silkworms grow, they pass through four stages known as molts, eventually attaching themselves to a twig and creating a cocoon by extruding two paired protein filaments that are glued together into a single fiber by a gummy substance of exceptional resilience. Each cocoon is composed of one extremely long braided fiber of incomparable strength and smoothness. If left alone, the larvae will mature into an adult moth, but on a silk farm the cocoons are collected and immersed in boiling water to dissolve the gummy substance. The fibers are then collected via a labor-intensive process known as "reeling," where they are floated in water and then woven together to create silk yarn, which is the basic commodity of silk manufacture.

Global demand for silk was estimated at about 320,000 tons annually in 1980 and increased to more than 535,000 tons by 2008,[21] representing the output of more than 630,000 hectares of mulberry plantations.[22] China dominates the industry, accounting for 70 percent of total global production and 90 percent of all exports, which were valued at $1.9 billion in 2005. Growth in exports averaged about 14.5 percent over the last decade, although projections have been revised downward due to the global recession of 2008 and 2009.[23] Japan once dominated world trade in silk production, but now produces less than 1 percent of this global commodity because it is unable to compete in what is an extremely labor intensive industry. Although China has developed methods to reduce labor costs, it too is beginning to face constraints and is seeing competition from silk producers in countries with lower labor costs, such as India.

There are about 68 species of mulberry (*Morus*). Although most are palatable to the silkworm, commercial operations depend largely on cultivars from two species: *Morus alba*, which predominates in temperate regions and *Morus indica*, which predominates in tropical regions, particularly India, Thailand, and Vietnam. Mulberries can be grown in pure stands, but are particularly well adapted to mixed agroforestry schemes and produce well when interplanted with other crops. Although adapted to a wide variety of climates and soils, optimal production requires adequate soil fertility and irrigation may be needed to maintain maximum production of leaves in the dry season.[24] Mulberry also has other uses in addition to sericulture: it produces an edible fruit and foliage that is high-quality forage for ruminants. If left alone, it will eventually form a small tree, but plants respond well to pruning, and the bush habit facilitates management and harvest. Tropical countries have a significant advantage in that they can grow multiple broods each year, while temperate producers are restricted to one or a few broods due to cold temperatures and the deciduousness of temperate species of mulberry.

Cambodian entrepreneurs are well positioned to enter the international silk market, both as a supplier of raw silk and as a supplier of fine silk fabrics to global markets. A silk-based production model would complement Cambodia's already robust textile industry based on cotton and synthetics and would allow manufacturers to diversify their product lines and add value to a raw material that can be produced nationally; this in turn would stimulate silk production and increase income for rural families. Finally, the Khmer tradition of fine handmade silk products provides an additional avenue for entry into the global textile market, particularly the high-value fashion boutiques of East Asia, Europe and North America.[25]

Rubber Trees

Rubber plantations (*Hevea brasiliensis*) were first established in the last decades of the nineteenth century in what was then known as British Malaya and the Dutch East Indies, with the French following suit in Indochina during the 1920s. The demand for rubber grew following the transportation revolution ushered in by the invention of the automobile and the demand for tires; as a consequence, rubber came to dominate the economies of these European colonies. The market for natural rubber was profoundly transformed by the Second World War, when both the Allied and German industries made a technological breakthrough and developed synthetic rubber to compensate for their lost access to natural rubber from Southeast Asia. Synthetic rubber made dramatic inroads as a feedstock for tire manufacturing and by 1960 dominated the global rubber market. Nonetheless, demand for natural rubber remains strong because of its superior elasticity, which makes it the preferred raw material for items like surgical gloves, condoms, and airplane tires. Even massed produced automobile tires contain as much as 40 percent natural rubber. Currently, natural rubber constitutes about 40 percent of the global demand for rubber products and its production has expanded annually by about 4.5 percent over the last decade, reaching 9.7 million tons in 2007. It is estimated that global demand will reach 18 million tons by 2018.[26]

Although global production of rubber is increasing, industrial plantations in Malaysia and Indonesia have been declining over the past two decades; this is due largely to the rise of the oil palm sector, whose greater profitability has led to the large-scale conversion of industrial rubber plantations to oil palm. Nonetheless, smallholders in both Malaysia and Indonesia continue to cultivate rubber and this has allowed Indonesia and Malaysia to maintain near dominant positions within the global rubber market. For small farmers, rubber is a more feasible crop when compared to oil palm, because it can be harvested and stored for an extended period of time on the farm. In contrast, palm fruits must be processed within 24 hours of harvest, which dictates that palm oil processing facilities be located immediately adjacent to plantations. Over the near term, rubber production probably will decline in both countries, as corporate and government policies focusing on increasing palm oil production by smallholders. Climate also influences the geographic segregation of oil palm and rubber plantations. Oil palm thrives only when annual precipitation exceeds 2,000 mm and rains must be evenly distributed throughout the year. Thailand is now the world's largest producer of natural rubber, where smallholders also dominate while efforts to expand production are underway in India, Vietnam, Laos, Myanmar and Southern China.

In Cambodia, oil palm can only be cultivated successfully on the narrow coastal plain situated between the Gulf of Thailand and the Cardamom Mountains, while rubber can be cultivated throughout the country. Cambodia currently has about 80,000 hectares of mature rubber plantations, up from a low of 50,000 hectares in 1990; over the last decade, the country renewed its existing plantations and added another 70,000 hectares. Importantly, approximately half are located on small- or medium-sized family farms. Cambodia currently ranks 16th out of the world's 23 natural rubber producing countries and is responsible for less than 0.5 percent of global production. Nonetheless, those exports were valued at $87 million in 2010 and created approximately 27,000 direct jobs for rural families, which translate into an additional 40,000 indirect jobs for the larger economy. The increasing global demand for natural rubber, and its accessibility to small farmers, provides Cambodia with an opportunity to increase market share and expand its rubber sector and some estimate that rubber cultivation could expand to 400,000 hectares by 2020. Rubber, particularly smallholder rubber, could be an important niche in a diversified agricultural economy throughout the settled areas of the Greater Cardamom Region.

Bioenergy

Cambodia has long depended on energy derived from plant biomass, primarily fuel wood and charcoal extracted from its forests. These fuels currently provide more than 80 percent of the nation's energy needs. A shift to a low-carbon economy would replace the country's dependence on forest biomass with a mixture of other energy sources: hydropower and natural gas if exploration efforts are effective, but also biofuels produced by Cambodia's agricultural sector. A biofuel is defined usually as any liquid fuel, such as ethanol and biodiesel that is derived from a cultivated crop; biomass energy is a broader term that also refers to organic waste, wood chips and charcoal. All of these forms of bioenergy represent opportunities for Cambodia to increase and diversify its agricultural production as part of a low-carbon economy — though all also bring their own environmental and social issues as well. Currently, there is a robust global debate as to the relative benefits of bioenergy, particularly biofuels, centering on three major issues:

(1) The energy needed to produce the biofuel is often only marginally less than the energy content of the fossil fuel it replaces;
(2) The conversion of land to biofuel production causes deforestation that essentially negates any emission reductions from the use of biofuels either directly or indirectly by displacing the cultivation of food crops;
(3) The competition for land can cause a displacement of food crops that can lead to local food shortages or food inflation, which exacerbate poverty.

These concerns do not mean that biofuels will not have a place in energy markets over the near to medium term; much to the contrary, as long as biofuels are cost-competitive with petroleum based fuels or are mandated by governments as part of national energy strategies, they will be cultivated and processed for sale in domestic and global markets.

Regardless of these limitations and risks, biofuels have the potential to reduce emissions within the confines of the existing industrial and transportation infrastructure and, consequently, provide a feasible way to reduce greenhouse gas (GHG) emissions over the short term. In countries like Cambodia, which imports 100 percent of the fossil fuels consumed by its transportation sector, liquid biofuels can provide a domestic energy source that would also create jobs in both agriculture and industry. There are many different bioenergy options and there will be both known and unknown tradeoffs when all of the economic benefits and environmental impacts are identified and evaluated; consequently, research is needed into species, cultivation methods and processing technology, in order to identify the best options in terms of climate change mitigation potential and the economic interests of Cambodian society.

The first modern biofuel plant in Cambodia was opened in November 2008, when a Korean company began producing ethanol from cassava starch 15 km north of Phnom Penh (Figure 7.3). Cassava roots contain more starch than any other food crop, which can be extracted using standard food processing technologies and fermented by a relatively simple process using yeast and bacteria to produce ethanol. From a ton of cassava with 30 percent starch content, a

Figure 7.3. Cambodia's first liquid biofuel factory opened in 2008 to process cassava starch into ethanol for export to overseas markets (top). Preliminary studies indicate that the sugar palm, Cambodia's national tree, has potential for the production of ethanol (bottom).

distillery can produce about 280 liters of 96 percent pure ethanol.[27] Thailand has pioneered the use of cassava starch as a feedstock for ethanol, and as of May 2010, had six processing plants producing approximately 1.8 million liters per day, about 60 percent of that nation's ethanol production.

Thailand discourages the export of ethanol as part of its national energy policy to reduce fossil fuel imports[28] and cassava ethanol has limited potential in international biofuel markets due to its relatively low energy efficiency, at least when processed using standard technology. The relative climate change benefit can be determined by conducting a Life Cycle Analysis (LCA), which is the standard methodology used for quantifying the environmental impact linked to the fabrication of a manufactured product. In the case of biofuels, a LCA is used to objectively calculate and compare the energy inputs and GHG emissions associated with the cultivation and transformation of different biofuel feedstocks. Regulatory agencies use LCAs when determining which biofuels to allow into their domestic markets; for example, the Renewable Energy Directive (RED) of the European Union mandates that starting in 2011, subsidies will be provided only to those fuels that have at least a 35 percent reduction in GHG emissions when compared to the fossil fuel they will displace — a threshold that will increase to 50 percent by 2017.[29] The directive explicitly requires the LCA to consider all the GHG emissions generated during cultivation of the crop, including: (a) from fossil fuels consumed by mechanized farm equipment, (b) fertilizer applications, including the manufacture of fertilizer, but more importantly the nitrous oxide that escapes into the atmosphere during its application, (c) the transport, distribution and processing of the biofuel after its harvest and (d) the carbon emissions that are linked to the conversion of high carbon landscapes, such as forest, to low biomass cropland. This last point is highly controversial, especially when the cultivation of the biofuel displaces other crops, leading to a phenomenon known as "indirect land-use change." In the case of traditional cassava ethanol, the LCA reveals an energy and GHG savings of only about 6 percent, which is due mainly to the high energy input required for its processing and transformation into ethanol; however, these calculations have assumed that this energy source is either fuel oil or coal, both of which are characterized by very high GHG emissions per unit of energy.[30]

The Phnom Penh ethanol facility would appear to be different, because the Korean investors have developed an innovative system that uses the waste water from the cassava processing facility to produce methane (CH_4), which is then captured and used to produce thermal energy for the ethanol distillery. Methane is a strong greenhouse gas with 21 times the atmospheric warming potential as CO_2,[31] and under the rules of the Clean Development Mechanism (CDM) the operators of the ethanol plant are entitled to claim an emission reduction credit for capturing and burning the methane. Even better, the use of methane, which is itself a form of bioenergy, to operate the cassava distillery dramatically changes the LCA rating of its ethanol product and brings cassava-derived ethanol into compliance with current standards for several major international markets. Over the medium term, the use of starch as a feedstock for the manufacture of ethanol biofuel will be replaced by "second generation" biofuel technologies currently under development that convert plant fibers composed of cellulose or lignocellulose (e.g., wood) to ethanol via enzymatic or chemical processes. Once that technology is perfected and integrated into world biofuel production systems, cassava will still retain its role as one of the planet's most productive food crops.

Sugar cane, which has significant agronomic potential in Cambodia, is recognized as the most efficient of the current generation of biofuels, with reductions in GHG emissions that exceed 100 percent. Its success is largely due to the research and development efforts of Brazil and India, where plant breeding programs have developed cultivars that maximize yield and sucrose content, while minimizing the need for fertilizer and pesticides. Simultaneously, the industry has increased the efficiency of the conversion process by using waste biomass, known as bagasse, to generate the thermal energy that powers the sugar mill and alcohol distillery. In some cases, this thermal energy is supplemented with methane captured from waste treatment ponds and used to co-generate electricity that is sold to third parties by way of the national energy grid. The most technologically sophisticated sugar cane refineries produce multiple types of sugar (e.g., sucrose, fructose, glucose, etc.), as well as ethanol distilled from a fermentation product of molasses that was once considered a byproduct of the sugar-making process, but which now constitutes a major revenue stream for integrated sugar-ethanol refineries. Modern sugar cane mills waste virtually nothing and manipulate the proportion of processed sugars and ethanol to maximize the profitability of an extremely successful business model that can be a cornerstone of a low carbon economy.

There are periodic reports in the Cambodian press of the pending development of sugar cane mills and associated plantations, including one announced for Koh Kong province in 2006. However, none of these proposed investments have materialized due to a variety of obstacles related to the complexities of establishing a vertically integrated plantation and processing facility. The profitability of a modern sugar mill and distillery complex is predicated on economies of scale and, consequently, requires a reliable source of feedstock to ensure that the mill and distillery operate at, or near to, full capacity. One way to mitigate the risk of inadequate feedstock supplies is to establish simultaneously an industrial-scale sugar cane plantation. Entrepreneurs who have obtained economic land concessions in Cambodia might be able to organize this type of investment package; however, the risk from land conflict may be impeding the development of this type of investment.[32]

An alternative business model would be to rely on small and medium-sized farmers as contract growers to supply feedstock to an independently owned and operated industrial sugar cane refinery. This would not be without risk, because it would entail organizing hundreds of individual producers with different skills, some of whom may not be inclined to engage in a new cultivation practice that they may view as a high risk venture when compared to traditional and more diversified agricultural systems. Such an arrangement has functioned well in other countries, however, including Thailand and India where land tenure systems are dominated by small landholders. In fact, Thailand has become one of the world's largest exporters of refined sugar, thanks to the efforts of tens of thousands of small farmers who produce more than 70 million tons of sugar cane on more than one million hectares of land. Thailand's system has evolved over decades and also includes hundreds of sugar refineries that range from small family-owned operations to large integrated sugar — ethanol facilities owned and operated by large industrial concerns. Sugar exports from Thailand have been projected at $1.5 billion in 2009 and are expected to increase 50 percent by 2011, while ethanol production is expected to grow from 2 to 4 million liters over the same time period.[33] Sugar cane has even held its own against cassava in Thailand's ethanol market, despite price considerations, but the competition between the two commodities has turned out to be more complex than originally thought, with cassava occupying only about 30 percent of the national ethanol market.

Cambodia currently has no industrial-scale production of sugar, and most domestic sugar is produced by the sugar palm (*Borrassus flabellifer*), which also happens to be the national tree of Cambodia (Figure 7.3). The potential of sugar palm as biofuel crop is both unknown and intriguing. Although studies are limited, sugar palms are prolific producers of sugar, yielding up to 146 kg per tree of soluble sugars per year. Extrapolated to a "per hectare" value, this number translates into approximately 17 tons of sugar per hectare, making it twice as productive as sugar cane.[34] There are several "sugar palm" species in Southeast Asia (e.g., *Nypa fruticans Arenga pinnata, Caryota urens*), all of which produce high levels of sugar and other products consumed by indigenous societies, including thatch, rough construction timber, and palm wine, which is fermented sugar syrup. The syrup harvested from the flowers of these trees could be fermented to produce industrial grade ethanol with very little additional processing. Unfortunately, the process of harvesting sugar palm is extremely laborious, as it requires an able bodied individual to climb the palm tree and "tap" the sap from the flowering stalks. Another hurdle would be the distillation phase of the palm wine, which would consume large amounts of thermal energy that might negate its potential to reduce GHG emissions. Nonetheless, its basic productivity warrants further research to determine the feasibility of a new production system, which might be based on short stature cultivars and a simplified tapping procedure. One obvious benefit for Cambodia would be the potential to combine rice farming with sugar palms as a complimentary crop and, in the process, provide small farmers with an additional source of income based on an industrial commodity with an international market.

Palms are members of a unique family of plants that have evolved into a remarkably diverse array of species. Their qualities and inherent productive capacity continues to surprise modern society as their utility is discovered — or rediscovered. The oil palm (*Elaeis guineensis*) is a case in point. It has been cultivated as a food crop for millennia in Africa and as a versatile industrial commodity for more than a century. In the past three decades, however, this remarkable species has come to dominate the international market for vegetable oils, currently accounting for more than 35 percent of all vegetable oils traded on global markets. Its extraordinary productivity is a consequence of its evolution in the high precipitation and energy environments of the tropics, and the morphological characteristics of its fruit that is characterized by unusually high fat content in its fleshy exterior tissues. Oil palm has come to dominate the plantation forest industry in Southeast Asia over the last two decades and, with the advent of biofuels, is poised to assume an even larger role as a plantation crop as it becomes adopted as a feedstock for the production of biodiesel (see Figure 1.6).

The often acrimonious debate about the utility or "appropriateness" of biofuels has led many environmental advocates to targeted palm oil as a source of GHG emissions, because its expansion has led to the deforestation of extensive tracts of tropical forest in Indonesia and Malaysia. In fact, most LCA analyses show that when oil palm plantations are established on forested landscapes, the GHG emissions linked to land-use change not only negates any potential benefit from the displacement of fossil fuels by biofuels, but in some cases can actually cause even greater GHG emissions when compared to fossil fuels. This has led environmental organizations to criticize corporations that have announced plans to use it as a feedstock for the production of biodiesel. While these accusations are based on documented cases of deforestation linked to the palm oil industry, not all palm oil plantations have been established as the direct consequence of deforestation. In some instances, they have been established on low biomass landscapes, such as pastures or abandoned crop land, and the subsequent LCA has demonstrated that the GHG emissions not only fall within the guidelines recently established by the European Union, but are nearly identical to the LCA values of ethanol produced by modern sugarcane technology.[35]

Because oil palm is adapted to equatorial regions with rainfall greater than 2000 mm per year and with no, or a very limited, dry season, it is not adapted to the greater part of Cambodia and can only be commercially cultivated on the Cambodia Coast, where extensive plantations have been established along Route 4 between Sihanoukville and Phnom Penh. The expansion of oil palm to nearby coastal areas is possible, but would require a parallel investment in a large processing facility, because fruits must be processed within a few hours to avoid spoilage. Production could be expanded at the existing facility, however, by involving adjacent land holders, but, in keeping with efforts to build a low-carbon economy, any expansion, whether for biofuel or cooking oil, should be targeted at low biomass landscapes and avoid the conversion of natural forests.

The large and growing market for biofuels in both the developed countries and emerging markets is driving biofuel producers to search for feedstocks that can be produced sustainably, which in the parlance of biofuels means without causing deforestation and without impacting local food supplies. One of the most popular candidates to fill this market niche is a small tree known as jatropha (*Jatropha curcans*), which has several attributes that make it an attractive biofuel feedstock. Its oil can be extracted easily from its seeds by a simple milling process that does not require large capital expenditures in processing plants; thus, it could be easily integrated into existing production systems in the developing world. Both the seeds and the vegetable oil are toxic, so the trade of this new commodity will not immediately lead to price hikes of basic food commodities.[*] In addition, jatropha is adapted to a broad range of ecological conditions from the seasonal tropics to semiarid areas of the subtropics; consequently, concern that its widespread cultivation will lead to new deforestation is minimized, because a great deal of land at these latitudes has already been cleared and many farmers would benefit from a perennial woody crop that might actually improve land-use practices.

Jatropha naturally occurs in semiarid ecosystems and has evolved characteristics that make it relatively tolerant of drought; similarly, it is adapted to soils that have low fertility, holding out the promise that it can be used to restore degraded landscapes. Even its toxicity is seen as an agronomic benefit, since that will protect the crop from many common invertebrate herbivores.[36] Jatropha actually grows naturally throughout much of Cambodia where farmers have cultivated it as a living hedge, known as "*lahong kwong*," and practitioners of traditional medicine have prescribed a concoction from its seeds as a remedy for intestinal parasites.

The miracle-like attributes of jatropha are clearly overstated, particularly those claims that it might produce relatively high yields on poor soils under drought stress, which are simply unrealistic. Nonetheless, jatropha could form the basis of a viable agricultural production system as part of integrated biofuel strategy. Of particular interest, are areas that receive between 1,000 and 1,500 mm of precipitation annually on soils of limited fertility appropriate for perennial tree crops, including the upland landscapes situated between the inundated rice plain surrounding the Tonle Sap and the Cardamom Mountains. Cambodia outlined an ambitious plan in 2007 to dramatically increase its agricultural output via jatropha and several assistance agencies and the government have initiated field trials to evaluate the feasibility of expanding jatropha cultivation for biofuel.[37] Because the oil can be extracted using small-scale seed grinders and filters, jatropha is amenable to a decentralized production model that would produce fuel for domestic consumption. Most studies have been based on small-scale production systems, which would allow rural communities to produce energy for local consumption. As of yet there has been no major move to initiate the large-scale cultivation of jatropha in Cambodia.

---

[*] Nonetheless, the expansion of biofuels may displace food crops and increase competition for land, which will have the same impact over the medium term.

Soil Organic Matter and Biochar

The most obvious and potentially most profitable way for Cambodia to link agricultural production with a low-carbon economy would be to increase the carbon content in the soils that underpin its farming sectors. Carbon exists in a variety of forms in the soil, including biomass (mainly roots) and soil organic matter (SOM), which is made up of decomposed stems, leaves and roots. Soil itself can be viewed as a type of ecosystem and is home to a complex assemblage of invertebrates such as insects, worms, and nematodes, as well as a multitude of microorganisms that are key mediators of the decomposition process. The end product of this complex decomposition process is humus, which is the most important constituent of soil organic matter.

Humus plays a key role in maintaining soil fertility, because it is composed of large complex organic molecules with many negatively charged sites that attract and bind the positively charged cations, which characterize many plant nutrients. Humus is relatively resistant to further decay and is a good candidate for sequestering additional carbon as a climate mitigation strategy. The key term is "relatively resistant," because humus is still a somewhat labile molecule, so maintaining high levels of humus requires maximizing the input of biomass into soils on an annual basis. Increasing soil organic matter is at the heart of almost all traditional and modern technologies for improving crop productivity and can be achieved by a number of agronomic practices, such as minimum tillage, planting cover crops, and manure application. Any and all of these practices, if applied consistently over many years, will increase carbon sequestration in the soil. Another simple method for improving soil organic matter is to restrict the use of fire to burn crop residues. Cambodian farmers routinely use fire when preparing fields thus removing organic matter that could be incorporated into soils and converted to humus.

Another approach for sequestering carbon in soils that is gaining attention in international forums is the storage of charcoal in soil — or *biochar* as it is known in this context. Charcoal is created by a process that, essentially, enhances the carbon content of firewood. Known as *pyrolysis*, it is a simple industrial process that chemically decomposes organic molecules by extreme heat, a process that differs from combustion because the wood is heated in an oxygen-deprived chamber. Large complex organic molecules, such as cellulose and lignin, are about 50 percent carbon and pyrolysis chemically transforms them into simpler molecules with a carbon content between 85 percent and 95 percent. Charcoal is considerably less biologically labile than humus, which is about 60 percent carbon, and will resist decay in soils over centuries. When charcoal is pulverized into small particles and incorporated into a soil medium, it becomes *biochar* and, importantly, begins to mimic the characteristics of humus by improving both soil structure and fertility. Interestingly, this technology was known to pre-Colombian civilizations who used charcoal to improve tropical soils on the main stem of the Amazon River;[38] known as *terra preta* (black earth), these soils are still more fertile and carbon rich than nearby non-transformed soils — even after hundreds of years of weathering in the hot humid environment of the Amazon. The positive impact of biochar on crop yields has also been demonstrated by agronomic research in both temperate and tropical ecosystems, including rice paddy systems in the Philippines, where it has increased yields by as much as 100 percent;[39] [40] the use of biochar is currently under evaluation in Cambodia.[41]

Biochar and its use as a soil amendment has many potential applications, including the conversion of carbon-rich waste from large urban centers, wood processing facilities and crop residue into charcoal. Defenders of fossil fuels have even suggested that it be linked with coal fired power plants to neutralize GHG emissions. The most ambitious proposals focus on biochar's potential as a geo-engineering solution for global warming and some proponents suggest growing trees for the express purpose of capturing carbon and sequestering it as biochar in the soils of terrestrial ecosystems.[*] Cambodia is in a unique position to test the economic feasibility of agricultural biochar, which would have synergistic benefits across different sectors of a low-carbon economy. In the energy sector, for example, Cambodia relies on firewood for almost 80 percent of its energy needs, and charcoal is more efficient than firewood as a primary energy source, especially if it is manufactured in modern kilns and consumed via energy efficient cook stoves. Since most fuel wood is extracted from native forests, any reduction in its use would contribute to the success of a national REDD+ policy.

---

[*] Most geo-engineering proposals are controversial and carry significant unknown environmental and social risks: for example, proposals to fertilize the ocean with iron to stimulate algal photosynthesis or to decrease incident solar radiation by manipulating cloud cover.

Even more emission reductions could be earned from international carbon markets if the fuel for making charcoal were obtained from crop residues or tree plantations. Tree plantations would reduce the take of firewood from natural forests, plus sequester additional carbon in the stems and roots of the trees themselves. Using crop residues would provide an additional GHG benefit, because rice paddies are a major source of methane (CH_4), which is emitted as an end product of decomposition in the anaerobic environments of inundated rice fields.[42]

As mentioned previously, methane has 21 times the GHG warming potential as CO_2 and accounts for between 10 and 20 percent of global methane emissions — and may represent as much as 5 percent of all global anthropogenic GHG emissions.[43] The quantity of methane emissions from rice paddies depends on multiple factors, particularly the duration of inundation that creates the anoxic conditions that leads to its formation, but also the quantity of organic matter from crop residue. Increasing soil organic matter in rice paddies through humus formation could increase methane emissions and cancel out the climate mitigation benefits of increasing soil carbon stocks — a classic example of a tradeoff between two potentially beneficial outcomes. The use of biochar as a soil additive, however, might resolve this paradox, particularly if the straw from the rice crop were used to manufacture biochar. Although charcoal has traditionally been made from woody plants, new technologies are available to produce charcoal pellets from herbaceous biomass; these pellets can be produced using small-scale processing facilities appropriate for rural Cambodia.

A biochar pilot project could take advantage of various synergies from different sectors of the Cambodian economy — rice, fuel wood, carbon markets — to make it attractive to a broad range of potential investors. As rice straw or tree farm sourced charcoal supplies are increased and supply the energy market, some can be diverted for use as a soil additive, which would create additional revenue from either compliance or voluntary carbon markets. The source of the emission reductions potentially could be certified either as a reduction of methane emissions from paddies or as carbon sequestration via biochar — or both. A biochar initiative undoubtedly holds much potential, but it must be implemented in a step-wise fashion to establish appropriate protocols and safeguards, as it might also radically alter the complex processes of soil chemistry and physics.[44] For example, how much biochar can maximize benefits; like anything, the potential to have "too much of a good thing" is a highly probable outcome. The evidence from the Amazon, while compelling, shows that creating black earth landscapes is not a simple process of dumping charcoal onto a soil surface; the Amazonian Amerindians also incorporated other soil amendments, including bones, shellfish, plant biomass and ceramics, in a process that occurred over centuries and is, as yet, not entirely understood. Soils are as diverse as the ecosystems they support and, what might function in a well drained sandy soil in the Central Amazon, would almost certainly act differently in the heavy clay soils of the Tonle Sap flood plain. Consequently, any endeavor must be well-researched and proceed in stages, or risk having the opposite effect and negatively impact the livelihoods of the farmers it was meant to help.

Eventually, carbon markets will recognize that soil carbon is of enormous value to the world, and schemes similar to REDD+ and the CDM AR will emerge to compensate individuals, communities, and nations that undertake efforts to sequester carbon in agricultural soils. When this occurs, carbon subsidies can be used to improve soil management techniques; Cambodia can and should move proactively to make this effort part of a strategy to create a low-carbon economy. As mentioned previously (see Chapter 6), the key step in any ecosystem-based carbon initiative is to establish baselines under existing practices and then initiate a monitoring program to track how stocks change over time and under different management regimes.

Fisheries and Aquaculture

Cambodia has been blessed with abundant natural fisheries, and the Cambodian government, aware of the importance of sustaining both the quality and quantity of the nation's freshwater resources, has made the sustainable management of these food resources a central part of its national development strategy.[45] Fish are essential to the Cambodian economy and its culinary traditions, and the demand for fish and quality fish products will increase as the Cambodian population grows and its consumers become more prosperous. However, there is overwhelming evidence that both the freshwater and marine fisheries are being exploited at near- or above-capacity, and that the continued increase in traditional fishing practices will degrade this resource, either by reducing the size of the overall catch or by reducing large predator species with high market value, thus changing the composition of fish communities — or both. In response, the Cambodian government has taken steps towards conserving fish stocks and seeking to identify and implement management practices that can protect endangered species. These include improving access for communities, reforming the regulatory and market framework to enhance benefits for farmers and fishermen. Most of these measures will also improve livelihoods. For example, participatory fisheries management initiatives can change behaviors that lead to over-fishing and help reduce post capture spoilage. The concessionaire system has also been reformed; commercial fishing lots have been transformed into community fish reserves with specific areas that have been set aside as fish sanctuaries. Moving forward, the government has determined that aquaculture will also play a key role in the future of the Cambodian fisheries industry.[46]

Aquaculture, or fish farming as a dedicated food production system, has become increasingly important globally over the past three decades as demand for high quality fish and fish products have sky-rocketed in both advanced and emerging economies. The increase in aquaculture is partly a response to overfishing and the decline in natural fish stocks, such as that documented in the Gulf of Thailand (see Chapter 2); however, it is also an efficient and cost effective way to produce animal protein needed to supply the increasing demand from both developing countries and emerging economies.* Aquaculture now accounts for almost 37 percent of all fish and seafood consumed in the world.[47] Asian nations lead the world in aquaculture production and innovation, with China at the head by an order of magnitude, although the overwhelming share of that massive production is used to satisfy an equally massive internal demand.[48] Two of Cambodia's neighbors, Thailand and Vietnam, have pioneered new aquaculture techniques over the past 20 years, generating important export revenues and contributing to rural economic development and poverty reduction. Aquaculture is still a small part of the Cambodian economy, but is growing rapidly and has expanded from less than 30,000 tons in 2007 to more than double that amount in 2009.[49] The recent

---

* Weight gain efficiency is measured by feed conversion ratio (FCR), which measures the amount of feed mass required to increase the body mass of the animal; these ranges from approximately 7:1 for beef cattle, 4:1 for pork and 2:1 for poultry; in contrast, aquaculture systems have values near 1:1. However, the FCR is an imprecise statistics that compares different types of feed from fish meal (salmon) to mixtures of grain and vegetable protein (poultry). The most efficient and sustainable aquaculture systems are based on herbivorous freshwater species fed grains and vegetable protein supplements similar to poultry feed.

interest of Cambodia's government to promote aquaculture is no doubt heavily influenced by its neighbor's experiences, which provide valuable lessons on how to expand production, but also on the environmental pitfalls that accompany high density production systems. Done with an eye to those lessons, the expansion of Cambodia's freshwater aquaculture and shrimp farming industries could form an important pillar of a low-carbon economy; and one which will simultaneously improve the nation's food security, reduce poverty and create an export sector with large growth potential.

Cambodian farmers have a long tradition of a quasi-natural aquaculture that is an intrinsic part of rice paddy farming, which is at the center of Khmer culture. Fish are ubiquitous throughout rice paddies and migrate into and out of the flooded rice fields from adjacent rivers and ponds in perfect synchrony with the monsoon. The presence of wild fish in rice paddies brings multiple benefits to family farmers, many of whom live on the edge of poverty. The economic value of the protein provided by fish is estimated to be as much as 50 percent of the value of the actual rice crop itself and accounts for almost 80 percent of the protein consumed annually by subsistence farmers. The fish catch from rice fields ranges between 25 and 80 kg per hectare and may be as much as 25 percent of the total freshwater fish harvest each year. The connection between rice and fish as a joint production system is reflected in the Cambodian proverb "plant rice and harvest fish;" however, the rice paddies also provide harvests of crabs, clams, snails, frogs and edible insects.[50] Fish also control pests, such as insects and weeds that would otherwise adversely affect rice plants, and are an integral component of the rice field nutrient cycling system. Most paddy fish are relatively small species, which allows them to survive the dry season in nearby wetlands and streams (i.e., pagoda ponds) from which they colonize the rice fields each year.

Farmers can influence the productivity of rice field fisheries by facilitating (or blocking) connections with natural waterways and through the use (and abuse) of pesticides. Intensifying rice production itself represents a threat to this traditional aquaculture system, because as wetland habitat is converted to rice fields, the number and size of natural refuges where fish can survive low-water periods is reduced. Consequently, any and all strategies to increase either rice yields or fish production should start with a commitment to protect the current system — which is extremely productive and inherently stable. Fortunately, within the current system there are several simple measures that can raise fish production, such as conserving and enhancing dry season ponds or supplementing the supply of fingerlings at the onset of the rainy season.[51] In the People's Republic of China, for example, scientists have modified the traditional rice-fish culture and increased fish populations in irrigated rice fields by eliminating carnivorous fish species and seeding herbivores, such as carp, into flooded paddies. Other practices include digging ditches to expand fish-favorable habitats, fostering the growth of azolla ferns and rotating fish cultivation and rice farming. These efforts require a high degree of technical assistance and significant labor investment and would only be applicable in a portion of Cambodia's rice fields; however, they could increase the fish harvest by as much as 40 percent and, because they provide a source of natural manure, rice yields by about 5 percent.[52]

Cambodia's fisherman and farmers have developed over the past centuries a long tradition of aquaculture on the Tonle Sap, where it is employed to store the amazing abundance of the dry season fishery for consumption later in the year. This traditional system employs shallow pens and cages uniquely suited to the lakes pronounced variation in seasonal water levels. During the dry season, juvenile fish are caught and placed in pens established in shallow water; these are transferred to floating cages with the onset of the monsoon rains and the subsequent rising water levels. On the Tonle Sap Lake, cages are attached to houses and entire villages will move inland as water rises, taking their captive fish with them. The preferred species are the striped catfish (*Pangasianodon hypophthalmus*) and the striped snakehead (*Channa striata*), but may include many other species. Initially, captive fish are fed the abundant small fish known as "*trey riel*" that characterize the Tonle Sap fisheries; however, as waters rise and fish populations disperse, *trey riel* become more difficult to catch, and captive fish must survive on a less nutritious diet of aquatic weeds, rice bran and kitchen waste.[53] Snakeheads are the most demanding captives, because they are strictly carnivorous and will feed only on live fish; in comparison, catfish are omnivorous bottom feeders and will accept a broad range of feed types.

In some respects, this production model can also be viewed as a human-mediated food chain that converts small fish into large fish, and as an adaptive response to the loss of large fish in a system that has long been subject to overfishing (see Chapter 2). However, the dependence on live fish as a feed source limits the potential expansion of this system. Unique to this system is the importance of *trey riel*, which would be considered "trash fish" in most other parts of the world with only limited direct market value. In Cambodia, however, *trey riel* are used to produce "*prahoc*," the fermented fish paste that is a staple of the Cambodian diet. Like the traditional aquaculture systems, *prahoc* originated as a means to preserve the abundance of the dry season fish harvest; it is a major source of protein for the rural poor, who use it in a wide variety of rice dishes, soups and other dishes.

To be successful, Cambodia's planned expansion of its aquaculture industry must rely on more sophisticated technologies than the traditional system and the use of *trey riel*. It will require an integrated approach with more sophisticated breeding techniques to increase hatching rates, improved nursery management to increase survival and growth, and feed supplements that increase yield at the final grow-out stage of production. Fortunately, much of this important research and development is being coordinated by the Mekong River Commission and is focused on species native to the Mekong River Basin.[54] For example, Vietnam has developed a massively successful production system based on the striped catfish (*Pangasianodon hypophthalmus*). In 2000, Vietnamese scientists developed an artificial propagation procedure for the catfish, which is known as *trey pra* in Cambodia, and revolutionized aquaculture in the Mekong Delta by providing growers with an essentially unlimited number of hatchlings. This success stimulated other innovations, such as the use of nurseries to ensure the survival of fingerlings before they were transferred to grow-out ponds or submerged cages in river channels. Fish were originally fed a farm-made concoction of crushed rice, waste from vegetable gardens, duck weed and locally harvested termites, supplemented with trash fish. However, producers have switched to locally produced commercial pellet feed, which provides a more nutritionally balanced diet and can be supplemented with imported soy instead of fishmeal.[55] Investments in processing facilities facilitated exports to sophisticated international markets and added commercial value to a basic commodity. In the first decade of the twenty-first century, the Vietnamese striped catfish aquaculture was the fastest growing food production system in the world; exports grew from virtually nothing in 2000 to more than \$1.5 billion by 2009 — an amazing accomplishment.

In spite of its remarkable success, environmental and social factors have created challenges that must be managed in order to ensure its sustainability. Competition has squeezed profit margins and small family producers are being displaced by larger enterprises that can meet the strict environmental and health standards imposed by consuming nations. Like any high-density animal production system, waste water treatment must be properly managed to assure the long-term health of the industry. Although waste water treatment is theoretically possible, it is not feasible (or at least commercially competitive) due to the very large volumes of water and the high cost of water treatment facilities. Consequently, the expansion of this model has been limited to the Mekong Delta where water volumes are so large that the effluents can be diluted to acceptable levels. If past experience is any guide, the dilution of waste will eventually be ineffective even in these large river systems and the long-term impact on water quality will prove to be unsustainable.*

The evolution of the freshwater aquaculture industry in Cambodia will be much less dramatic, except for certain limited areas along the Mekong and Bassac rivers that might be able to duplicate Vietnam's striped catfish production model. In the Cardamom Region, the tributaries of the Tonle Sap could support closed pond aquaculture cultivation — but waste water must be managed sustainably to protect downstream water supplies. One method would be to recycle pond effluents via rice fields, as is done in neighboring Thailand, which removes between 25 to 30 percent of nitrogen and phosphorus from catfish pond effluents; not surprisingly, rice yields were boosted to approximately the same levels as those attained with standard fertilization.[56]

Approximately 40 percent of global aquaculture occurs in marine or coastal waters, with the most important species being shrimp and prawns, in terms of both volume and economic value. Shrimp and prawns accounted for approximately \$14 billion dollars in trade in 2008, with China being the largest producer of farmed shrimp at 25 percent of global production; Thailand was the next biggest exporter with 18 percent global production and exports valued at \$2.5 billion annually; Vietnam trailed with 5 percent of global production and exports approximately equal to stripped catfish at \$1.2 billion annually.[57] The expansion of the shrimp farming industry in Cambodia can either contribute to a green economy or do the opposite and destroy mangrove habitats and degrade important natural fisheries including wild shrimp (see Chapter 5). The outcome depends on whether the industry adopts an integrated management regime that can avoid or mitigate the environmental impacts of intensive shrimp production

* Engineering schools used to teach that water quality could be dealt with via water flows management, rather than the removal of noxious effluents and was epitomized by the saying "the solution to pollution is dilution."

facilities, while simultaneously increasing production and maximizing economic returns.

Such a sustainable management regime would include establishing shrimp farms on appropriate sites and managing water flows, both of which are essential for maintaining high quality ecological conditions and avoiding disease outbreaks with cataclysmic economic consequences. Advanced shrimp production systems avoid the brackish environments of converted mangrove swamps, as these develop toxic acid-sulfate soils over the mid-term that compromise ecological conditions and affect shrimp health. Instead, technologically advanced shrimp farms are now placed on upland landscapes where the physical and chemical properties of ponds can be closely monitored to ensure maximum productivity. It is now recognized that maintaining healthy shrimp populations will allow their natural immunological defenses to successfully combat a wide range of pathogens.[58] Sophisticated shrimp farmers never exploit wild shrimp populations to stock production ponds, because these often introduce pathogens. Instead, they use genetically improved and certified brood stocks that ensure high productivity and reduce the cost and environmental risk of using antibiotics to control disease outbreaks. Better quality feeds increasingly avoid fish meal — an essential step in an industry that is overly dependent on trash fish from over exploited freshwater and marine ecosystems.[59] Best practices, however, go beyond shrimp production and environmental issues to ensure that processing facilities are also efficient and safe, both for the workers and for food safety reasons. Social issues should also be taken into consideration, especially in regions like the Cambodian Coast where communities have poorly defined legal rights to land and other natural resources.

The standards of a sustainable shrimp industry have now been identified by a multi-stakeholder process and are organized around eight major themes: farm siting, farm design, water use management, brood stock and larvae, health management, feed, food safety and social responsibility.[60] Walmart, the world's largest retail company, which accounts for approximately 3 percent of total shrimp sales in the United States, has announced that all future suppliers of shrimp must adhere to best practices guidelines by the end of 2009.[61] One of the largest suppliers in Thailand immediately took steps to ensure that it could meet these requirements; however, the announcement has been met with criticism by some producers, because of the scale of the changes and the time-line under which they must be adopted. In the current Thai shrimp sector, about 80 percent of shrimp farms are small operations run by families who lack the capital to make the required upgrades or who cannot afford to pay for the certification process that overseas markets will soon require. Consequently, the drive for sustainability may be accompanied by a consolidation process, where small producers are displaced by large, vertically integrated companies that can more efficiently combine all aspects of the production chain to comply with certification standards.

The extraordinarily rapid development of the aquaculture industry globally over the past two decades bodes well for the potential to expand this sector in the Greater Cardamom Region, where the region's abundant freshwater resources would provide the foundation for enormous growth, and the biological diversity of the Mekong contains a treasure trove of species with specialized niche markets in Asia and the rest of the world. The traditional knowledge of the Khmer people also confers a unique advantage in developing aquaculture systems; few people on Earth know more about the care and feeding of fish than the Khmer. This knowledge can be used to enhance the rice – fish culture, as well as to develop dedicated aquaculture production systems founded on appropriate technology and wise water management. The recent success of the striped catfish production model in the Mekong Delta highlights the potential for the region's biodiversity to contribute to the economic development of the nation, while the pitfalls from an unsustainable shrimp farming model demonstrate the importance of responsible management.

A Diversified and Green Tourism Model

Tourism has a relatively benign image as far as its environmental and social impact, at least when compared with other "traditional" industries such as oil and gas, mining and manufacturing, or agricultural and forestry — all of which are considered mainstays of a modern economy. In part, this is because it is a "service industry" rather than an "extractive industry" or "heavy industry," terms that are loaded with images of smokestacks and blighted landscapes. In contrast, tourism is associated with images of wooded hills and pristine beaches, and even urban destinations are typically clean and well ordered. However, most educated people, and certainly all experienced travelers, know that tourism comes with a fairly heavy environmental footprint, which can include increased sewage effluents into overstretched waste treatment facilities, the conversion of natural habitat, and the massive emission of green house gases, to name three of the most obvious.

Efforts to support biodiversity conservation have led to the development of what is now termed "ecotourism," a specialized sector of the tourist industry that has experienced phenomenal growth since its inception about 20 years ago. The principal motivation for individuals who purchase an ecotourism package is the desire to visit a pristine or an exotic, "wild" natural habitat. This desire predates the modern tourist industry and many of the proto-tourists of the eighteenth and nineteenth centuries were also seeking adventures in wilderness areas. The current concept of ecotourism goes beyond merely enjoying nature; it also requires both tourist and host to limit their impact on the environment, to provide an educational experience and to support the economic development of local communities.

The success of ecotourism and the general push for more sustainable enterprises in general has led to initiatives within the broader tourism industry to improve environmental performance: terms like "green tourism" or "sustainable tourism" are often heard these days, referring to initiatives that integrate the economic, environmental and social dimensions of tourism into their management philosophy. Guidelines and standards to certify tourism enterprises for "best practices" are currently under development, the most detailed of these having been developed by a multi-stakeholder process that lists the "principals and criteria" a business must adopt to become certified.[62] These include sustainable concepts in the design and operation of hotel and resort facilities, which is particularly relevant to the Cambodian Coast and the Cardamom Mountains where investments in hotel infrastructure are at an early stage.

These criteria start with a commitment to obey all laws and pertinent regulations, including those dealing with health, labor and the environment, as well as all national and local zoning ordinances — especially those related to protected areas or cultural heritage sites. The design of the hotel or resort is particularly important, since the physical infrastructure is the most conspicuous part of the facility and can either add to the beauty of a locality — or be an eyesore. A commitment to reduce green house gas emissions is a common component and includes the use of renewable energy, as well as new technology and common sense practices that promote efficiency and minimize the consumption of energy and water. Pollution control should be of paramount importance in pristine areas and include efforts to minimize the use of harmful substances, such as pesticides, swimming pool disinfectants, and cleaning materials, as well as to reduce noise, light, and runoff — measures that will not only preserve the environment, but improve the experience of the hotel guest. Where chemicals are required, they should be properly managed following procedures for each individual compound, including a solid waste management plan and water treatment facility. If this infrastructure and basic services do not exist near the resort, the investor must develop them and make them available to surrounding communities at a reasonable cost.

Tourist operations should also maximize benefits for nearby communities. This includes hiring local residents for both menial and managerial positions, which may require providing specialized training and providing equal opportunities for women. The exploitation of children and adolescents should not be tolerated, and all service providers should be encouraged to adopt these criteria to bring supply chains into compliance with sustainability standards. One of the most important standards, however, is for companies to adopt a code of conduct for interacting with local communities, to ensure the consent of the full community and not some self-appointed representative. Particular attention must be paid to issues concerning natural resources, including land and water use, but also fishing rights and forest access; for example, traditional paths that might connect villages or provide access to the sea. Finally, as it is not in the financial interest of resort operators to operate a luxurious facility adjacent to impoverished communities, they should proactively assist in the development of basic health and education services.

The Cambodian tourist industry has been a leader in implementing the concepts of sustainable tourism as part of its efforts to both promote and manage the growth of tourism at Angkor Wat. This leadership was motivated by the early recognition of the unique cultural importance of the temple complex and its designation as a World Heritage Site. One strategically important decision was to create a core area where the construction of hotels is prohibited and where food service facilities, transportation and guide services are regulated to maintain order and ensure quality. This policy benefits local residents by allowing local women to provide food and beverages services at strategically located food stalls within the outer walls of the central temple complex. Although crowds are often large and the mixture of ethnic and economic groups is remarkably diverse, the management and the flow of visitors are characterized by dignity and respect for what is essentially a religious temple. The larger complex of temples conserves large areas of forest habitat, while the surrounding landscapes are covered by rice fields and traditional farmsteads where the local inhabitants live and work. Efforts to provide tour guides with language training and to educate them on the archeological and historical significance of the temples have improved the tourist experience and are reaping massive dividends in word-of-mouth marketing around the world.

The experience of Angkor Wat should serve the Cambodian government well in organizing the already vibrant tourist industry on the Cambodian Coast and emerging ecotourism opportunities in the Cardamom Mountains. Chief among these lessons should be the recognition that land-use zoning is an essential component in planning future growth, and that some prime real estate should be set aside from development (see Chapter 8). Fortunately, large areas in the mountains and the coast have already been set aside as protected areas, but within those areas there remains an urgent need to zone future land use. Efforts to develop a large beach resort complex within Bokum Sakor National Park, for example, must be carefully considered and operated according to a philosophy that is appropriate for a national protected area dedicated to biodiversity conservation. The restoration of the casino and hotel complex at Bokor Hill Station provides an opportunity to integrate a tourist facility into its natural surroundings in a way that adds value to the resort as a tourist destination while conserving the natural habitats of the Elephant Mountains.

Land-use planning should also manage the diverse nature of international tourism. Casinos, golf courses, beach resorts, restaurants, bars and night clubs, forest lodges, trekking expeditions and cultural experiences in local communities, scuba diving and snorkeling on coral reefs — these are all part of a diversified tourist economy. Success in attracting different tourists will depend on providing them with the experience they seek; at the same time, mixing tourists with different tastes can create negative impressions that will impede return visits. For example, nature lovers looking for a tranquil spot to explore a coral reef would not be keen on sharing a beach with a water sports enthusiast fond of loud music. Nonetheless, everybody probably appreciates a good meal, a cold drink and quite possibly a night out on the town. Obviously, cost and budget can distinguish between different tourist clienteles; however, it is a mistake to cater preferentially to the affluent, because greater gross monetary flows are almost always associated with the middle class, and even budget (backpacker) tourists have a place in a diversified tourist model, as they seek out the most economical options and patronize service providers at the bottom end of the economic scale.

Current tourist investments are concentrated on the Cambodian Coast, particularly near Sihanoukville where the international airport facilitates access, but there is growing interest in the coastal areas on the Bokum Sakor peninsula and virtually all of the offshore islands. Nature tourism in the Cardamom Mountains will likewise become attractive to investors and tourists as the hydropower dams are completed and access roads connect them to the coast. The cool mountain air will be a welcome respite after a few days walking around in the suffocating heat at Angkor Wat or in the intense sun on the nearby beaches. The dams themselves can become tourist destinations, as they are in Arizona, Brazil and China, where hotel lodges around the reservoirs provide yet another option to the region's tourist offerings. Eventually, improved roads will connect the communities of the Tonle Sap rice plain with Koh Kong City and open up local commercial traffic and another set of tourism opportunities, including adventure tours for those inclined to trek to a megalithic site in the vast wilderness of the Cardamom Plateau.

Whether tourism development occurs in an orderly and sustainable fashion depends in part on the government and its ability to impose an integrated coastal management plan on its foreign and domestic investors, many of whom will have radically divergent visions as to what constitutes an aesthetically pleasing development. The adoption of certification standards to promote green tourism provides another option; although that route needs to include middle class investors who play a large and strategically important role in the development of the tourist industry. An important strategy will be to engage local and regional governments in creating and enforcing standards for the tourism industry, as they understand the communities and issues in their individual areas best, and investors can look to them for guidance.

The Rest of a Green Economy

The previous sections describe how four strategic sectors of Cambodia's current economic model can be transformed into the pillars of a green economy: forestry, agriculture, fisheries and tourism. What makes these sectors relevant to resolving the Cardamom Conundrum are their dependence on renewable natural resources and the multiple opportunities they provide to stimulate economic growth based on environmentally and socially sound production models — the essence of sustainable development. This diversity of opportunities is itself a sound development strategy, because just as biological diversity lends resilience to a natural ecosystem, a diversity of production models lends stability to the economy of a nation. Cambodia cannot rely on just four sectors for its future growth and development, however, no matter how robust and sustainable. In order to thrive, the country must foster innovation across all sectors of its economy and integrate them into regional and global markets that are likewise experiencing profound change. The four natural resource sectors highlighted in this chapter can form the foundation of a low carbon economy, but to be successful, Cambodia must also transform and "green" the other sectors of its national economy.

The most important sector to engage in a low emissions development strategy is the financial services industry, because of the increasing importance of international carbon markets and their potential to fund climate change mitigation and adaptation. This combination of public and private resources, which is now being referred to as "climate finance," includes the potential revenue flows from REDD+ and CDM, but also other types of development assistance and investment capital with even greater potential for influencing global trade. The early stages of climate finance are being organized via traditional bilateral and multilateral agencies,[63] but private capital is expected to become the primary source for financing carbon abatement investments in the near future. The development of a cap-and-trade system for carbon credits may, or may not, emerge over the medium term; but this controversial market mechanism is not the only option available for influencing business decisions and reducing greenhouse gas emissions. Moving forward, climate finance will include a broad range of financial instruments that support climate change mitigation and adaptation, including risk insurance, loan guarantees, subsidized debt, and public-private partnerships, as well as cash payments for carbon credits. Consumers across the World are demanding that the production of the goods and services they consume are truly sustainable; the financial industry is responding to these very clear market signals by developing standards for their own industry,[64] because finance impacts all other sectors of the economy, as they change others will follow.

The low emission development strategies described in this chapter will require the active intervention of the Cambodian financial services industry. To be successful, Cambodian bankers will need to understand the requirements of international carbon markets, the nature and degree of risk of Cambodian projects, and the potential tools, financial and otherwise, available for managing that risk. Opportunities for accessing climate finance in Cambodia can only be achieved if the appropriate enabling conditions are created to attract investment; for example, a well informed financial services sector can play a key role in regulatory reform and providing objective input when evaluating infrastructure options. The cost of finance is always a key consideration when weighing investments and the primary role of climate finance is to change that calculus in favor of green development options.[65]

Investments in renewable energy will be in the forefront of any strategy to mitigate the impact of climate change; in this and previous chapters, the potential from hydropower and biofuels were identified as attractive green options — if they are developed according to sustainable criteria. Cambodia also has tremendous potential to exploit solar power, which can play an important role in rural areas that lack electrical transmission lines and, eventually, set the stage for an alternative national energy strategy. Transmission lines will expand into the countryside over the next decade, a process that will coincide with a reduction in the price of photovoltaic technology. In this alternate energy scenario, households and businesses across the country would be equipped with solar panels that feed power back to the grid when the sun is shining. Solar power can meet demand during the long sunny days of the dry season when hydropower facilities are under stress; similarly, hydropower can fill the inherent gaps in solar energy systems at night and on cloudy days. The integration of solar and hydropower can minimize the risk of energy shortages from drought and long cloudy periods, creating a robust green energy model.

This energy model will not be attractive to power companies, however, particularly when viewed from the perspective of maximizing the return on investment and the challenges in organizing a "distributed generation system." The development of this integrated system requires a specific regulatory framework that motivates power companies to purchase electricity from its clients and specialized technology that allow for the two way flow of energy in a system that is referred to as a "smart grid."[66] Lowering the capital cost of acquiring such a system is precisely the role of climate finance and regulatory reform is an example of creating the enabling conditions to access those resources. Fortunately, the potential for creating a smart grid and a distributed energy generation system is easier in a country like Cambodia, which does not have a legacy industry and the associated infrastructure investments in traditional fossil fuel generating systems.

Investments in renewable energy should be complimented by investments in energy efficiency. Cambodia is currently characterized by a severe lack of power generation; consequently, electricity is expensive and saving energy is standard practice in most homes and businesses. This situation will change when the hydropower facilities under construction in the Cardamom Mountains are incorporated into the national energy grid and the cost of electric power generation declines. At that point, Cambodia will face a situation more common to most advanced and developing economies where growth in electricity demand is met with a sequential increase in energy generation. Another option is to invest in what is referred to within the power industry as "end-use efficiency," and ask companies to assist their clients consume less energy by changing habits, design buildings with passive climate control systems, or purchase more efficient home appliances.

Promoting end-use efficiency requires exactly the same regulatory framework needed to motivate an electric utility company to purchase power from its clients via a distributed energy system. In a traditional regulatory system, companies seek to sell as much energy as they can generate and have no incentive either to promote energy conservation or purchase surplus energy their customers may generate via solar panels. The role of a regulator in a traditional system is to ensure that utilities do not overcharge their customers, but those systems do not provide an incentive mechanism for promoting energy conservation — much to the contrary, low prices have the opposite effect. Fortunately, an alternative regulatory framework has been pioneered and tested by the State of California, which simultaneously promotes energy efficiency and ensures low rates for consumer, while rewarding companies for both outcomes. Known as "decoupling," the regulating authority in this system disassociates a power company's sales revenue from its profit margin and, instead, ensures that its "rate of return" (another word for profit) is matched to revenue targets, while "adjusting" electricity rates according to the interaction of supply and demand and the actual cost of energy production. As in all regulated energy markets, a utility company is allowed a comfortable profit, but in this scheme they are rewarded on how well they meet their customer's energy needs, rather than the total amount of energy they can produce and sell. Because improving energy efficiency is almost always less expensive than in investing in new infrastructure, companies have a strong incentive to assist their customers become more efficient.[67]

The switch to a decoupled regulatory framework in 1984 has allowed California to stabilize energy consumption per capita and save hundreds of billions of dollars, all while enjoying three decades of unparalleled economic growth. The concept of decoupling has been embraced by the United Nations Environment Program (UNEP) and is being adopted by dozens of countries and municipalities across the globe, including China and the European Union.[68] In the case of Cambodia, the adoption of decoupled regulatory framework could have a significant impact on its overall development trajectory. It would foster a domestic energy industry that benefits rural citizens, an innovative concept attractive to climate financiers, as well as avoid the importation of coal that would negatively impact both human health and the nation's balance of payments. It might also allow Cambodia to forgo the construction of one, or both, of the large hydropower facilities proposed for the main stem of the Mekong River. Both of these projects, one near Stung Tren and the other at Sambor near Kratie, are highly controversial due to their potential impact on the natural fisheries of the Mekong River and Tonle Sap. The electricity demand that would be met by these two facilities could be reduced over the short term by improving end-use efficiency, and eventually replaced by photovoltaic technology, which is projected to become cost competitive with coal and large-scale hydro by 2020.[69] The adoption of a regulatory framework that rewards efficiency can have a similarly transformative impact on the construction sector. Designing homes, offices, factories and hotels so that they are energy efficient can yield massive savings in air conditioning costs, improve the standard of living, and improve the cost competiveness of manufacture goods.

An efficient transportation sector is an essential component of all economies, green or otherwise. Transportation costs can make or break the competitive position of an enterprise or production model, particularly bulk transport and logistical systems that connect factories and farms to national, regional and global markets. Improving efficiency and reducing greenhouse gases in these systems yield obvious economic benefits. Poorly conceived, unregulated, and chaotic urban transport systems can be massive consumers of fossil fuels, add costs to business transactions, and significantly degrade the quality of urban life. Cambodia currently has a motorcycle model of urban and rural transport, which provides Cambodians with a remarkably level of mobility at a very reasonable cost; in addition, many tourists find the hoards of motorcycles exotic and consider it to be part of the charm of Phnom Penh. But as the country becomes more affluent, motorcycles will be replaced by automobiles and very soon (probably within a couple of years) Phnom Penh will be characterized by traffic gridlock, not unlike that currently suffered by megacities such as Jakarta, Bangkok and Manila.

Mass transit is the only rational long-term option for Phnom Penh and city planners should be preparing its existing infrastructure by planning for dedicated bus lanes, while undertaking feasibility studies for elevated light rail. In the countryside, trucks and buses provide a cost-effective transportation system; however, the basic road infrastructure is still rudimentary and will very quickly become saturated. Investments in both secondary and primary road systems remain a priority, particularly as the production of agricultural commodities multiplies, which require low cost bulk transport systems. Fortunately, Cambodia has already embarked on a renovation of its legacy rail network that can form the backbone for a renovated and energy efficient terrestrial transportation system, which can be integrated into the regional rail network and facilitate commerce with neighboring countries.

The adoption of a low emissions developments strategy that spans all sectors of the national economy will yield multiple benefits for the manufacturing sector. As discussed in Chapter 5, the textile industry has carved out a special market niche for their products due to their commitment to improving the social welfare of their workers. A similar opportunity is available for companies that demonstrate they have a comparable concern for the environment. Discerning consumers in advanced economies are seeking "sustainability" in the clothes they wear and the food they eat; consequently, retailers across the globe are seeking to improve the environmental and social standards of their supply chain. This has led to the proliferation of certification initiatives, such as the Consumer Goods Forum, which brings together more than 650 enterprises in 70 countries with combined sales estimated at more than $3 trillion.[70] In the process, these same companies are discovering the benefits of a green economy and that embracing sustainability not only opens markets, but also improves efficiencies within the enterprise and creates a positive attitude in the work force, both of which can substantially improve profit margins.

Some industrial sectors are inherently energy dependent or are particular damaging to the environment; these types of enterprises are not amenable to being "reformed" into green production strategies, but they can take important steps to reduce their environmental footprint. For example, responsible firms in resource extraction industries have developed "best practices," which mitigate unavoidable impacts and reduce risk, while improving productivity. Cambodia is not a major participant in the non renewable resource sector, but efforts are underway to discover and exploit hydrocarbons, while mining companies have initiated exploration and development in various parts of the country (see Chapter 5). The exploitation of non renewable natural resources is a normal part of economic development, particularly for a country that needs public revenues to invest in development or, in the case of Cambodia, natural gas to fuel its economy.

It is not necessarily a good development strategy for a country to participate in all stages of a development trajectory based on resource extraction and transformation, particularly those stages where minerals are processed into intermediate industrial commodities, such as steel and aluminum or refined petroleum products and petrochemicals. These so-called "heavy industries" are characterized by massive capital outlays and economies of scale; enterprises must compete in highly competitive global markets that are dominated by vertically integrated transnational corporations and countries that often resort to subsidies and unfair trading practices to protect their market positions. Another option, and one that is particularly strategic for a small nation seeking to excel as a green economy, is to use the revenues earned from the exploitation of its mineral resources to capitalize a sovereign wealth fund dedicated to developing the human resources that are essential to a modern economy.

Cambodian families understand this fundamental fact and are singularly intent on educating their children, as evidenced by the high levels of primary school attendance and the proliferation of private academies and universities. This common-sense concern is shared by the elected officials of Cambodia, who dedicate approximately 10 percent of total government expenditures to public education.[71] Most of this is dedicated to teaching and school infrastructure, but there are still gaps in the educational system, particularly at the university level and in teacher preparation and support. The Cambodia state has many demands on its budget and must carefully allocate scarce resources to infrastructure and productive activities if it is to succeed. The example of other "knowledge based" economies should be instructive; for example, Singapore allocates 20 percent of its national budget to education. The leaders of that nation, which must organize an economy in the absence of any significant natural resources, are keenly aware that the real key to developing a globally competitive economy, green or otherwise, is to nurture a well educated and highly motivated population.

CHAPTER 8

The Way Forward

In the past, environmental advocates have argued for conservation largely on moral grounds, citing society's obligation to protect our nature heritage; however, recent advances in the science of economics has shown that there are exceptionally strong bottom-line arguments for nature conservation.[1] The resolution of the Cardamom Conundrum depends on Cambodia's capacity to capitalize on the value of its natural resources and develop a green economy that creates real economic opportunity for its citizens. Such a strategy can begin with a commitment to conserve forests landscapes, but this must be coupled with other low emission development strategies that span other sectors of the domestic economy, particularly agriculture, fisheries, and tourism, but also energy, manufacturing, transportation and construction. Like any coherent development strategy, adequate health services and educational opportunities must be extended to all Cambodian citizens, in order to provide a labor force commensurate with the demands of the global market place. A green development paradigm is ideally suited for the Greater Cardamom Region because of its rich natural resources, but it is equally appropriate for Cambodia as a whole and for many other countries in the developing world.

Building a green economy and investing in the sustainable development strategies outlined in the previous chapter will depend to a large part on how Cambodia allocates the use of its geographic space and how it manages its natural resources. This demands good governance and decision-making processes that includes all constituencies, some of which will have radically different views as to what is in the best interest of the country. Land-use decisions need to be made not on short-term perspectives that respond to the interest of a single group, but on a strategic analysis that considers broader perspectives. Land that is best suited for a specific economic activity should be accurately identified and zoned for that purpose. This should incorporate an effective conservation strategy, but nature conservation needs to be conceived within a multiple-use framework that accommodates land-use options backed by powerful constituencies and supported by strong traditional economic logic. The decisions to limit virtually any specific economic activity will generate tension, as people and industries will act to protect their own economic interests. Consequently, economic analyses must be an essential part of the land-use planning process and must clearly detail the economic benefits that low-carbon enterprises and conservation initiatives bring to society. This includes the long-term value of a renewable resource, such as soil fertility or timber, as well as the positive economic contributions that well-managed landscapes provide on a daily basis, such as water for energy and irrigation, and fish for human consumption.

Viewed from this context, Cambodia's current conservation strategy is deeply flawed and probably unviable, because it was designed without an assessment of future development options and based on a process that did not consult important constituencies. Consequently, landscapes are now being modified in ways that are incompatible with conservation objectives. New communities are being established in frontier areas and within protected areas, while agro industrial investors are converting natural habitats on surrounding landscapes. Timber exploitation has been put on hold for almost a decade, but a plan to exploit the forest estate will soon be developed and implemented. Environmental advocates no doubt will rally to defend the integrity of the protected area system; however, failure to address the deficiencies in the current system will lead to its gradual breakdown. Worse still, the social dynamic linked to the failure of the protected area system will foster a fundamental disrespect for the rule of law and impede the emergence of an effective governance system that is absolutely essential for effective natural resource management.

Conservation organizations and their allies in civil society and local governments have the opportunity, however, to lead a reevaluation of current conservation and development strategies in the Greater Cardamom Region. If organized and led strategically, it can set in motion a process that transforms Cambodia and establishes the foundation for a low-carbon sustainable economy — rather than just a suite of protected areas. This process can, and should, bring more actors into an expanded discussion about what constitutes a sustainable society. It must include common sense solutions like community management of forests and fishery resources, while employing sophisticated spatial planning tools and, most importantly, identify and promote innovative production strategies that foster economy growth and create jobs. The planning must be sufficiently flexible to include the long-term impacts that might occur due to changing climate, as well as the incremental change that will accompany Cambodia's inevitable transition into a modern urban society. Fortunately, this revitalized planning process can be aligned with current government policies to decentralize natural resource management decisions to the provincial and district level, while a broader participatory process can build the support that is essential for a successful conservation strategy. In the end, an expanded collaborative and consultative process will lead to greater understanding of the economic value of the region's natural resources and the contribution of nature conservation to the long-term and sustainable economic growth of the nation.

Monetize the Value of Forest Carbon

Forest conservation can be the starting point for Cambodia's low-carbon economy. The economic resources from the international carbon markets and the impending REDD+ framework can initiate the green economic model and provide seed money for other low-carbon initiatives, particularly in agriculture. These initiatives will require a coherent institutional structure that goes beyond creating protected areas, so that revenues are used to promote forest management, foster economic opportunity and create jobs. In its simplest form, REDD+ can be viewed as payment for slowing or stopping deforestation. This approach, however, treats REDD+ revenues as a rent or royalty paid to individuals or communities who agree to change their land-use and natural resource extraction practices. This course of action might very well alleviate poverty, particularly in rural areas, but it would not reduce the underlying causes of poverty or create the conditions for long-term sustainable change. A more innovative strategy would be to treat REDD+ revenues as a source of investment capital and invest them in the region's productive capacity. This would require the support of key stakeholder groups, so that everybody involved understands the link between forest conservation, as well as agreeing to use those financial resources to boost productive enterprises. Being realistic, REDD+ revenues are not likely to provide sufficient investment capital to change the way people do business, but if they were leveraged with other public and private financial resources, they could enhance their impact and stimulate investments in low-carbon production systems with strong global and regional markets. As shown in Chapters 6 and 7, these projects could actually create additional carbon assets that could be monetized in future carbon markets for linked to agricultural production systems.

Development assistance is currently being channeled toward proto-REDD+ pilot projects in dozens of countries and several pilot projects under development in Cambodia, two of them in the Greater Cardamom Region. The Greater Cardamom Region is probably too large and diverse to organize a single all-encompassing REDD+ project and a suite of distinct REDD+ initiatives may be a more efficient strategy for attracting investors and implementing successful projects that address the needs and interests of different stakeholder groups. Some might be community forests, for example, while others focus on protected areas. Some may be geographically defined, based on specific upland or lowland landscapes and communities, while others address the underlying drivers of supply and demand that lead to deforestation and forest degradation. For example, the strategies needed to avoid deforestation in hilly, remote areas will be radically different from the tactics needed to promote the conservation of forest corridors and reserves on plains populated by farming settlements.

The most obvious way to organize a suite of REDD+ initiatives would be adopt the investment principle that investors who assume greater risk are rewarded with a larger economic return. Following this logic, REDD+ revenues should be distributed according to the probabilities of future deforestation, with greater revenues flowing to buffer zone areas where population density is high and increasing and where the need for intervention is more acute (Figure 8.1). Following this logical framework, the core area of the Central and Southern Cardamom Protected Forests and the former logging concessions of the Forest Estate could be aggregated into a single REDD+ initiative managed under the aegis of the Forestry Administration (FA), while the buffer zones surrounding this core area might best be coordinated by provincial and community authorities, who have administrative authority over these more populated landscapes. Within multiple-use protected areas, the Ministry of the Environment (MOE) could assume a leadership role together with local governments, while the FA should assume that responsibility in areas allocated as production forest.

Because the REDD+ mechanism has yet to be formally approved by the UNFCCC*, its impact on the private sector has been limited.

---

* United Nations Framework Convention on Climate Change.

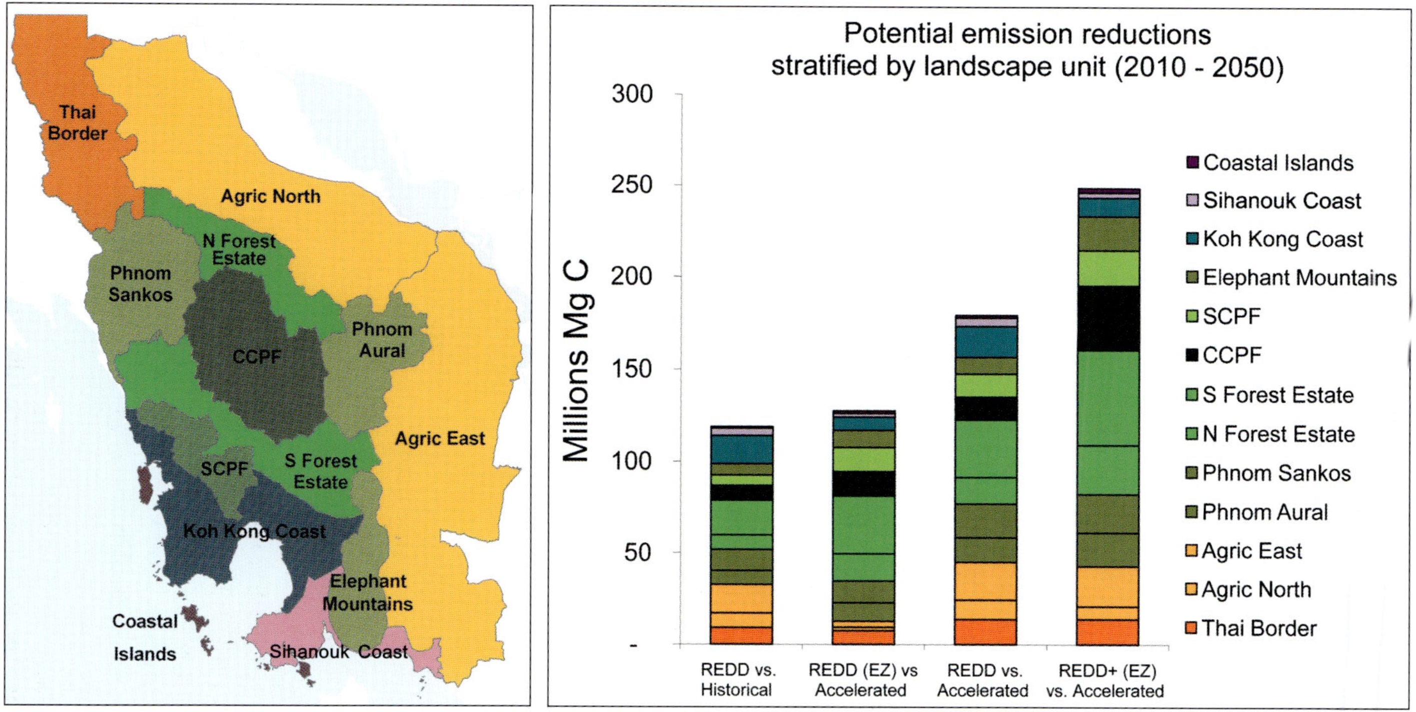

Figure 8.1. The distribution of potential REDD revenues in different units of the Greater Cardamom region; the dimension of the potential revenues varies depending upon the baseline used to calculate reductions (historical vs. accelerated) and the strategy used to reduce deforestation and forest degradation (REDD, REDD with Economic zoning (EZ) and REDD+) with efforts to increase carbon stocks in standing forest.

Once a formal mechanism is approved, however, there will be a boom in the demand for carbon credits generated via REDD+ initiatives, and the marketplace almost immediately will classify carbon credits associated derived from forest conservation and management according to risk, quality and cost. Countries with good baseline data and monitoring systems that validate emission reductions or sequestration offsets will have a competitive advantage, as will those countries with a coherent strategy for managing revenues and a governance system that honors the commitments that underlie the carbon credits. Although markets will primarily be seeking value in terms of carbon volumes, countries can add value to their carbon credits by conserving biodiversity or improving livelihoods and, in the process, make them more attractive to corporate clients with multiple agendas.

The greatly expanded supply and demand of a fully functional global REDD+ mechanism will also focus attention on the transaction costs for carbon credits. As markets have learned from the CDM mechanism, these costs can limit the ability of suppliers to scale up and meet demand; more importantly, these costs will subtract from a country's potential to use REDD+ revenues for productive investment. A truly competitive market should seek better and more reliable carbon credits, particularly from initiatives that change behaviors at the national or sub national scale. Although many initiatives will be geographically defined to focus on specific landscapes and communities, others must be targeted at the underlying drivers of deforestation and forest degradation, such as reducing the demand for fuel wood or improving agricultural yields. To validate the carbon credits linked to that type of policy initiative will require demonstrating real changes in deforestation rates and levels of forest degradation across broad landscapes. In other words, show that a change in cooking fuels has led to a reduction in forest degradation or that better farming practices are linked to a slowdown in land-clearing.

A comprehensive national monitoring program will be essential for documenting both deforestation and forest degradation, as well as validating new carbon sinks on productive landscapes. At the same time, Cambodia will need to expand REDD+ initiatives from the project scale to a national scale. A national REDD+ strategy cannot be based, realistically, on a bundle of projects; other mechanisms must be developed for distributing revenues and providing positive outcomes. This does not mean that projects should be eliminated; much to the contrary, there are many investors that will specifically seek project-based carbon credits. Nonetheless, countries like Cambodia need a different type of revenue sharing mechanism that allows local governments, and through them communities, to benefit from REDD+. If properly organized it can strengthen the role of the state and, in the process, improve security for carbon investors.

To be successful, Cambodia's REDD+ system must instill confidence in both international markets that invest in the credits and domestic stakeholders who will share the revenues from the credits. This will require that it be both transparent and equitable. There are two basic and complimentary approaches:

- A "*book and claim*" system would distribute revenues to stakeholders via the state, essentially paying carbon royalties to local governments and state agencies responsible for meeting land management goals.
- A "*book and trade*" system would establish a market mechanism to manage contractual agreements between individuals and communities who create carbon credits and the individuals or institutions seeking to purchase them.

Both systems could be administered by the same national authority, which should also assume responsibility for establishing emission reduction targets from land-use and formulating policies to reach targets. One option would be for this agency to act as the central clearinghouse for carbon revenues, allocating a portion of the revenues to a market mechanism and distributing the rest to local governments where people are reducing emissions or sequestering carbon. Conversely, the state could allow a market mechanism to develop between international buyers and domestic sellers of carbon credits and then garner a portion of those revenues to distribute to local governments, in order to ensure equity and promote low carbon strategies more widely. Regardless of how the system is structured, Cambodia will need to develop capacity in preparing baselines and monitoring carbon stocks, particularly in regional and local governments that must oversee the process.

Care will also need to be taken that governmental agencies and service providers, be they consultants or NGOs, do not appropriate excessive revenue, which would deprive stakeholders of the resources necessary to change their behavior. Competition among service providers will be essential, as will a streamlined book and claim system that does not become an obstacle to market transactions. For example, some environmental NGOs might seek to ensure that all probability of leakage is eliminated and that all projects provide 100 percent "additionality".* However, the adage that perfection is the enemy of the good should be a guidepost, and proponents and participants must understand that a certain amount of leakage and "non additionality" can be tolerated if they bring about expanded carbon flows into the system.

Stabilize the Agricultural Frontier

One of Cambodia's biggest challenges will be to slow and eventually stop deforestation; this will require an integrated strategy that identifies appropriate land-use and establishes a regulatory system with clear, enforceable rules governing land-use. It will also require market-based incentives that reward wise land-use decisions. Spatial planning can demarcate where agriculture is appropriate and where it is not, and improved governance and law enforcement, particularly at the local level, can motivate inhabitants to respect land-use regulations and stop illegal land use. These tactics will only be successful, however, if the benefits from improved land-use and forest conservation are tangible and equitably distributed among communities and the individuals who will be affected by these decisions. Funds from REDD+ and other emission reduction mechanisms must be invested in production systems that improve people's livelihoods, and the law enforcement system that enforces land-use regulation must be accompanied by a legal system that also protects the property of communities and individuals alike.

Conservation organizations and government ministries are fully aware that communities need to be integrated into conservation planning and natural resource management decisions, a step essential to ensure buy-in for protected areas and to create awareness of the benefits from sustainable production schemes. Cambodia's government and its partners are beginning to test these concepts in communities throughout the region. Law enforcement programs have shown good results when combined with initiatives to improve the livelihoods of communities in or adjacent to protected areas. For example, this approach has been integral in suppressing deforestation along a 40-kilometer stretch of Route 138 on the border between the Bokum

* Additionality and Leakage are terms used by the UNFCCC to describe situations where GHG emissions are not really reduced because they are simple displaced to another site (Leakage) or where the actions taken to generate carbon credits would have happened anyway and, thus, are not additional.

Sator National Park and the Southern Cardamom Protected Forest.* This represents a remarkable conservation achievement, as it is the only paved highway in Cambodia bordered by lush rainforest. Hopefully, it will soon be a major attraction where tourists can comfortably visit an intact primary rain forest, without elaborate travel arrangements involving small boats, airplanes or 4-wheel drive vehicles that such destinations often entail. With only modest investments, some enterprising tourist operator could develop a zoological garden or environmental center that caters to busloads of arm-chair explorers while protecting the integrity of what is essentially a wilderness landscape.

Crucial to these conservation results have been the actions undertaken to help communities improve their food security. For example, impoverished farmers in the Sovanna Baitong commune in Koh Kong Province are being provided with traditional farm implements such as plows, as well as drip irrigation technology and seed for high-yield crop varieties. In this instance, crop diversification is being promoted to diversify diets and increase household income, but is accompanied by market analysis to ensure that farm products can be commercialized in local and national markets. Like most farmers in Cambodia, these smallholders do not have clear title to their land and, consequently, face the additional uncertainty of losing their small plots to unscrupulous land speculators. Efforts are thus underway to help these communities formalize their legal rights and create a stable and respectful social dynamic where the rule of law brings tangible benefits, as well as restrictions on land-use.[2]

Stabilizing the agriculture frontier will require that positive experiences like those in Sovanna Baitong are replicated in thousands of communities across Cambodia. In 2006, two Cambodian ministries and several conservation organizations came together to create the Cardamom Biodiversity Conservation Corridor Project,[3] whose goal was to consolidate the protected area network and improve habitat connectivity outside of the protected areas.† The primary objective of

---

* Wildlife Alliance works with the Forest Administration and the Ministry of the Environment to combat illegal logging, poaching and forest fires with extensive ground patrols and aerial surveillance. <http://www.wildlifealliance.org/forest-rangers/> [accessed 15 July 2011].

† Partners are four NGOs: Conservation International (CI), Flora and Fauna International (FFI), Wildlife Alliance, and The World Conservation Union (IUCN) and two government ministries: Ministry of Environment and Ministry of Agriculture, Forestry and Fisheries (MAFF).

this initiative was to improve the livelihoods of local communities by expanding economic opportunities based on the precepts of sustainable development, but it brought additional benefits by improving the exchange of ideas and the coordination of programs among government ministries and conservation organizations.

Revenues from REDD+ or similar emission reduction mechanisms can provide the financial resources to improve and expand this type of community assistance and help small farmers overcome the constraints that have impeded many past efforts to reduce poverty. This carbon-based assistance cannot be a mere replication and amplification of past development assistance, however, and must bring a new feature to development projects. Recipients will have to demonstrate that investments mitigate the emission of greenhouse gases, which requires that communities understand that there is an explicit *quid pro quo* that links the arrival and continuity of the benefits with a commitment to respect the forest. That link is relatively easy to establish in pilot projects, where the interaction with individuals occurs on a personal scale and creates a bond that facilitates effective communication; however, it will be much more difficult to inculcate into state institutions who will lead these initiatives as they are scaled up to include communities across Cambodia. Finally, farmers are conservative by nature and subsistence farmers are particularly risk adverse, because if their crops fail, the survival of their families is at stake. Thus, any and all interventions to stop or slow deforestation must produce obvious and risk-free benefits for the small farmer.

Expand Community-Based Management

Along with improving the livelihood of communities, engaging community participation is vital — perhaps the most essential element — in any strategy to preserve a shared natural resource. Conservation incentive agreements have proven to be an effective tool in building local support for protected areas and can help traditional communities improve their livelihoods. These efforts are exemplified on the southern border of the Central Cardamom Protected Forest, where the Chumnoab community in the Areng Valley have recolonized their ancestral lands after nearly two decades of life in a resettlement village.[4] Upon their return to their ancestral village, this *Khmer Daeum* community adopted a rudimentary form of slash-and-burn agriculture and cleared approximately 40 hectares of forest each year. Fortunately, most families were able to supplement their livelihoods by subsistence hunting and fishing;

however, some individuals engaged in illegal logging and the trade of endangered species. Dialogue with the community led to a "conservation incentive agreement" where the village as a whole agreed to change farming practices and collaborate in efforts to stop illegal hunting.

In this instance, the change in farming practices was relatively easy to implement. The Areng Valley has substantial areas of flat topography and abundant water resources, conditions that are ideal for rice paddies, which was the villager's preferred farming technology before their forced exile in the early 1980s. Following abandonment, the heavy clay soils of the valley floor developed a thick hard pan that impeded their recuperation as functional rice paddies. The conservation agreement assisted the inhabitants of Chumnoab reestablish their water buffalo herd that enabled them to rehabilitate their rice paddies using traditional soil tillage practices. As mentioned in Chapter 7, growing rice in flooded paddies is the most sustainable and reliable form of tropical agriculture and the recuperation of this age old farming technique significantly improved the food security of the community. Villagers also agreed to participate in routine patrols to monitor and police the forest reserve and, working with forest rangers, have removed hundreds of traps and snares from the forest, as well as dozens of chains saws and small arms (Figures 5.9 and 6.11). Military police frequently accompany forest rangers and their presence has led to the confiscation and destruction of illegal drugs valued in the hundreds of millions of dollars.[5]

Other initiatives have created new economic opportunities for communities by generating cash income. The community-based tourism project in the Chipat village adjacent to the Southern Cardamom Protected Forest provides wilderness trekking and mountain bike expeditions featuring spectacular scenery and wildlife watching and, if tourists are willing to spend several days in the wilderness, an opportunity to visit a remote archeological site in the Central Cardamom Protected Forest. Less ambitious tourists are offered tours on the Stung Proac River and can experience life in a typical rural village with rustic lodging and meals provided by Khmer families. Benefits to locals are maximized because there are no intermediaries, so 80 percent of all revenues go directly to the families that provide the services.[6]

Community participation also allows conservationists to draw on the skills of traditional communities that have been acquired over decades and centuries of living in a natural ecosystem and depending on a natural resource — knowledge that can improve the management of a protected area. This common sense concept underpins the Man and

Biosphere (MAB) Program of the United Nations Educational, Science and Cultural Organization (UNESCO), which recognized Cambodia's Tonle Sap Lake and floodplain as a UNESCO Biosphere Reserve in 1997. The Tonle Sap is an extraordinarily productive ecosystem upon which virtually the entire population of Cambodia depends, either directly or indirectly, as a food resource. Recognition of its importance to human well-being has motivated multiple efforts to organize and improve the management of its fisheries.

In a well documented example of the potential of community participation, a consortium of NGOs and government agencies supported a community-based management system in the Tonle Sap following the fisheries reform law of 2000, when approximately 50 percent of the commercial fishing concessions were converted to community fishing lots. They chose as their model community Kampong Phluk, a small village on the northern shoreline that had managed to preserve a long tradition of resource management. In this village, a council of village elders stratified its territory and organized its inhabitants, so that wood gathering or fishing was permitted only under a defined set of conditions, in specific areas, and during certain times of the year. For example, dead wood could be collected in flooded forests near the villages, while harvesting from live trees was allowed in other areas according to a rotational system that allows trees to recuperate. Interestingly, villagers were aware that moderate levels of wood harvesting encourage the formation of shrubby vegetation that improves the quality of fish habitat in flooded forests; thus their wood gathering supported their fishing activities. Fishing rules defined the types of allowable fishing gear and the season in which certain gear was allowed. Internal enforcement is largely voluntary, but villagers actively participated in patrolling duties to keep outsiders from infringing upon the rules and their territory.[7]

This and other examples of community participation have been replicated and incorporated into a community-based fisheries management model that is being promoted by the Fisheries Action Coalition Team (FACT), an innovative network that has assisted more than 400 communities in the Tonle Sap and on the Cambodian Coast to confront the challenges of managing a shared natural resource. Key to these efforts has been the role of spatial planning in deciding where certain activities are allowed and the ability for communities to regulate the behavior of their members within these areas. Perhaps most importantly, FACT has taken a concept developed and perfected in pilot projects and expanded it to a larger initiative with a national scope. Community fisheries are now an example of a grass roots movement that has successfully engaged both national and local authorities, who provide legal support and police action to enforce the regulations and control the actions of individuals both inside and outside the community.[8]

Redefine Cambodia's Protected Area System

Efforts to involve forest and fishing communities in conservation, like those described above, have been very successful and bring important lessons for future initiatives. However, there has been virtually no dialogue about land-use or conservation with other individuals or influential interest groups from agroindustry or the mineral and hydropower sectors. This lack of dialogue exacerbates the potential for conflict within or adjacent to protected areas, because regulations governing natural resource use and protected area management can be vague or, in some cases, contradictory with a range of other government policies that promote development.[9,10,11,12] Many of the individuals associated with these sectors neither understand nor appreciate the goals of the conservation community, and are completely unaware of how environmental economics has emerged over the last decade to incorporate economic criteria as a justification for conservation. Many enterprises have committed significant financial resources in projects that have been identified as national priorities and approved by government authorities and, not surprisingly, feel aggrieved when they are criticized in the press based on partial or even erroneous information.[13] The failure to design and implement a broader and more participatory process is, in the end, detrimental to conservation and to the consolidation of Cambodia's protected area system.

The Greater Cardamom Region's unique and remarkable biodiversity enjoys legal protection thanks to the creation of five national parks, four wildlife reserves, three multiple-use areas and three protected forests, which include habitats as varied as the flooded forests of the Tonle Sap, the rainforests and pine savannas of the Cardamom Plateau, the mangrove swamps and beaches of the Cardamom Coast, and the coral reefs and sea grass beds of the Gulf of Thailand. This remarkable network of protected areas has value much greater than the sum of its parts, however, because it provides the fundamental units of a conservation corridor (see Figure 1.5). Spanning multiple watersheds, it provides strategic links between terrestrial, freshwater, coastal and marine ecosystems. Even more importantly, the protected area network spans a broad range of altitudes and enjoys a precipitation regime moderated by mountainous topography, making it particularly resilient to the potential impacts of climate change (see Figures 2.3 and 2.4).

As good as it is, however, portions of the protected areas are suffering significant degradation and intense pressure from land speculators and settlers. The system needs to be reformed and reconfigured. Land that has been deforested must be excised immediately and protected areas facing severe encroachment should have their boundaries modified to reflect the likely future distribution of communities and farmland. The protected area system was created by western consultants during the early stages of Cambodia's reconstruction; most of these individuals were conservation biologists who lacked training in spatial planning methodologies. Moreover, the design of the protected area system was carried out without an extensive consultation process that included all stakeholders and without an open and honest discussion that evaluated development options and priorities. Consequently, the protected area system of Cambodia, which is relatively well designed to protect biodiversity, reflects the priorities of only one major stakeholder — conservation biologists.

A broader and more effective participatory process could bring significant new opportunities and improve the long-term probability of organizing an effective conservation corridor. Such an endeavor would be challenging and risk having significant areas excised from the system, but the payoff would come by clarifying the concept of what a protected area is in the public imagination and educating the local population and elected officials about the economic opportunities of natural landscapes. For example, few probably understand why the conservation of the forests in the upper watersheds of the Cardamom Mountains is also an investment in the long-term utility of the nation's strategic hydropower facilities. Conservation science has advanced in the nearly 15 years since Cambodia's protected area system was formulated. Much more emphasis is placed on the protection of coastal and wetland habitats and the need to establish physical linkages between terrestrial, freshwater and marine ecosystems. The issue of climate change, while not new, is far more important and provides not just a rationale for design, but also the potential to access significant new financial resources via international carbon markets.

Many conservation scientists agree that it is necessary to revisit the fundamental assumptions underlying the current protected area system and to organize an open multi-stakeholder process to reassess its spatial extent, the designations as parks, reserves, and multiple use areas, and the government's framework for managing these areas — including the logic of having two parallel systems with overlapping mandates.* A redesigned protected areas system could be smaller, better and more secure.

A redesign of the protected area system would also streamline and reinforce Cambodia's financial commitment to these lands. The best plan in the world is useless without the financial resources to implement it, and the financial solvency of the nation's existing protected area system is very much in doubt. When the protected area system was created in the 1990s, the government was almost totally reliant on international support and agreed to establish the existing protected area system without much internal debate. Short-term financing was made available, and conservation organizations stepped in to provide strategic planning, training, infrastructure development, and community engagement, while the Cambodian state assumed responsibility for paying salaries and financing the administrative structure that underpins the system. Support came from a variety of public and private sources, but was seldom sufficient to cover the needs of any single protected area; looking forward, support from outside donors is certainly not sustainable over the mid-term. The government has kept its commitment to the institutions that manage the protected area system and the corresponding protected forest and fishing sanctuaries; however, budgets have not kept pace with the increasing threats that these areas are facing, nor will the national treasury be able to cover the activities and investments currently provided by NGOs and their donors.

Bilateral and multilateral assistance allocated to the environment is being channeled increasingly toward climate change issues, making support for protected area management and biodiversity conservation scarcer. Not surprisingly, the prospect of carbon revenues is viewed by many conservation organizations as a solution to the perennial issue of budget shortfalls and the fatigue of fund raising.[14] However, access to these revenues, if and when they materialize, will be sharply contested, and the greater part of these funds will probably be channeled to communities and institutions on the margins of protected areas and not to the protected areas themselves. Fortunately, there are

---

* The Ministry of Agriculture, Fishing and Forestry administers protected forest and fish sanctuaries, while the Ministry of the Environment oversees national parks, wildlife reserves and multiple use areas; see Chapter 4.

other potential revenue sources, such as the payment for ecosystem services scheme linked to watershed management and hydropower facilities described in Chapter 6 and initiatives like the Global Conservation Fund (GCF) that create endowments, trusts and other financial instruments to support protected areas in many parts of the world. As important as external support may be, nothing can replace the long-term commitment of the government to finance the protected area system and this will occur only when Cambodian society fully recognizes the benefits from conservation as an integral and essential component of a green economy.

A redesign of the protected area system in Cambodia should also take place in coordination with other development initiatives, particularly efforts to improve forestry and fisheries management, promote sustainable agricultural and attain energy independence. To be effective, it must be conducted within the context of a larger land-use planning process, which would promote the creation of corridors that link protected areas and avoid conflicts over minerals or high value real estate needed for tourism enterprises. A visionary approach would seek public participation via parallel efforts to strengthen democracy and promote decentralization; local governments can provide legitimacy to protected area management decisions and are critical for ensuring sustainable land-use on the landscapes surrounding the protected areas.

Spatial Planning, Strategic Environmental and Social Assessment

Strategic planning with a strong spatial component has been employed in both developed and developing countries for decades. Land-use planning, regional planning and agro-ecological zoning are all based largely on spatial information, while urban planning and integrated coastal zone management use spatial data in specialized situations. In recent years, the need for more integrated planning has led to a specialized methodology known as a strategic environmental and social assessment (SESA). This methodology originally came about due to the recognition that the commonly used environmental impact analyses (EIA) were too narrowly focused in their spatial, temporal and thematic dimensions and, consequently, only generated recommendations on how to mitigate the negative impacts of a specific investment. SESAs are intended, instead, as a more proactive and flexible decision-making tool that can identify the best available investment options for communities or nations, particularly when these are engaged in long-range strategic planning. SESAs can be organized according to sector or geographic criteria, or include both simultaneously, but their objective is to envision alternative future scenarios similar to those presented in the first chapter of this book so that societies fully understand their development options.

The SESA has been adopted as a key planning tool by all multilateral assistance agencies to manage and evaluate their investment portfolios in developing countries. It is typically used in the transportation, energy and tourism sectors, but should also be used to organize regional and national planning processes for land-use and natural resource management. The SESA has been adopted as the major planning tool by a number of countries with widely different political systems, including the United Kingdom and the People's Republic of China.[15,16] The Global Environmental Outlook (GEO) program of United Nations Environment Program (UNEP) uses similar methodologies to monitor the state of the environment at local and regional scales, incorporating information from social and economic studies to identify the drivers of change, as well as to provide recommendations on how different policy options can modify outcomes.[17] In Southeast Asia, the Mekong River Commission (MRC) is conducting a SESA to evaluate the risks and opportunities of different hydropower development options in the lower Mekong river basin,[18] with the collaboration of the governments of Cambodia, Thailand, Laos and Vietnam.

A SESA typically starts by compiling all available physical and biological information, which is used to stratify a region into subunits according to natural resources; of particular importance is data relating to soil type and rainfall patterns that determine a landscape's suitability for agriculture and forestry, while information on geology and hydrology is essential for identifying areas appropriate for mineral extraction and hydropower development. Census tract information and economic data analyses are required to site areas appropriate for urban expansion and industrial development, or preferred routes for transportation corridors and energy grids. Natural assets with intrinsic economic value, such as wetlands that maintain fisheries or forest slopes that control erosion, are identified so they can be set aside within protected areas or as environmental easements.* A SESA should also

---

*An easement is a limited right to use land that belongs to another; for example, water, sewage, and electricity utilities hold easements across private property to build infrastructure. A conservation easement is a legal right to restrict future development, even though the owner may continue to reside on and make some use of the property.

include both land-use and land-tenure data, in order to protect the pre-existing rights of long-established communities and assist them develop their lands according to sustainable paradigms. The technical aspects of a SESA are the responsibility of a team of specialists that might include geographers, geologists, biologists, agronomists, foresters, fisheries experts, sociologists, and economists, which integrated into a synthetic report which include a series of thematic documents and maps.

The technical aspects of a SESA are imbedded within a participatory process that is as important as the spatial plan and the land-use recommendations it makes. Typically, a spatial plan starts out as a proposal that attempts to adjudicate land among competing interests; eventually, consensus is reached or, more likely, an imperfect political process leads to compromise. A poorly managed political process can render the most technically competent land-use plan meaningless — if it is perceived as unfair, is poorly understood, or fails to take into account the interests of an important segment of the population or economically influential interest group. Unlike an EIA consultation, where people are provided the opportunity to "comment" on a draft, participants are sought out and proactively engaged in a SESA process as the analyses and report writing are underway; this ensures that their knowledge is integrated with the technical information and their perspectives are incorporated into the future scenarios models that are one of the major outputs of an SESA.

The parties engaged in the SESA process are generally referred to as "stakeholders," and in Cambodia they are as varied as the nation itself, including both large and small landholders who wish to protect their prerogatives to develop properties, as well as entrepreneurs seeking to avoid restrictions or regulations that might limit their ability to develop a mine or exploit timber. Corporations and banks whose headquarters are far removed from the actual landscapes of the Greater Cardamom Region are also legitimate stakeholders, particularly if they are investing in hydropower dams or resort complexes. These interests are counterbalanced by grassroots organizations from civil society, many of which are supported by international nonprofit organizations dedicated to fighting poverty and promoting social justice. Local environmental organizations promote the conservation of nature, particularly endangered species, and these also usually have foreign sponsors motivated by global concerns about species extinction and climate change. In addition to this cast of characters, there are other interested parties motivated by a range of interests, which in Cambodia's case includes national ministries with jurisdiction over existing land units or specific sectors of the national economy (mining, energy, fisheries, forestry, tourism, or construction). All of these — individuals and institutions, public and private, national and international — are legitimate stakeholders and should be sought out as part of the SESA consultation process. But the most important stakeholders are the Cambodian people: the inhabitants of the forest, the farmers of the rice plain and agricultural frontier and the fishermen of the Tonle Sap and Cambodian Coast, but also the urban residents in regional cities and in the capital city of Phnom Penh.

A lack of dialogue leads to lost opportunities. For example, conservation organizations may invest unknowingly in projects on landscapes slated for conversion due to their inherent agricultural potential or mineral value, while missing opportunities to improve land-use and habitat connectivity within mining or economic land concessions. Corporations can learn from civil society, particularly in the areas of environmental sustainability, which can open markets, improve efficiencies and boost profitability, as well as protect the long-term productivity of agricultural properties or the commercial appeal of tourist resorts. In almost all cases, an improved consultation process should be part of a due diligence process that reduces risk from conflict, delays and other complications that can lead to project failure. Good due diligence is in the interest of the investor, the government and the effected population.

The spatial planning aspect of a SESA is particularly important in the frontier areas of the Greater Cardamom Region because of the rapid change they are undergoing. The next few years represent, perhaps, the last chance to maintain some level of connectivity between larger blocks of natural forest habitat within the core areas of national parks, wildlife reserves and protected forests. Spatial planning on the agricultural frontier can identify areas suitable as conservation easements, where clearing or intensive logging should be prohibited on steep slopes or rocky soils susceptible to erosion. These smaller, less conspicuous patches and strips of habitat are of great long-term importance because they provide the "connectivity" that can transform a group of individual protected areas into a conservation corridor. Forests along rivers, referred to as riparian corridors, are also especially important because they also channel the provision, regulation, and quality of water from natural areas to communities in the downstream portion of the watershed. They are often the most difficult component of a spatial plan to adjudicate and set-aside as a conservation easement, because of the understandable desire of families to live near a reliable source of water. The SESA process can identify the competing demands for resources and provide solutions that benefit everybody; in this case a prohibition of settlements on rivers could be linked to the provision of secure water supplies for those communities affected by an easement.

Spatial planning can identify and conserve patches of natural habitat on agricultural landscapes and the SESA process can help educate local communities and improve their support for the conservation of natural areas. Once the link between nature conservation and improved livelihoods has been established on the landscapes where people reside, it is much more likely society will understand similar linkages at larger scales and involving protected areas. Moreover, if the aspirations of the various stakeholder groups are at least partially satisfied, then a broad consensus can be forged that leads to a more harmonious development dynamic, facilitating investment, economic growth and poverty reduction. The growing importance of the international carbon markets provides a new area for SESA analysis, which could be used to identify opportunities and strategies for maximizing carbon stocks and for providing a forum to discuss how REDD+ revenues might be distributed and invested.

The recommendations and policy options identified by a SESA must ultimately be formalized by a legal framework that commits society to the recommendations and agreements that emerge from the technical and political components of the process. This may take the form of regulatory framework that mandates land-use by zoning laws or concessionary agreements, although a system based on taxes and subsidies may be more effective in some societies. In either case, both national and local government authorities must demonstrate willingness to adopt the recommendations made in the SESA and the rest of the stakeholders must accept the rules that emanate from the government as part of a social compact that defines a functional society.

Governance and Decentralization

Good governance is a necessary precondition for economic growth and poverty reduction and will be the decisive factor in whether or not Cambodia will be able to carry out the innovations and decisions necessary to create a low-carbon economy. Good governance requires viable and coherent policy options, a competent bureaucracy dedicated to serving the public, rules people can understand, and a transparent process where citizens can act with confidence they will be treated fairly and allowed to participate in a process that actually leads to positive outcomes. Good governance is largely the responsibility of individuals, mainly elected officials and government authorities, who have the power to create the conditions where citizens trust state institutions. Poor governance takes many forms, but is characterized by arbitrariness, corruption, and indifference on the part of the public officials; not surprisingly, citizens react by ignoring the rules and acting in their own narrow interest. The so-called "natural resource curse" common to many resource rich developing countries is essentially a failure of governance, where individuals maintain power by using the revenues generated by the exploitation of their country's natural resources to placate its citizenry. Fortunately, the world has many examples of countries who have broken out of the cycle of poverty and natural resource exploitation and many — if not all — of these successful societies have come to recognize the benefits of environmental and social sustainability.

The government of Cambodia has embraced democracy and a market economy as two major pillars of its "rectangular strategy" that underpins its national development plan (see Chapter 4). An important part of this strategy is a decentralization process that would transfer the administration of key social services, such as health and education, to local governments as a means to increase the efficiency and effectiveness of their delivery to the rural poor.* Cambodia's decentralization plan also calls for a "desconcentration" of natural resource management decisions, which could improve the management of the nation's fisheries and forest resources, as well as protect the legitimate rights of its rural and coastal inhabitants. The Cambodian government has been taking determined, if gradual, steps to empower local institutions to assume these responsibilities, starting with elections in 2002 for members of commune councils and followed by a major administrative reform act in 2008, which clarified the administrative responsibilities at the provincial, district and commune level and allocated financial resources to the decentralization process.[19]

---

* The decentralization process is coordinated by the National Council for Democratic Development, an interministerial body led by the Ministry of the Interior (MOI) and the Ministry of Economics and Finance (MEF), but other agencies are also developing action plans for delegating specific tasks and responsibilities to local government, including the Ministries of Agriculture, Forestry and Fisheries, Rural Development, and Water Resources and Meteorology.

As might be expected, there are many challenges in implementing a decentralized governance system in a country emerging from decades of central planning and, before that, a colonial system where important decisions regarding land-use, infrastructure and industrial development were even further removed from the local community. Because of this cultural legacy and the recognition that new local and regional entities have little technical capacity or administrative experience in executing projects, the national government is proceeding in a gradual, stepwise fashion. The possibility that an important and vital democratic reform could be perverted into a source of corruption and inefficiency has led to a climate of caution, with systematic investments in capacity training. Cambodia is a small country whose population shares the same language and culture; consequently, the ethnic divisions that drive and complicate many decentralization processes are largely absent and national agencies are able to view the process more as a question of hierarchical organization, rather than as a competition for power among competing political groups or regional movements. The challenges are rather to avoid corruption, elite capture, or the creation of a new bureaucracy that consumes resources meant for services or infrastructure. To be successful, the move towards decentralization also must be accompanied by the creation of bottom-up decision-making processes to complement the top-down policies that will continue to flow from the national government.

Decentralization will be particularly beneficial in monitoring land-use decisions and engaging local communities in low-carbon production strategies. Because access to fisheries and forest resources are fundamentally important issues to every rural Cambodian (and about 80 percent of all Cambodians are rural inhabitants), the involvement of commune councils in allocating these resources would make the decentralization process relevant to the lives of average Cambodian citizens. Given their current lack of legal authority to govern these issues, commune councils are unable to respond to complaints from local communities about the unsustainable exploitation of these resources by outsiders, thus reducing their credibility among their constituents. Changing that dynamic will require both bottom-up and top-down solutions: residents must provide commune councils legitimacy via the democratic election process, and the efficacy of their decisions will depend, in part, on the response of law enforcement authorities to uphold them.

Allocating resource management responsibility to local government would also help Cambodian citizens become more invested in both the decentralization process and natural resource management. By participating in a local electoral process that actually determines resource access, citizens would experience the benefits of democracy in a way that is much more relevant than voting for a minister of parliament; thus, faith in democracy is strengthened. Similarly, if rural Cambodians benefit from the distribution of natural resources, they are more likely to respect the rules; thus, sustainability is improved. Forest and fisheries tend to be shared resources that require consensus among users; access to a shared resource is best mediated by local government, but only one that is viewed as legitimate by its constituents.

Decentralization would also improve the allocation and zoning of land-use in Cambodia, particularly the distribution of public lands. Participatory spatial planning is integral to improved land management at every level of government: national, provincial, district and commune, as well as urban areas. This will require the development of a statutory and regulatory framework that transcends and links the different administrative units. At the national level, land-use is the responsibility of the Ministry of Land Management, Urban Planning and Construction, which is the convening body of the Council of Land Policy Strategy (see Chapter 4). This organization pursues three fundamental and linked objectives: (1) advance the effort to compile a national land registry, (2) promote the wise use of land and other natural resources and (3) ensure that the distribution of state land is transparent and equitable.[20] These objectives illustrate the urgency of advancing the decentralization process. The overwhelming proportion of land holdings are small family plots; consequently, the land registry must be locally accessible. More importantly, a land registry can bring revenues to local government; in part, because taxes on landholdings are typically allocated to local governments, but also a land registry can be used to facilitate the distribution of revenues from carbon mitigation mechanisms similar to REDD+.

Finally, efforts to decentralize government functions will provide much needed opportunities to improve forest management and facilitate interactions between the Forest Administration (FA) and the private firms that will harvest timber according to the new quota system (see Chapters 5 and 7). Delegating or sharing this responsibility to district and provincial governments is a logical option for the FA, because the commune is the smallest subnational administrative body and the logical unit for organizing and overseeing community forests. Its expanded role in forest management and the development of community forest programs could be funded through REDD+ revenues, providing local governments with increased revenue to enforce and protect the forests.

Sooner Rather than Later

The development options outlined in this chapter are not radically different than many of the recommendations purveyed by development specialists around the world for decades. If there is a novel element, it is the potential to use carbon markets to provide a sustainable source of financing for development and highlighting the benefits that certain development options bring in the context of climate change mitigation and adaptation, which in the case of land-use are almost always synergistic. Unfortunately, however, carbon markets do not yet exist, or at least do not channel the financial resources at the scale needed to motivate a change in global development paradigms. Worse still, the reality of climate change is viewed as an uncertainty in the minds of too many influential individuals in government and business, while it is just totally meaningless to the average Khmer farmer struggling to exist on a dollar a day. How can individuals and institutions be motivated to pursue options that are in the vanguard of conventional thinking? Why should Cambodia be a leader in the effort to combat the impacts of climate change? Might it be better to see how the global policy debate on global warming unfolds before adopting policies that improve resilience to climate change? The simple answer to these questions is because it is in Cambodia's national interest to pursue a green economy "sooner rather than later".

Regardless of what eventually transpires in international policy forums, pursuing a green economic model is good for Cambodia. International markets for carbon credits may remain anemic for the foreseeable future, but other types of climate finance are beginning to emerge and consumers across the world are demanding carbon neutral production schemes. Sustainability is no longer limited to niche products or used as public relations gimmicks by corporations to protect their image; the concept of sustainability, including efforts to reduce the emission of greenhouse gases, is being mainstreamed into global supply chains. Cambodia will benefit by becoming a global leader in efforts to formulate and test novel types of climate finance and then to use those resources to increase efficiencies, improve natural resource management, and increase carbon stocks on both natural and productive landscapes. Early action will be rewarded by investors seeking to link their business enterprises with concrete examples of climate change mitigation, while investments in ecosystem based adaptation will bring long-term benefits in the form of food security and reliable water supplies. The example of Costa Rica can be exemplary. A small country provided leadership by testing new concepts in nature conservation; that early action led to increased growth in the tourist sector, which attracted foreign capital that became diversified over time to include biotechnology, pharmaceuticals, information technology and consumer products, as well as improving the competitive position of its agricultural sector by adding value to commodities such as coffee, bananas and pineapples. Being a leader reaps benefits and provides an additional justification for acting "sooner rather than later."

The best arguments for pursuing sustainable development options have nothing to do with the eventual outcome of global policy debates. One of the central tenets of sustainability as a public policy is to maximize benefits for as many people as possible, for as long as possible — preferably in perpetuity. Development strategies based on sustainability will improve the lives of the rural poor; however, by increasing equity and fostering community welfare, they will also create the social and economic conditions amenable for long-term investors. Taking immediate action to alleviate poverty is a moral imperative, but using the profit motive to drive development is also the most effective pathway to reduce poverty. Many, if not all, of the recommendations or options described in this book represent a "no regrets" approach to climate change mitigation and adaptation. For example, the adoption of good land use planning and sound watershed management practices will yield economic benefits in the short, medium and long term, regardless of the rate and degree of global warming. These are often described as "best practices" and are fundamentally good for business and it makes sense to adopt them "sooner rather than later."

What gets lost in the public debate about the cost and benefits of adopting environmental criteria, be they voluntary, regulatory or market driven, is that too many people focus on tradeoffs and the perception that conservation and development are mutually exclusive. This is a false debate. The differences are sometimes an issue of priority and timing, which can be overcome if the parties can identify common goals that bring benefits to all stakeholders. For example, environmental economists identify the pursuit of short-term economic gain as the culprit driving environmental degradation; they contend that investments should be evaluated over longer time frames when the advantages of sustainable options become apparent. Most private investors, on the other hand, are obligated to focus on maximizing the return over the short term and are acutely aware of how small differences in the rate of return can lead to very large differences when compounded over time. Are these two concepts fundamentally different? Obviously not! What can unite these two seemingly different investment philosophies is an appreciation and valuation of risk. The risk of catastrophic failure in the first instance and the risk of a lost opportunity in the second, both parties are interested in reducing risk and, presumably, would be willing to collaborate in order to identify that risk. Perhaps the most important future role of climate finance will be to help investors manage risk and shift investment decisions towards longer horizons that incorporate sustainable criteria. In this instance, "sooner" means adopting good practices that will pay off "later."

Decisions to fast track some green development options can postpone others characterized by a high environmental cost, even those that may seem unavoidable over the long term. For example, the conversion of land with arable soils is seen as inevitable in order to feed a growing population. However, investments in technology can increase crop yield or otherwise intensify production systems on existing landscapes and waterways; thus future options may alleviate the need for habitat conversion. Increased prosperity and educational opportunity may reduce rural population growth; urbanization and industrialization may lead to opportunities for conservation or, unlikely as it may seem, even habitat restoration. From this perspective, investing in agricultural and aquaculture technology "sooner" may allow Cambodian society to conserve more natural habitat "later" without compromising the welfare of its rural population. Similar arguments can be made for just about any development option with irreversible environmental impacts, particularly the large-scale hydropower projects on the Mekong River; these should be delayed for as long as possible. The development of affordable solar technology and smart energy grids provides a plausible option for the country, but the emergence of that technology to reach the economy of scale may take time. In this instance, investing in energy efficiency "sooner," can buy time for technology to provide an alternative option "later."

Perhaps one of the most important issues related to the allocation of priority investments is how that might impact the distribution of the costs and benefits of the various options, including those linked to conventional development pathways and the green options described in this and the previous chapter. Traditional forest and fishing communities suffer most from deforestation and the disruption of freshwater systems, because the impact on their livelihoods is direct and difficult to mitigate, even when genuine efforts are made to implement environmental and social action plans to manage those impacts. Even in rural communities that are able to adapt to change, some families will fail to make the transition to new production strategies; adapting to change is difficult in any society, but it is even more challenging when it is imposed by outside forces. Green development options will also bring change, particularly if communities are asked to assume a greater role in natural resource management. They will start with the advantage of traditional knowledge, but to prosper their children must have access to educational opportunities and extension programs that enhance their productive capacity. Thus, the term "sooner rather than later" might refer to the investments in initiatives that assist rural communities adapt to the changes that will accompany Cambodia's development trajectory, regardless of the pathway that is chosen and so that they can effectively participate in the decision on which pathway to elect.

Like all nations, Cambodia, via its political and societal processes will determine its future development trajectory and the first chapter of this book describes the contrasting outcomes that might result from two essentially positive, but radically different philosophical orientations, as well as a negative scenario characterized as Business as Usual. Either one of the positive outcomes, characterized as the Utilitarian or Utopian, will require technically competent Cambodian institutions that employ both regulatory tools and market mechanisms to manage the economy. Neither one of those positive scenarios is likely to unfold, however, unless the Cambodian state firmly embraces the rule of law and the need for strong independent institutions that respond to the needs of the Khmer people. The thesis elaborated in the various chapters of this book contends that the best option for the Khmer people is to pursue a sustainable development paradigm that creates a green economy based on democratic principles. The eventual outcome is not, however, likely to be a replica of any of the scenarios described in the first chapter, which are actually caricatures based on stereotypical views of the world. The eventual outcome is impossible to predict, obviously, but it is a safe wager to assume that much will depend upon the capacity of the political and economic elite of Cambodia to envision a future that allows them to benefit from a green economy. This book was written in an effort to facilitate an understanding of that future, describe what that future might look like, and identify some of the decisions that need to be made "sooner rather than later."

Cambodia is to be admired for its ability to overcome such tremendous adversity in organizing and undertaking what amounts to a national renaissance. Its commitment to the preservation and enrichment of its own cultural heritage has underpinned its progress, which has been fortified by its willingness to engage the international community. Cambodia has one of the most open and honest societies in Asia; its brutal past has, perhaps, disabused it of hubris and opened the eyes of its citizens and leaders to the possibility of change. Change in Cambodia is a constant: new roads, new bridges, new hospitals, new schools, new crops, new settlements, new cell phone towers are evident in every corner of rural Cambodia. In the cities, more motorcycles, more cars, more billboards, more cell phones, more fashion, and more people. This willingness to embrace new ideas has created a unique opportunity to organize a green economy. Dynamic, as only an Asian nation can be in the first decade of the twenty-first century.

ES

)aeum: *Khmer de l'Origine. Société*
orêt, *de l'Écologie à l'Histoire'.* Presses
ent, Paris, 1997, Cited by Ironsides
ear (Por) people in Cambodia, NGO
forum.org.kh/Land/Docs/Indigenous/
)11].

Bohay and D.W. Holdsworth. 2004.
Micro-computed tomography of a 500 year old tooth. *Journal of the Canadian Association of Radiology* 55: 242–5.

3. M. Osborne. 2000. *Southeast Asia: An Introductory History*, Sydney: Allen & Unwin.
4. Kampot Pepper Farmers' Associations. <http://www.kampotpepper.biz/en/context.html> [accessed 15 June 2011].
5. D. Chandler. 2008. *A History of Cambodia*, Boulder Colorado: Westview Publishers.
6. P. Short. 2004. *Pol Pot: Anatomy of a Nightmare*, New York: Holt & Co.
7. B. Kiernan. 2004. *How Pol Pot Came to Power: Colonialism, Nationalism, and Communism in Cambodia, 1930–1975*, New Haven: Yale University Press.
8. M. Vickery. 1984. *Cambodia: 1975–1982*, Boston, Mass: South End Press.
9. Rithy Panh. 2003. *La Machine De Mort Khmere Rouge*, Documentary film <http://www.festival-cannes.com/en/archives/ficheFilm/id/4075477/year/2003.html> [accessed 15 June 2011].
10. Thet Sambath and R. Lemkin. 2009. *Enemies of the People*, Old Street Films, 24th International Documentary Film Festival Amsterdam. <http://www.idfa.nl/industry/Festival.aspx> [accessed 15 June 2011].
11. Beauty and Darkness: Cambodia in Modern History. <http://www.mekong.net/cambodia/> [accessed 15 June 2011].
12. *Land Mine Monitor Report — Cambodia*. <http://www.icbl.org/lm/2006/cambodia.html> [accessed 15 June 2011].
13. F. Momberg and H. Weiler. 1999. *Conservation Status of the Cardamom Mountains in Southwestern Cambodia, Preliminary Studies*, Phnom Penh: Fauna & Flora International.

14. S. Milne and J. Brunner. 2005. Conservation and Care in Cambodia's Cardamom Mountains, Conservation International, Woodrow Wilson Centre for International Scholars, May 09 2005, <http://www.wilsoncenter.org> [accessed 15 June 2011].
15. Japan International Cooperation Agency — JICA. 2007. The Master Plan Study of Hydropower Development in Cambodia, JICA & Ministry of Industry, Mines and Energy, Kingdom of Cambodia, prepared by Nippon Koei Co., Ltd. (Tokyo), presentation from workshop held on 13 November 2007, Phnom Penh.
16. Coordinating Committee for Geoscience Programs in East and Southeast Asia — CCOP. 2011. Exploration Promotion Forum: Cambodia. <http://www.ccop.or.th/epf/cambodia/cambodia_status.html> [accessed 15 June 2011].
17. G. Wright, D. Moffatt, and J. Wager. 2004. Tonle Sap Basin Profile, Cambodia National Mekong Committee, Asian Development Bank, Document TA4212-CAM.

Chapter 2

1. Save Cambodia's Wildlife — SCA. 2006. Atlas of Cambodia, National Poverty and Environment Maps, Phnom Penh: Save Cambodia's Wildlife. <http://www.cambodiaatlas.com/index/> [accessed 15 June 2011].
2. I. Douglas. 2005. The Mekong River Basin. In *The Physical Geography of Southeast Asia*, ed. A. Gupta, pp. 193–218. Oxford: Oxford University Press.
3. Cambodian National Petroleum Authority — CNPA. 2009. Petroleum Geology of Cambodia. <http://cnpa-cambodia.com/pages/petroleum-exploration.php.> [accessed 15 June 2011].
4. C.D. Woodroffe. 2005. Southeast Asian Deltas. In *The Physical Geography of Southeast Asia*, ed. A. Gupta, pp. 219–36. Oxford: Oxford University Press.
5. Y. Zhang, T. Li, B. Wang and G. Wu. 2005. Onset of the Summer Monsoon over the Indochina Peninsula: Climatology and Interannual Variations. *Journal of Climate* 15: 3206–21.
6. M.H. Glantz, ed. 2001. *Once Burned, Twice Shy: Lessons Learned from the 1997–98 El Niño*. Tokyo: United Nations University Press.
7. United Nations University — UNU. 2000. Reducing the Impact of Environmental Emergencies through Early Warning and Preparedness, The Case of the 1997–98 El Niño. Country Reports, Vietnam. Boulder: Center for Atmospheric Research (NCAR) <http://www.unu.edu/env/govern/ElNino/CountryReports/inside/vietnam.html> [accessed 15 June 2011].
8. Y. Zhang, T. Li, B. Wang and G. Wu. 2002. Onset of the Summer Monsoon over the Indochina Peninsula: Climatology and Interannual Variations. *Journal of Climate* 15: 3206–21.

9. Save Cambodia's Wildlife — SCA. 2006. *Atlas of Cambodia.*
10. Asia Sustainable and Alternative Energy Program — ASTAE. 2001. The Wind Energy Resources Atlas of Southeast Asia. Washington, DC: World Bank. <http://siteresources.worldvbank.org/EXTEAPASTAE/Resources/wind_atlas_complete.pdf> [accessed 15 June 2011].
11. M. Kummu and J. Sarkkula. 2008. Impact of the Mekong River flow alteration on the Tonle Sap flood pulse. *AMBIO* 37: 185–92.
12. V. Kite. 2000. Modeling the Mekong, Hydrological simulation for environmental impact studies. *Journal of Hydrology* 253: 1–13.
13. M. Saburo, M. Keskinen, P. Sokhem, and M. Nakamura, 2005. *Tonle Sap, Experience and Lessens Learned Brief (Lake Basin Management Initiative).* International Waters Learning Exchange and Resource Network — IW: LEARN. <http://iwlearn.net/iw-projects/Msp_112799491968/experience-notes-and-lessons-learned/laketonlesap_2005.pdf/view> [accessed 15 June 2011].
14. G. Wright, D. Moffatt and J. Wager. 2004. Tonle Sap Basin Profile, Cambodia National Mekong Committee, Asian Development Bank, Document TA4212-CAM. <http://www.adb.org/Projects/Tonle_Sap/tonle-sap-basin-profile.pdf.> [accessed 15 June 2011]
15. Lecompte, H., H. Humbert, H. and F. Gagnepain. 1907–1951. *Flore Général de lIndchine*, Vols. 1–9, Paris: Masson.
16. Dy Phon, P. 1982. Végetacon du cambodge: endemisme et affiones de sa floreavec les regiones voisines. *Comptes Rendus des Seances de la Societe de Biologie et de Ses Filiales* 58: 135–44.
17. R.J. Whittaker, K.J. Willis and R. Field. 2001. Scale and species richness: towards a general, hierarchical theory of species diversity. *Journal of Biogeography* 28: 453–70.
18. Cambodia Tree Seed Project (2003) Forest Gene Conservation Strategy, Part A. Conservation of Forest Genetic resources, Cambodian Tree Seed Project (Annex 1 — Flora of Cambodia), Forest Administration, Kingdom of Cambodia & DANIDA. <http://www.treeseedfa.org/doc/ForestGene/> [accessed 15 June 2011].
19. R. Corlett. 2005. Vegetation. In *The Physical Geography of Southeast Asia*, ed. A. Gupta, pp. 105–20. Oxford: Oxford University Press.
20. P. Ashton. 1988. Dipterocarp biology as a window to the understanding of the tropical forests structure. *Annual review of Ecology and Systematics* 19: 347–70.
21. G. Kaury-Lechion and L. Curfat. 1998. Biogeography and Evolutionary Systematics of Dipterocarpaceae In *A Review of Dipterocarps, Taxonomy, Ecology and Silviculture*, ed. S. Appanah and J.M. Turnbull, pp. 5–44. Bogor: Center for International Forestry Research — CIFOR.
22. International Union for Conservation of Nature — IUCN. 2010. IUCN Red List of Threatened Species. <http://www.iucnredlist.org/> [accessed 15 June 2011].

23. World Wildlife Fund — WWF. 2007. Cardamom Mountains Rain Forests. *Encyclopedia of Earth*. <http://www.eoearth.org/article/Cardamom_Mountains_rain_forests> [accessed 15 June 2011].
24. Cambodia Tree Seed Project (2003) Forest Gene Conservation Strategy.
25. R. Hiramatsu. 2007. Open woodland patches in an evergreen forest of Kampong Thom, Kratie, Monolkiri. In *Forest Environments in the Mekong River Basin*, ed. H. Sawada, pp. 201–13. Tokyo: Springer.
26. Meng Monyrak. 1998. Bamboo resources, conservation and utilization in Cambodia. In *Bamboo — Conservation, Diversity, Ecogeography, Germplasm, Resource Utilization and Taxonomy*, ed. Rao, A.N. and R.R. Ramanatha. Proceedings of training course cum workshop 10–17 May 1998, Kunming and Xishuanbanna, Yunnan, China. The International Plant Genetic Resources Institute – IPGRI. <http://www2.bioversityinternational.org/publications/Web_version/572/> [accessed 15 June 2011].
27. Global Rattan Programme. 2010. International Network for Bamboo and Rattan. <http://www.inbar.int/index.ASP.> [accessed 15 June 2011].
28. M. Auliya. 2003. *Hot trade in cool creatures: A review of the live reptile trade in the European Union in the 1990s with a focus on Germany*. Brussels: TRAFFIC.
29. J.C. Daltry and W. Wüster. 2002. A new species of Wolf Snake (Serpentes: Colubrida, Lycodon) from the Cardamom Mountains, Southwestern Cambodia. *Herpetologica* 58: 498–504.
30. C. Campbell, C. Poole, W. Giesen and C. Valb-Jorgemnso. 2006. Species diversity and ecology of Tonle Sap Great Lake, Cambodia. *Aquatic Science* 68: 355–73.
31. Ibid.
32. Ibid.
33. Flora and Fauna International — FFI. 2008. Green-blooded, blue-boned frog discovered in Cambodia. Mongobay <http://news.mongabay.com/2008/1218-frog_ffi.html> [accessed 15 June 2011].
34. Mohd Nazip Suratman. 2008. Carbon Sequestration Potential of Mangroves in Southeast Asia. In *Managing Forest Ecosystems: The Challenge of Climate Change*, ed. F. Bravo, V. LeMay, R. Jandland and K. von Gadow, pp. 297–315. *Managing Forest Ecosystems, Volume 17*. Springer-Netherlands.
35. Ramsar Convention on Wetlands — RAMSAR (2010), The Annotated Ramsar List: Cambodia, <http://www.ramsar.org/profile/profiles_cambodia.htm> [accessed 15 June 2011].
36. M.D. Amarasinghe, D. Varunthat, W. Taparhudee, R. Yoonpundh and S. Jumnongsong. 2009. Mangroves: Mangroves of Nakhon Si Thammarat Province in southern Thailand: Species diversity, community structure and current status. *Aquaculture Asia Magazine*. <http://www.enaca.org/modules/news/article.php?storyid=1822> [accessed 15 June 2011].
37. H.O. Arceo, C. Cheung, K. Vathana and H. Vibol. 2002. *Marine Protected Areas in Southeast Asia*. ASEAN Regional Centre for Biodiversity

Conservation <http://www.arcbc.org.ph/arcbcweb/publications/mpa.html> [accessed 15 June 2011].

38. Global Wildlife Conservation — GWC. 2008 Southeast Asian Field Expeditions. <http://globalwildlife.org/field.southeast.php> [accessed 15 June 2011].
39. Ibid.
40. National Oceanic Atmosphere Administration — NOAA. 2008. Gulf of Thailand LME (Large Marine Ecosystem), *Encyclopedia of Earth*. <http://www.eoearth.org/article/Gulf_of_Thailand_large_marine_ecosystem> [accessed 15 June 2011].
41. D. Pauly and R. Chuenpagdee. 2003. Development of Fisheries in the Gulf of Thailand: Analysis of an Unplanned Experiment. In *Large Marine Ecosystems of the World, Volume 12: Trends in Exploration, Protection and Research*, ed. G. Hempel and K. Sherman. Amsterdam: Elsevier.
42. K. Adulyanukosol. 2002. Report of dugong and seagrass survey in Vietnam and Cambodia. Phuket Marine Biological Center. <http://www.sirenian.org/andulyanukosolfinalreport.pdf> [accessed 15 June 2011].
43. Tuoch Bunthang. 2002. Strengthening Sea Turtle Conservation in Cambodia Indian Ocean and South-East Asia Conservation and Management of Marine Turtles and their Habitats. *IOSEA Marine Turtle Memorandum of Understanding*. <http://www.ioseaturtles.org/eleclib/Cambodiaturtles.pdf> [accessed 15 June 2011].
44. Kim Sour, Chou Loke Ming and Karenne Tun. 2004. Status of coral reefs in Southeast Asian countries: Cambodia. In *Status of Coral Reefs in East Asian Seas Region*, pp. 17–32. Japan Wildlife Research Center and World Fish Centre. Tokyo: Ministry of Environment, Japan.
45. J. Kleypas. 2010. Coral Reefs. *Encyclopedia of Earth* <http://www.eoearth.org/article/Coral_reef?topic=49570> [accessed 15 June 2011].
46. Nagelkerken, I.G. van der Velde, M.W. Gorissen, G.J. Meijer, T. Van't Hof, and C. den Hartog. 2000. Importance of mangroves, seagrass beds and the shallow coral reef as a nursery for important coral reef fishes. *Estuarine, Coastal and Shelf Science* 51: 31–44.
47. Intergovernmental Panel on Climate Change. 2007. Ocean acidification due to increasing atmospheric carbon dioxide. In *The Physical Science Basis, IPCC Fourth Assessment Report: Climate Change 2007*. <http://www.ipcc.ch/publications_and_data/ar4/wg1/en/ch10s10-4-2.html> [accessed 15 June 2011].
48. Kim Sour *et al.* 2002. Status of coral reefs in Southeast Asian countries.
49. J.V. van Bochove, M. Sandström and P. Raines. 2009. Cambodia reef conservation project, Coral Cay Conservation. <http:// www.coralcay.org> [accessed 15 June 2011].
50. Kim Sour *et al.* 2002. Status of coral reefs in Southeast Asian countries.
51. Ibid.

52. Marine Conservation Cambodia. 2010. <http://www.marineconservation cambodia.org/> [accessed 15 June 2011]

Chapter 3

1. Save Cambodia's Wildlife — SCA. 2006. *Atlas of Cambodia, National Poverty and Environment Maps*, Phnom Penh: Save Cambodia's Wildlife.
2. Ibid.
3. R.R. Russell. 1987. Cambodia: A Country Study. Washington: Government Printing Office for the Library of Congress. <http://countrystudies.us/cambodia> [accessed 25 June 2011].
4. J. Ironsides. 2005. *Overview of the distribution of Pear (Por) people in Cambodia.* Phnom Penh: NGO Forum. <http://www.ngoforum.org.kh/Land/Docs/Indigenous/Overview.htm> [accessed 25 June 2011].
5. Ministry of Planning — MOP. 2006. A Poverty profile of Cambodia 2004, Phnom Penh: Royal Government of Cambodia. <http://www.mop.gov.kh/Others/PovertyProfile/tabid/191/Default.aspx> [accessed 25 June 2011].
6. Save Cambodia's Wildlife — SCA. 2006. *Atlas of Cambodia.*
7. Ministry of Planning — MOP. 2006. A Poverty profile of Cambodia 2004.
8. The World Factbook. 2011. Distribution of family income — Gini index Washington, DC: Central Intelligence Agency. <https://www.cia.gov/library/publications/the-world-factbook/fields/2172.html#41> [accessed 25 June 2011].
9. M. Osborne. 2004. *Southeast Asia: An Introductory History* (9th Edition). New York: Allen & Unwin.
10. S.S. Dy and A. Ninomiya. 2003. Basic Education in Cambodia: The impact of UNESCO on policies in the 1990s. Education Policy Analysis Archives 1: 48. <http://epaa.asu.edu/epaa/v11n48/> [accessed 25 June 2011].
11. National Institute of Statistics — NIS. 2008. General Population Census of Cambodia 2008. Phnom Penh: Ministry of Planning, Kingdom of Cambodia. <http://www.nis.gov.kh/nis//census2008//Census.pdf> [accessed 25 June 2011].
12. Ministry of Education, Youth and Sport — MEYS. 2005. *Education Strategic Plan 2006–10*, Kingdom of Cambodia. <http://www.moeys.gov.kh/en/education/esp06-10/forward.htm> [accessed 25 June 2011].
13. Ibid.
14. M. Osborne. 2004. *Southeast Asia: An Introductory History.*
15. The World Factbook. 2009. Distribution of Family Income — Gini Index. Washington, DC: Central Intelligence Agency. <https://www.cia.gov/library/publications/the-world-factbook/fields/2172.html#41> [accessed 25 June 2011].

16. Presidents Emergency Plan for AIDS Relief — PEPFAR. 2008. Country Profile: Cambodia. <http://www.pepfar.gov/press/81877.htm> [accessed 25 June 2011, site discontinued].
17. United Nations Programme on HIV/AIDS — UNAIDS. 2010. Cambodia. <http://www.unaids.org/en/regionscountries/countries/cambodia/> [accessed 25 June 2011].
18. Population Reference Bureau — PRB. 2002. Fewer Malaria Cases in Cambodia. <http://www.prb.org/Articles/2002/FewerMalariaCasesinCambodia.aspx> [accessed 25 June 2011].
19. World Health Organization — WHO. 2007. Malaria Epidemiology. <http://www.wpro.who.int/sites/mvp/epidemiology/malaria/> [accessed 25 June 2011].
20. M. Enserink. 2010. Malaria: signs of drug resistance rattle experts, trigger bold plan. *Science* 322: 1776.
21. Kaiser Global Health. 2009. Recorded TB Cases increasing in Cambodia, health officials report <http://www.globalhealthreporting.org/article.asp?DR_ID=56264> [accessed 25 June 2011].
22. R.R. Russell (1990), *Cambodia: A Country Study*.
23. Ibid.
24. National Committee on Planning and Development — NCPD. 2005. A Review of Migration in Cambodia. <http://www.ncpd.gov.kh/> [accessed 25 June 2011].
25. R.R. Russell (1990), *Cambodia: A Country Study*.
26. B. Maltoni. 2007. Migration in Cambodia: Internal vs. External Flows, 8th Asia Pacific Migration Research Network, Conference on "Migration, Development and Poverty Reduction", in Fuzhou (China), 25–29 May 2007. <http://apmrn.usp.ac.fj/conferences/8thAPMRNconference/> [accessed 25 June 2011].
27. Ibid.
28. US Census Bureau. 2011. International Database: Cambodia. <http://www.census.gov/ipc/www/idb/country.php> [accessed 25 June 2011]
29. Save Cambodia's Wildlife — SCA. 2006. *Atlas of Cambodia*.
30. Ministry of Planning — MOP. 2006. *A Poverty profile of Cambodia 2004*.

Chapter 4

1. Mekong Express. 2011. Cambodia Investment Law. <http://www.mekongexpress.com/cambodia/index.htm> [accessed 30 June 2011].
2. Royal Government of Cambodia — RGC. 2009. National Strategic Development Plan: Update 2009–2013, For growth, employment, equity and efficiency to reach Cambodia Millennium Development Goal. <http://www.mop.gov.kh/Home/NSDP/NSDPUPDATE20092013/tabid/206/Default.aspx> [accessed 30 June 2011].

3. Ministry of Land Management Urban Planning and Construction — MLMUPC. 2010. <http://www.mlmupc.gov.kh/mlm/index.php?page=01&lang=e> [accessed 30 June 2011].
4. Royal Government of Cambodia — RGC. 2009. Declaration of the Royal Government on Land Policy. <http://www.mlmupc.gov.kh/mlm/documents/document_464.pdf> [accessed 30 June 2011].
5. Ministry of Agriculture, Forestry and Fisheries — MAFF. 2010. Economic Land Concessions. <http://www.elc.maff.gov.kh/overview.html> [accessed 30 June 2011].
6. Ministry of Agriculture, Forestry and Fisheries — MAFF. 2010. Agriculture Sector: Strategic Development Plan 2006–2010. <http://www.maff.gov.kh/eng/> [accessed 30 June 2011].
7. Ministry of Water Resources and Meteorology — MOWRAM. 2010. <http://www.mowram.org/temp/en/mission.php> [accessed 30 June 2011].
8. Cambodia National Mekong Committee — CNMC. 2010. <http://www.cnmc.gov.kh/> [accessed 30 June 2011].
9. Ministry for Rural Development — MRD. 2010. <http://www.mrd.gov.kh/> [accessed 30 June 2011].
10. The World Bank. 2010. The SEA Toolkit, Strategic Environmental Assessment — SEA, World Bank Environment Strategy. <http://web.worldbank.org/WBSITE/EXTERNAL/TOPICS/ENVIRONMENT/0,,contentMDK:20885949~menuPK:549265~pagePK:148956~piPK:216618~theSitePK:244381,00.html> [accessed 30 June 2011].
11. Council for the Development of Cambodia — CDC. 2010. The Cambodian NGO Database, Council for the Development of Cambodia, Cambodian Rehabilitation and Development Board. <http://cdc.khmer.biz/Reports/reports_by_Donor.asp> [accessed 30 June 2011].
12. International Development Statistics — IDS. 2010. Online databases on aid and other resource flows, Development Co-operation Directorate-Development Assistance Committee (DCD-DAC, Organization of Economic Cooperation and Development (OECD). <http://www.oecd.org/dac/stats/idsonline> [accessed 30 June 2011].
13. Ministry of Commerce, People's Republic of China. 2007. <http://english.mofcom.gov.cn/aarticle/newsrelease/commonnews/200701/20070104301533.html> [accessed 30 June 2011].
14. The Korean Embassy in Cambodia. 2007. <http://khm.mofat.go.kr/eng/as/khm/mission/greetings/index.jsp> [accessed 30 June 2011].
15. Korea Eximbank. 2007. Annual Report, Export-Import Bank of Korea. 2007. <http://www.koreaexim.go.kr/en/file/etc/annualreport2007.pdf> [accessed 30 June 2011].
16. Council for the Development of Cambodia — CDC. 2010. The Cambodian ODA Database, Council for the Development of Cambodia, Cambodian Rehabilitation and Development Board. <http://cdc.khmer.biz/index.asp> [accessed 30 June 2011].

17. Cambodian Chamber of Commerce. 2010. <http://www.ccc.org.kh/> [accessed 30 June 2011].

Chapter 5

1. Investment News in Cambodia. 2007. Cambodia's Open Door Investment Policies, Bangkok Post, June 2, 2007. <http://daracambodia.wordpress.com/2007/06/04/cambodias-open-door-policy-is-a-boon-for-foreign-investors-to-the-detriment-of-cambodias-own-industry/> [accessed 5 July 2011].
2. Leopard Cambodia Funds. 2009. <http://www.leopardasia.com/index.php?action=cambodia_fund> [accessed 5 July 2011].
3. Ministry of Mining, Industry and Energy — MIME. 2009. Cambodia Power Sector Strategy, 1999-2016. Phnom Penh: The Royal Government of Cambodia, Ministry Of Industry, Mines And Energy, accessed at <http://www.recambodia.org/index.html> [accessed 5 July 2011].
4. Ibid.
5. Ibid.
6. M. Keskinen. 2008. Water resources development and impact assessment in the Mekong Basin: Which way to go? *Ambio* 37: 193–8.
7. Japan International Cooperation Agency — JICA. 2007. The Master Plan Study of Hydropower Development in Cambodia, JICA & Ministry of Industry, Mines and Energy, Kingdom of Cambodia, prepared by Nippon Koei Co., Ltd. (Tokyo), presentation from workshop held on 13 November 2007, Phnom Penh.
8. Much Chhun Horn. 2005. Renewable Energy in Cambodia. In 2nd Seminar on Harmonized Policy Instruments for the Promotion of Renewable Energy and Energy Efficiency In the ASEAN Member Countries, 6–7 September 2005, Kuala Lumpur, Malaysia. Hydro Electricity Department, Ministry of Industry, Mines and Energy, Kingdom of Cambodia. <http://www.serd.ait.ac.th/cogen/62/events/klumpur_sep05/country_presentations/cambodia.pdf> [accessed 5 July 2011].
9. Forest Administration — FA. 2008. The Central Cardamom Management Plan, Ministry of Agriculture, Forestry and Fisheries (MAFF) and Conservation International (CI-Cambodia).
10. World Commission on Dams, Final Report.World Commission on Dams. 2000. Dams and Development: A New Framework. London: Earthscan Publications Ltd. <http://www.dams.org/report/> [accessed 5 July 2011].
11. Japan International Cooperation Agency — JICA. 2007. Master Plan Study of Hydropower.
12. Ministry of Planning — MOP. 2006. A Poverty profile of Cambodia 2004. Phnom Penh: Ministry of Planning, Royal Government of Cambodia.
13. The World Bank. 2006. Cambodia. Halving poverty by 2015? Poverty Assessment 2006. Report No. 35213-KH. <http://web.worldbank.

org/WBSITE/EXTERNAL/COUNTRIES/EASTASIAPACIFICEXT/CAMBODIAEXTN/0,,contentMDK:20815621~pagePK:141137~piPK:141127~theSitePK:293856,00.html> [accessed 5 July 2011].

14. J.P. Mund and N. Bunthan. 2006. Present situation and future perspective of Cambodian Agriculture. In Proceedings of the Conference on International Agricultural Research for Develop-Tropentag 2007, Utilisation of diversity in land use systems: Sustainable and organic approaches to meet human needs, edited by Eric Tielkes, Göttingen: Cuvillier Verlag, <http://www.tropentag.de/2007/proceedings/proceedings.pdf> [accessed 5 July 2011].
15. Ibid.
16. Chay Sakun. 2011. Economic Land Concession Granting in the Kingdom of Cambodia, Presentation by the Deputy Director, Department of Planning, Statistics and International Co-operation of the Ministry for Agriculture Forestry and Fisheries (MAFF) at the Investment Seminar on Cambodia on 28 April 2011, ASEAN-Korea Centre.
17. Forest Administration — FA. 2008. Cambodian Forest Cover, Forest Cover Map: 2002–2008. Ministry of Agriculture, Forestry and Fisheries, Kingdom — MAFF, Phnom Penh: Royal Government of Cambodia.
18. Ministry of Land Management, Urban Planning and Construction — MLMUPC. 2003. General Department of Cadastre & Geography. <http://www.cadastraltemplate.org/countrydata/kh.htm> [accessed 5 July 2011].
19. Ministry for Agriculture Forestry and Fisheries — MAFF. 2010. Overall Status of Economic Land Concession in Cambodia. <http://www.elc.maff.gov.kh/en/news/12-elc-status.html> [accessed 5 July 2011].
20. Ibid.
21. Chay Sakun. 2011. Economic Land Concession.
22. Ministry for Agriculture Forestry and Fisheries — MAFF. 2010.
23. Office of the United Nations High Commissioner for Human Rights — OHCHR. 2009. Land and Housing Rights in Cambodia. Parallel Report 2009 Phnom Penh: Special Representative of the Secretary-General for Human Rights in Cambodia. <http://www2.ohchr.org/english/index.htm> [accessed 5 July 2011].
24. The World Bank. 2009. Land Management and Administration Project. Ministry of Land Management, Urban Planning and Construction (MLMUPC), The World Bank, GMZ (Germany) and Ministry of Foreign Affairs (Finland). <http://siteresources.worldbank.org/INTCAMBODIA/Resources/Cambodia_Project_Updates.pdf> [accessed 5 July 2011].
25. Forest Administration — FA. 2008. *Cambodian Forest Cover Map*.
26. B. Khlok, K.S. Heov, K. Hansen and C. Sloth. 2006. Timber Productivity in Selected Cambodian Forest Types. *Cambodia Development Resource Institute — Policy Brief*, No. 3. <http://www.cdri.org.kh/webdata/policybrief/nre/nrepbE01.pdf> [accessed 5 July 2011].

27. M. Osborne. 2000. *The Mekong: Turbulent Past, Uncertain Future*. Sydney: Allen & Unwin.
28. M. Cleary. 2005. Managing the Forest in Colonial Indochina. *Modern Asian Studies* 39: 257–84.
29. P. Dauvergne. 1997. *Shadows in the Forest: Japan and the Politics of Timber in Southeast Asia*, Cambridge, Mass.: Massachusetts Institute of Technology.
30. P. Hurst. 1990. *Rainforest Politics: Ecological Destruction in South-East Asia*. London: Zed Books.
31. P. Dauvergne. 1990. *Shadows in the Forest: Japan and the Politics of Timber in Southeast Asia*. Cambridge, MA: Massachusetts Institute of Technology.
32. International Timber Trade Organization — ITTO. 2005. Cambodia, Status of Tropical Forest Management. <http://www.itto.or.jp/live/Live_Server/1237/Cambodia.e.pdf> [accessed 5 July 2011].
33. Ibid.
34. K.S. Heov, B. Khlok, K. Hansen and C. Sloth. 2006. Forest Management options in Cambodia *Cambodia Development Resource Institute — Policy Brief*, No. 4. <http://www.cdri.org.kh/webdata/policybrief/nrepolicyb.htm> [accessed 5 July 2011].
35. Ibid.
36. Save Cambodia's Wildlife — SCA. 2006. Atlas of Cambodia, National Poverty and Environment Maps, Phnom Penh: Save Cambodia's Wildlife, Cambodia. <http://www.cambodiaatlas.com/index/> [accessed 5 July 2011].
37. J. Fichtenau, Ly Chou Beang, Nup Sothea and Dy Sophy. 2002. An Assessment of ongoing Community Forestry Initiatives in Cambodia — Implications for the development of a forestry extension strategy, Cambodian-German Forestry Project, Phnom Penh: MAFF and GTZ
38. Food and Agriculture Organization —FAO. 1997. The role of wood energy in Asia, Food and Agriculture Organization. <http://www.fao.org/docrep/W7519E/w7519e00.HTM> [accessed 5 July 2011].
39. Food and Agriculture Organization — FAO. 2008. Yearbooks of Fishery Statistics Summary tables of Fishery Statistics: Capture — Aquaculture — Commodities Fisheries and Aquaculture Department. <ftp://ftp.fao.org/FI/STAT/SUMM_TAB.HTM> [accessed 5 July 2011].
40. S. Ling. 1977. *Aquaculture in South-east Asia: A Historical Overview*. Seattle: University of Washington Press.
41. The Wall Street Journal. 2007. The New Wal-Mart Effect: Cleaner Thai Shrimp Farms (July 24, 2007) <http://online.wsj.com/article_email/SB118523019620675464-lMyQjAxMDE3ODI1NTIyMzUwWj.html> [accessed 5 July 2011].
42. B.W. Szuster. 2003. Shrimp Farming in Thailand's Chao Phraya River Delta: Boom, Bust And Echo, International Water Management Institute, Colombo, Sri Lanka.

43. I. Valiela, J.L. Bowen and J.K. York. 2001. Mangrove forests: One of the world's threatened major tropical environments. *Bioscience* 51: 807–15.
44. I. Nagelkerken, S. Kleijnen, T. Klop, R.A.C.J. van den Brand, E.C. de la Morinièrel, G. van der Velde. 2001. Dependence of Caribbean reef fishes on mangroves and seagrass beds as nursery habitats: a comparison of fish faunas between bays with and without mangroves/seagrass beds. *Marine Ecology Progress Series* 214: 225–35.
45. E. Baran and J. Hambrey. 1999. Mangrove Conservation and Coastal Management in Southeast Asia: What Impact on Fishery Resources? *Marine Pollution Bulletin* 37: 431–40.
46. P.J. Mumby, A.J. Edwards, J.E. Arias-González, K.C. Lindeman, P.G. Blackwell, *et al.* 2004. Mangroves enhance the biomass of coral reef fish communities in the Caribbean. *Nature* 427: 533–6.
47. B. Szuster. 2006. Coastal Shrimp Farming in Thailand: Searching for Sustainability. In *Environment and Livelihoods in Tropical Coastal Zones: Managing Agriculture Fisheries Aquaculture*, ed. C.T. Hoanh, T.P. Tuong, J.W. Gowing and B. Hardy. Comprehensive Assessment of Water Management Series, No. 2. Oxford: CABI B International.
48. E. Barbier and M. Cox. 2002. Economic and Demographic Factors Affecting Mangrove Loss in the Coastal Provinces of Thailand, 1979–1996. *AMBIO: A Journal of the Human Environment* 31: 351–7.
49. Mangrove Action Project. 2010. Protection of mangrove forests: a nascent idea in Cambodia. <http://mangroveactionproject.org/search?SearchableText=Cambodia> [accessed 5 July 2011].
50. B. Brady. 2008. Tourism moves into mainstream. *Phnom Penh Post* (07 March 2008), available at <http://khmerization.blogspot.com/2008/03/tourism-moves-into-mainstream.html> [accessed 5 July 2011].
51. A. Levy and Cathy Scott-Clark 2008. Country for sale. *The Guardian* (April 26, 2008) <http://www.guardian.co.uk/world/2008/apr/26/cambodia> [accessed 5 July 2011].
52. Investment News in Cambodia. 2008. French company to invest $250 million for resort development in Ream (June 16, 2008) <http://daracambodia.wordpress.com/?s=Cambodia> [accessed 5 July 2011].
53. The Royal Group. 2011. <http://www.royalgroup.com.kh/our-portfolio/property-development.php> [accessed 5 July 2011].
54. Bloomberg News (2008) Royal Group Seeks $2 Billion for Cambodian Resort, Bloomberg.com <http://www.bloomberg.com/apps/news?pid=20601087&sid=apnLDDcRdAow&refer=home> [accessed 5 July 2011].
55. G. Bristol. 2007. Cambodia: The struggle for tenure, Enhancing Urban Safety and Security. Global Report on Human Settlements 2007. <http://www.unhabitat.org/grhs/2007> [accessed 5 July 2011].
56. International Trade center — ITC. 2001–2005. International Trade Statistics by Country and Product Group, United Nations Committee for

Trade and Development (UNCTAD) and World Trade Organization (WTO). <http://www.intracen.org/menus/countries.htm> [accessed 5 July 2011].

57. Coordinating Committee for Geoscience Programs in East and Southeast Asia — CCOP. 2011. Exploration Promotion Forum: Cambodia <http://www.ccop.or.th/epf/cambodia/cambodia_status.html> [accessed 5 July 2011].
58. Ibid.
59. Fox Business. 2011. Chevron Executive: Speeding Up Thailand Platong II Gas Project. <http://www.foxbusiness.com/markets/2011/01/18/chevron-executive-speeding-thailand-platong-ii-gas-project/#ixzz1QnUUM82j> [accessed 5 July 2011].
60. J.C. Wu. 2007. The Mineral Industry of Cambodia, 2007 Minerals Yearbook, U.S. Geological Survey (USGS), U.S. Department of the Interior. <http://minerals.usgs.gov/minerals/pubs/country/2007/myb3-2007-cb.pdf> [accessed 5 July 2011].
61. Science Centric. 2007. Understanding the origin of rubies and sapphires to improve prospecting strategies. <http://www.sciencecentric.com/news/article.php?q=07120330> [accessed 5 July 2011].
62. Extractive Industries Transparency Initiative — EITI. 2010. A global standard for transparency in oil, gas and mining. <http://eitransparency.org/> [accessed 5 July 2011].
63. International Trade center — ITC. 2008. Cambodia, International Trade Statistics by Country and Product Group, United Nations Committee for Trade and Development (UNCTAD) and World Trade Organization (WTO). <http://www.intracen.org/tradstat/sitc3-3d/er116.htm> [accessed 5 July 2011].
64. International Labor Organization — ILO (2007) International trade agreements and the Cambodian garment industry. <http://www.betterfactories.org/content/documents/International%20trade.pdf> [accessed 5 July 2011].
65. National Institute of Statistics — NIS. 2008. Ministry of Planning (MOP), Phnom Penh: Royal Government of Cambodia <http://www.statsnis.org/SURVEYS/depth-cips04/lfe-cips/summary_lfe.htm> [accessed 5 July 2011].
66. The Association for Women's Rights in Development — AWID. 2008. Overtime work for women in Cambodia. <http://www.awid.org/eng/issues-and-analysis/library/overtime-work-for-women-in-cambodia.> [accessed 5 July 2011].
67. Asia Times Online. 2007. Coal clouds over climate change talks (7 December 2007) <http://www.atimes.com/atimes/Southeast_Asia/IL05Ae01.html> [accessed 5 July 2011].
68. Intergovernmental Panel on Climate Change — IPCC. 2007. The AR4 Synthesis Report: Contribution of Working Groups I, II and III to the Fourth Assessment Report of the Intergovernmental Panel on Climate

Change, edited by Pachauri, R.K. and A Reisinger. Geneva: <http://www.ipcc.ch/publications_and_data/ar4/syr/en/contents.html> [accessed 5 July 2011].

69. R. Wassmann, Nguyen Xuan Hien, Chu Thai Hoanh and Phuc Tuong. 2004. Sea Level Rise Affecting the Vietnamese Mekong Delta: Water Elevation in the Flood Season and Implications for Rice Production. *Climate Change* 66: 89–10.
70. A.S. Kiem, H. Ishidaira, H.P. Hapuarachchi, M.C. Zhou, Y. and K. Takeuchi. 2008. Future hydroclimatology of the Mekong River basin simulated using the high-resolution Japan Meteorological Agency (JMA) AGCM. *Hydrological Processes* 22: 1382–94.
71. R.G. Ashrit, K.R. Kumar and K.K. Kumar. 2001. ENSO-monsoon relationships in a greenhouse warming scenario. *Geophysical. Research Letters* 28: 1727–30.
72. A.S. Kiem *et al.*, 2008. Future hydroclimatology.
73. C.H. Hoanh, K. Jirayoot, G. Lacombel and V. Srinetr. 2010. Impacts of climate change and development on Mekong flow regime. First assessment — 2009. MRC Technical Paper No. 29. Mekong River Commission, Vientiane, Lao PDR. <http://www.mrcmekong.org/download/free_download/Technical_paper29-revisedNov10.pdf> [accessed 5 July 2011].
74. M.H. Glantz, ed. 2001. *Once Burned, Twice Shy: Lessons Learned from the 1997–98 El Niño*. Tokyo, Japan: United Nations University Press.
75. Ashrit *et al.* 2001. ENSO-monsoon relationships.
76. K. Ashok, Z. Guan and T. Yamagata. 2001. Impact of the Indian Ocean Dipole on the relationship between the Indian monsoon rainfall and ENSO. *Geophysical Research Letters* 28: 4499–502.
77. T.J. Overpeck and J.E. Cole. 2007. Climate change: Lessons from a distant monsoon. *Nature* 445: 270–1.

Chapter 6

1. K.M. Chomitz and K. Kumari. 1996. The Domestic Benefits of Tropical Forests: A Critical Review Emphasizing Hydrological Functions. World Bank Policy Research, Working Paper No. 1601. <http://ssrn.com/abstract=636137> [accessed 7 July 2011].
2. M. Kummu, D. Penny, J. Sarkkula and J. Koponen. 2008. Sediment: Curse or blessing for Tonle Sap Lake? *AMBIO: A Journal of the Human Environment* 37: 158–63.
3. Anonymous. 2004. Critical issues in Brazil's energy sector. Baker Institute, Rice University. <http://www.rice.edu/energy/publications/docs/BrazilEnergySector_MainStudy.pdf> [accessed 7 July 2011]
4. A. Palmieri, F. Shah and A. Dinar. 2001. Economics of reservoir sedimentation and sustainable management of dams. *Journal of Environmental Management* 61: 149–63.

5. R. White. 2001. Evacuation of sediments from reservoirs. London: Thomas Telford. <http://www.icevirtuallibrary.com/docserver/fulltext/eosfr.29538.fm.pdf> [accessed 7 July 2011].
6. Chunghong Hu. 1995. Controlling reservoir sedimentation in China. *Hydropower & Dams* March: 50–2.
7. J. Liu, B. Liu and K. Ashida. 2002. Reservoir Sedimentation Management in Asia. In Proceedings of the Fifth International Conference on Hydro-science and Engineering (ICHE); Warsaw <http://www.iche2002.pl/default.html> [accessed 7 July 2011].
8. J. Fan and G.L. Morris. 1992. Reservoir Sedimentation. II: Reservoir desiltation and long-term storage capacity. *Journal of Hydraulic Engineering* 118: 370–84.
9. Intergovernmental Panel on Climate Change — IPCC. 2007. The AR4 Synthesis Report: Contribution of Working Groups I, II and III to the Fourth Assessment Report of the Intergovernmental Panel on Climate Change, edited by Pachauri, R.K. and A Reisinger. Geneva: <http://www.ipcc.ch/publications_and_data/ar4/syr/en/contents.html> [accessed 7 July 2011].
10. Y. Malhi. 2010. The carbon balance of tropical regions: 1990–2010. *Current Opinion in Sustainability Science*, 2: 237–44. <http://www.elsevier.com/wps/find/journaldescription.cws_home/718675/description#description> [accessed 7 July 2011].
11. Terrestrial Carbon Group — TCG. 2008. How to include terrestrial carbon in developing nations in the overall climate change solution. <http://terrestrialcarbon.org> [accessed 7 July 2011].
12. P.A. Enkvist, T. Nauclér and J. Rosander. 2007. A cost curve for green-house gas reduction, MacKinsey Quarterly, McKinsey & Company. <http://www.mckinseyquarterly.com/A_cost_curve_for_greenhouse_gas_reduction_1911#> [accessed 7 July 2011].
13. A. Angelsen. 2007. Forest cover change in space and time: combining the von Thunen and forest transition theories, World Bank Policy Research Working Paper 4117. <http://siteresources.worldbank.org/INTKNOWLEDGEFORCHANGE/Resources/491519-1199818447826/wps4117.pdf> [accessed 7 July 2011].
14. G.A.B. da Fonseca, C.M. Rodriguez, G. Midgley, J. Busch, L. Hannah, *et al.* 200. No Forest Left Behind. Public Library of Science 5: e216 <http://www.plosbiology.org/article/info:doi/10.1371/journal.pbio.0050216> [accessed 7 July 2011].
15. Intergovernmental Panel on Climate Change — IPCC. 2003. Good practice guidance for land use, land-use change and forestry (GPG-LULUCF). IPCC National Greenhouse Gas Inventories Programme, Hayama, Kanagawa, Japan: Institute for Global Environmental Strategies. <http://www.ipcc-nggip.iges.or.jp/public/gpglulucf/gpglulucf_contents.html> [accessed 7 July 2011].

16. Reuters. 2010. Factbox: Voluntary carbon market growth stalls in 2009 (Jun 16, 2010). <http://www.reuters.com/article/idUSTRE65F33D20100616> [accessed 7 July 2011].
17. PointCarbon. 2010. Value of global carbon markets up 5% in H1 2010 (07 Jul 2010). <http://www.pointcarbon.com/aboutus/pressroom/pressreleases/1.1460926> [accessed 7 July 2011].
18. Forest Administration — FA (2008) Cambodian Forest Cover, Forest Cover map change 2002–2008. Forestry Administration, Ministry of Agriculture, Forestry and Fisheries. Phnom Penh: Kingdom of Cambodia.
19. C. Campbell, W. Poole, W. Giesen and V. Valb-Jorgemnso. 2006. Species diversity and ecology of Tonle Sap Great Lake, Cambodia. *Aquatic Science* 68: 355–73.
20. Marine Conservation Cambodia. 2010. Cambodian Marine fish species list. <http://www.marineconservationcambodia.org/> [accessed 7 July 2011].
21. National Institute of Statistics — NIS. <http://www.ins.gov.kh> and <http://www.fao.org/fishery/countrysector/FI-CP_KH/en)> [accessed 7 July 2011].
22. S.T. Touch and B.H. Todd. 2003. The inland and marine fisheries trade of Cambodia. Phnom Penh: Royal Government of Cambodia and Oxfam, America.
23. Food and Agriculture Organization — FAO. 2007. Fishery and aquaculture country profiles, Cambodia, Fisheries and Aquaculture Department. <http://www.fao.org/fishery/countrysector/FI-CP_KH/en> [accessed 7 July 2011].
24. Save Cambodia's Wildlife — SCA. 2006. *Atlas of Cambodia, National Poverty and Environment Maps*. Phnom Penh: Save Cambodia's Wildlife, Cambodia. <http://www.cambodiaatlas.com/index/> [accessed 7 July 2011].
25. Global Wood, Industry News and Markets. 2009. International Log and Sawnwoods Prices. <http://www.globalwood.org> [accessed 7 July 2011].
26. Neth Top, M Mizoue N.S. and S. Kai. 2004. Estimating forest biomass increment based on permanent sample plots in relation to woodfuel consumption: a case study in Kampong Thom Province, Cambodia. *Journal of Forest Resources* 9: 117–23.
27. Global Wood, Industry News and Markets. 2009.
28. NationMaster.com. 2009. Energy Statistics: Fuel wood consumption by households (based on UN database). <http:// www.nationmaster.com/> [accessed 7 July 2011].
29. International Network Bamboo and Rattan — INBAR. 2007. Database on bamboo and rattan trade. <http://www.inbar.int/trade/main.asap> [accessed 7 July 2011].
30. Ibid.
31. H. Jin. 2003. Chinas experience in trade statistics on bamboo and rattan. *Journal of Bamboo and Rattan* 2: 417–27.

32. P. Vantomme, A. Markhula and R.N. Leslie. 2002. Nonwood forest products in 15 countries of Tropical Asia, EC-FAO Partnership Program, Food and Agriculture Organization. <http://www.fao.org/DOCREP/AB598E/AB958E00.htm#TOC> [accessed 7 July 2011].
33. Hang Suntra. 1995. Non-wood forest products in Cambodia. In *Beyond Timber: Social, Economic, and Cultural Dimensions of Non-Wood Forest Products in Asia and the Pacific*, ed. P.B. Durst and A. Bishop. Food and Agriculture Organization (FAO). Bangkok. <http://wwwfao.org/docrep/X5336e00.htm> [accessed 7 July 2011].
34. Prom Tola and B. McKenney. 2003. *Trading Forest Products in Cambodia: Challenges, Threats, and Opportunities for Resin*, Working Paper 28. Phnom Penh: Cambodia Development Resource Institute. <http://www.cdri.org.kh/webdata/workpap/wp28-abs.htm#_ftn1> [accessed 7 July 2011].
35. Agarwood. 2008. Wikipedia. <http://www.wikipedia.org/wiki/Agarwood> [accessed 7 July 2011].
36. Tola and McKenney. 2003.*Trading Forest Products in Cambodia.*
37. K. Sophanarith, N. Sasaki & M. Koike. 2008. Assessment of non-timber forest products in Phnom Kok community forest, Cambodia. *Asian Europe Journal* 6: 345- 354.
38. Flora and Fauna International — FFI. 2009. Huge seizure of ecstasy oil in Cardamom Mountains. <http://www.fauna-flora.org/docs/Media_Release-Huge_seizure_of%20_Ecstasy_Oil'_in_Cardamom_Mountains.pdf> [accessed 7 July 2011].
39. Dy Phon, P. 2001. Dictionary of Plants used in Colombia, cited in D. Ashwell and N. Watson. 2008. An overview of the use and the trade of plants and animals in traditional medicine systems in Cambodia. TRAFFIC, Southeast Asia.
40. D. Ashwell and N. Watson. 2008. An overview of the use and the trade of plants and animals in traditional medicine systems in Cambodia, TRAFFIC Southeast Asia. <http://www.traffic.org/home/2008/6/30/healthy-living-wildlife-use-in-traditional-medicines-in-camb.html> [accessed 7 July 2011].
41. S. Hout, A. Chea, S.S. Bun, R. Elias, M. Gasquet, P. Timon-David, N. Balansard and G. Azas. 2006. Screening of selected indigenous plants of Cambodia for antiplasmodal activity. *Journal of Ethno-Pharmacology* 107: 12–18.
42. Ashwell and Watson. 2008. An overview of the use and the trade of plants and animals.
43. Ibid.
44. New Scientist, Environment. 2009. Bear tapping: A bile business. <http://www.newscientist.com/article/mg20227061.400-bear-tapping-a-bile-business.html> [accessed 7 July 2011].
45. Ashwell and Watson. 2008. An overview of the use and the trade of plants and animals.

46. G. Hardin. 1968. The Tragedy of the Commons. *Science* 162: 1243–8.
47. D. Normille and M. Enserink. 2003. SARS in China, Tracking the routes of a killer. *Science* 301: 297–9.
48. Phnom Penh Post. 2008. Tourism moves into mainstream (07 March 2008) <http://www.phnompenhpost.com/index.php?option=com_jcs&view=jcs&layout=form&Itemid=254> [accessed 7 July 2011].
49. K.G. Hortle. 2007. Consumption and the yield of fish and other aquatic animals form the lower Mekong Basin. *MRC Technical Paper No 16*, Vientiane, Laos: Mekong River Commission.
50. The Association of Buddhists for the Environment — ABE. 2007. <http://www.sanghanetwork.org/index.php?option=com_content&task=view&id=105&Itemid=76> [accessed 7 July 2011].

Chapter 7

1. Climate Fund Update. 2011. <http://www.climatefundsupdate.org/> [accessed 15 July 2011].
2. F.E. Putz and M.A. Pinard 1993. Reduced-Impact Logging as a Carbon-Offset Method. *Conservation Biology* 7: 755–7.
3. F.E. Putz, M.A. Pinard, T.S. Fredericksen, and M. Peña-Claros. 2004. Forest science and the BOLFOR experience: Lessons learned about natural forest management in Bolivia. In *Working Forests in the Neotropics: Conservation through Sustainable Management?*, ed. D.J. Zarin, J.R.R. Alavalapati, F.E. Putz and M. Schmink, pp. 64–96. New York: Columbia University Press.
4. S. Siebert. 2004. Demographic effects of collecting rattan cane and their implications for sustainable harvesting. *Conservation Biology* 18: 424–31.
5. United Nations Framework Convention on Climate Change — UNFCCC. 2011. CDM in Numbers. <http://cdm.unfccc.int/Statistics/index.html> [accessed 15 July 2011].
6. Council for the Development of Cambodia — DCD. 2009. The ODA Database. Phnom Penh: Cambodian Rehabilitation and Development Board. <http://cdc.khmer.biz/Reports/reports_by_sector.asp> [accessed 15 July 2011].
7. Save Cambodia's Wildlife — SCA. 2006. Atlas of Cambodia, National Poverty and Environment Maps, Phnom Penh: Save Cambodia's Wildlife. <http://www.cambodiaatlas.com/index/> [accessed 15 July 2011].
8. Food Security and Nutrition — FSN. 2009. Ministry of Agriculture, Forestry, and Fisheries (MAFF), Phnom Penh: Kingdom of Cambodia. <http://www.foodsecurity.gov.kh/RiceYield.aspx> [accessed 15 July 2011].
9. M.K. Papademetriou, F.J. Dent and E.M. Heath. Eds. 2000. *Bridging the Rice Yield Gap in the Asia-Pacific Region*, Food and Agriculture Organization (FAO), Regional Office for Asia and the Pacific, Bangkok, Thailand. <http://www.fao.org/docrep/003/X6905e00.HTM> [accessed 15 July 2011].

10. K. Kyuma. 2005. Paddy soils around the world. In *Rice is life: scientific perspectives for the 21st century*, ed. K. Toriyama, H.L. Heong and B. Hardy. Manila: *International Rice Research Notes*. <http://www.irri.org/publications/wrrc/index.htm> [accessed 15 July 2011].
11. Papademetriou, Dent and Heath (eds.). 2000. *Bridging the Rice Yield Gap*.
12. Mekong River Commission — MRC. 2007. The MRC Initiative on Sustainable Hydropower (ISH). <http://www.mrcmekong.org/ish/ish.htm> [accessed 15 July 2011].
13. M. Roberts and Anthea Long. 2005. Drip Irrigation Trials in Prey Veng and Svay Rieng, International Development Enterprises <http://www.ide-cambodia.org/download/DripPilotProject-FinalReport-20061121.pdf> [accessed 15 July 2011].
14. Food and Agriculture Organization — FAO. 2003. Selected Indicators of Food and Agriculture Development in Asia-Pacific Region 1992-2002, RAP PUBLICATION: 2003/10, <http://www.fao.org/DOCREP/004/AD452E/ad452e00.htm#Contents> [accessed 15 July 2011].
15. Food and Agriculture Organization — FAO. 2006. Starch market adds value to cassava, Spotlight: *Issues in World Agriculture*. <http://www.fao.org/Ag/magazine/default.htm> [accessed 15 July 2011].
16. J. Huang, S. Seng, A. Kang, P. Chiv, S. Suon, and Y. Tian. 2000. Cassava Production and Utilization in Cambodia. Presented at the 7th Regional Cassava Workshop (28 October–1 November 2002 in Bangkok, Thailand. International Center for Tropical Agriculture (CIAT)). <http://webapp.ciat.cgiar.org/asia_cassava/workshop_pdf/> [accessed 15 July 2011].
17. A. Markandya and S. Setboonsarng. 2008. Organic Crops or Energy Crops? Options for Rural Development in Cambodia and the Lao People's Democratic Republic, Discussion Paper No: 101. <http://www.adbi.org/discussion-paper/2008/04/11/2523.organic.crops.energy.crops/
18. Green Power Group. <http://greenpowergroupltd.com/index.php> [accessed 15 July 2011].
19. Ministry for Agriculture Forestry and Fisheries — MAFF. 2010. Overall Status of Economic Land Concession in Cambodia, <http://www.elc.maff.gov.kh/en/news/12-elc-status.html> [accessed 15 July 2011].
20. Chan Vuthy. 2007. Cambodia's export diversification and value addition. Presentation available at the Economic Institute of Cambodia. <http://www.eicambodia.org/library.php> [accessed 15 July 2011].
21. Food and Agriculture Organization — FAO. 2011. Commodities by Country, Production, FAOSTAT. <http://faostat.fao.org/site/339/default.aspx> [accessed 15 July 2011].
22. M.D. Sanchez. 2000. Mulberry for Animal Production, FAO Animal Production and Health Paper #147, Food and Agriculture Organization. <http://www.fao.org/docrep/005/x9895e/x9895e00.HTM> [accessed 15 July 2011].

23. Food and Agriculture Organization — FAO. 2011.
24. M.D. Sanchez. 2000. Mulberry for Animal Production.
25. Golden Silk — Cambodian Silk Forum 2011. <http://www.cambodiansilk.net/index.html> [accessed 15 July 2011].
26. Malaysian Rubber Board. 2009. Natural Rubber Statistics. <http://www.lgm.gov.my/nrstat/NRStatisticWorld.aspx> [accessed 15 July 2011].
27. Food and Agriculture Organization — FAO. 2006. Starch market adds value to cassava. *Spotlight: Issues in world agriculture/2006*. <http://www.fao.org/ag/magazine/0610sp1.htm> [accessed 15 July 2011].
28. Global Agricultural Information Network — GAIN. 2009. GAIN Report Number: TH9082 Thailand Biofuels Annual, USDA-Annual 2009, Foreign Agricultural Service, United States Department of Agriculture (USDA). <http://www.usdathailand.org/upload/TH9082 percent20Biofuel percent20Annual.pdf> [accessed 15 July 2011].
29. European Council — EC. 2009. Directive 2009/28/EC of the European Parliament and of the Council of 23 April 2009 on the promotion of the use of energy from renewable sources. <http://www.r-e-a.net/document-library/thirdparty/rea-and-fqd-documents/REDDoc_090605_Directive_200928EC_OJ.pdf/view> [accessed 15 July 2011].
30. T.L.T. Nguyen and S.H. Gheewala. 2008. Life Cycle Assessment of Fuel Ethanol from Cassava in Thailand. *International Journal of LCA* 13: 147–154.
31. Intergovernmental Panel on Climate Change — IPCC. 2007. The AR4 Synthesis Report: Contribution of Working Groups I, II and III to the Fourth Assessment Report of the Intergovernmental Panel on Climate Change, edited by Pachauri, R.K. and A. Reisinger. Geneva: <http://www.ipcc.ch/publications_and_data/ar4/syr/en/contents.html> [accessed 15 July 2011].
32. Ministry for Agriculture Forestry and Fisheries — MAFF. 2010. Overall Status of Economic Land Concession in Cambodia. <http://www.elc.maff.gov.kh/en/news/12-elc-status.html> [accessed 15 July 2011].
33. Bangkok Post. 2009. Ethanol makers seek cassava Sugar surge pushes up molasses (2/09/2009). <http://www.bangkokpost.com/business/economics/23109/ethanol-makers-seek-cassava> [accessed 15 July 2011].
34. Borin Khieu. 1998. Sugar palm (*Borrassus flabellifer*): potential feed resource for livestock in small scale farming systems. *World Animal Review* 91-1998/2. <http://www.fao.org/docrep/W9980T/w9980t00.htm> [accessed 15 July 2011].
35. H.K. Gibbs, M. Johnston, J.A. Foley, T. Holloway, C. Monfreda, N. Ramankutty and D. Zaks. 2006. Carbon payback times for crop-based biofuel expansion in the tropics: the effects of changing yield and technology. *Environmental Research Letters* 3:03400. <http://iopscience.iop.org/1748-9326/3/3/034001> [accessed 15 July 2011].

36. R.E.E. Jongschaap, W.J. Corre, P.S. Bindraban, and W.A. Brandenburg. 2007. Claims and facts on *Jatropha curcas*. Global *Jatropha curcas* evaluation, breeding and propagation programme. Plant Research International Report 158, Wageningen, The Netherlands.
37. Cambodia Bio-energy. 2007. Development Promotion Project Study Report, Engineering and Consulting Firms Association, Japan Development Institute (JDI), Kimura Chemical Plants Co., Ltd.
38. C.M. Mann. 2005. 1491: New Revelations of the Americas before Columbus. New York: Knopf.
39. S.M. Haefele, C. Knoblauch, M. Gummert, Y. Konboon and S. Koyama. 2009. Black Carbon (Biochar) in Rice-Based Systems: Characteristics and Opportunities, Chapter 26. In *Amazonian Dark Earths: Wim Sombroek's Vision*, ed. W.I. Woods *et al.* Netherlands: Springer Science + Business media BV.
40. A. Masulili, W.H. Utomo and M.S. Syechfani. 2010. Rice Husk Biochar for Rice Based Cropping System in Acid Soil. *Journal of Agricultural Science* 2: 31–38.
41. Biocharm Project. 2010. Chapter 3. SME Cambodia biochar production. <http://biocharm.wordpress.com/biochar-production/biochar-sme> [accessed 15 July 2011].
42. Intergovernmental Panel on Climate Change — IPCC. 2007.
43. Ibid.
44. Ernsting A. and R. Smolker. 2009. Biochar for Climate Change Mitigation: Fact or Fiction? <http://www.globalbioenergy.org/bioenergyinfo/bioenergy-and-climate-change/detail/en/news/10791/icode/> [accessed 15 July 2011].
45. Royal Government of Cambodia — RGC. 2009. National Strategic Development Plan: Update 2009–2013, For growth, employment, equity and efficiency to reach Cambodia Millennium Development Goal.
46. Ibid.
47. Food and Agriculture Organization — FAO. 2010. World fisheries production, by capture and aquaculture, by country (2008), Fishery Statistics Programme, <http://www.fao.org/fishery/statistics/programme/3,1,1/en> [accessed 15 July 2011].
48. Ibid.
49. Mekong River Commission — MRC. 2009. Propagating indigenous fish species is a profitable business in Cambodia, Catch and Culture Fisheries Research and Development in the Mekong Region, Volume 15, No. 2. <http://www.mrcmekong.org/Catch-Culture/vol15_2Aug09/indigenous-fish-species.htm> [accessed 15 July 2011].
50. Gregory R. and H. Guttman. 2010. Capture and culture rice field fisheries in Cambodia. In *Rice Production in Cambodia*, ed. H.J. Nesbitt. Manila: International Rice Research Institute.
51. Ibid.

52. Cai Renkui, Ni Dashu, and Wang Jianguo. 1995. Rice–Fish Culture in China: The Past, Present, and Future, Document 6. In *Rice–Fish Culture in China*, ed. K.T. MacKay. Manila: International Development Research Centre. <http://www.idrc.ca/en/ev-9299-201-1-DO_TOPIC.html> [accessed 15 July 2011].
53. D. Lamberts. 2001. Tonle Sap Fisheries: A Case Study on Floodplain Gillnet Fisheries, RAP Publication 2001/11, Asia-Pacific Fishery Commission, Food and Agriculture Organization (FAO) Regional Office for Asia and the Pacific, Bangkok, Thailand. <http://www.fao.org/docrep/004/ab561e/ab561e09.htm#bm9> [accessed 15 July 2011].
54. S. Ingthamjitr. 2009. Recent developments in promoting the farming of indigenous species, Catch and Culture, Fisheries Research and Development in the Mekong Region, Volume 15, No.2. <http://www.mrcmekong.org/Catch-Culture/vol15_2Aug09/recent-developments.htm> [accessed 15 July 2011].
55. Nguyen Thanh Phuong and Dang Thi Hoang Oanh. 2009. Striped catfish aquaculture in Viet Nam: A decade of unprecedented development. In *Success Stories in Asian Aquaculture*, ed. Sena S. De Silva and F.B. Davym. Network of Aquaculture Centres in Asia-Pacific (NACA)/ International Devlopment Research Center Springer (IRDC)/Springer. <http://www.idrc.ca/en/ev-142350-201-1-DO_TOPIC.html> [accessed 15 July 2011].
56. C.K. Lin and Yang Yi. 2003. Minimizing environmental impacts of freshwater aquaculture and reuse of pond effluents and mud. *Aquaculture* 226: 57–68.
57. Food and Agriculture Organization — FAO. 2010 International trade in fishery commodities by principal importers and exporters, Fishery Statistics programme. <http://www.fao.org/fishery/statistics/programme/3,1,1/en> [accessed 15 July 2011].
58. L. Lebel, N. Hoang Tri, A. Saengnoree, S. Pasong, U. Buatama and Le Kim Thoa. 2002. Industrial Transformation and Shrimp Aquaculture in Thailand and Vietnam: Pathways to Ecological, Social, and Economic Sustainability? *AMBIO: A Journal of the Human Environment* 31: 311–23.
59. R.L. Naylor, R.J. Goldburg, J.H., Primavera, N., Kautsky, M.C. Beveridge, J. Clay, C. Folke, J. Lubchenco, H. Mooney and M. Troell. 2000. Effect of aquaculture on world fish supplies. *Nature* 405: 1017–24.
60. Food and Agriculture Organization — FAO. 2006. International Principles for Responsible Shrimp Farming. Bangkok, Thailand: Network of Aquaculture Centres in Asia-Pacifific (NACA). <http://www.thefishsite.com/articles/245/international-principles-for-responsible-shrimp-farming> [accessed 15 July 2011].
61. Mongobay. 2007. Wal-Mart demand drives "greener" shrimp farms (July 24, 2007). <http://news.mongabay.com/2007/0724-shrimp.htm> [accessed 15 July 2011].

62. Global Sustainable Tourism Criteria. 2010. <http://www.sustainabletourismcriteria.org> [accessed 15 July 2011].
63. Climate Fund Update. 2011. <http://www.climatefundsupdate.org/> [accessed 15 July 2011].
64. Equator Principals. 2002. A financial industry benchmark for determining, assessing and managing social & environmental risk in project financing. <http://www.equatorinitiative.org/> [accessed 15 July 2011].
65. Y. Glemarec. 2011. Catalyzing Climate Finance: A Guidebook on Policy and Financing Options to Support Green, Low-Emission and Climate-Resilient Development. New York: United Nations Development Programme (UNDP).
66. Department of Energy — DOE. 2011. The SMART GRID: an introduction. Exploring the imperative of revitalizing America's electric infrastructure. <http://www.oe.energy.gov/DocumentsandMedia/DOE_SG_Book_Single_Pages(1).pdf> [accessed 15 July 2011].
67. M.P. Totten, T.J. Killeen and T.A. Farrell. 2010. Non-dam alternatives for delivering water services at least cost and risk. *Water Alternatives* 3: 207–230. <http://www.water-alternatives.org> [accessed 15 July 2011].
68. The Regulatory Assistance Project — RAP. 2011. Decoupling/Financial Incentive, Best Practices for the Energy Sector. <http://www.raponline.org/Home.asp> [accessed 15 July 2011].
69. Europan Energy Research Councel — EREC. 2010. Renewable Energy Technology Roadmap: Up to 2020. Brussels: Renewable Energy House. <http://www.erec-renewables.org/.../erec.../EREC-Technoloy_Roadmap_def1.pdf> [accessed 15 July 2011].
70. Consumer Goods Forum. 2009. <http://www.ciesnet.com/1-wweare/index.asp> [accessed 15 July 2011].
71. Khmer Weekly. 20010. National Assembly adopted 2011 national budget (November 29, 2010). <http://khmerweekly.com/2010/11/29/national-assembly-adopted-2011-national-budget/> [accessed 15 July 2011].

Chapter 8

1. The Economics of Ecosystems and Biodiversity — TEEB. 2010. The Ecological and Economic Foundation. <http://www.teebweb.org/> [accessed 25 Jan. 2012].
2. Wildlife Alliance. 2010. Sustainable Livelihoods. <http://www.wildlifealliance.org/sustainable-agriculture/> [accessed 25 Jan. 2012].
3. Asian Development Bank — ADB. 2007. Cardamoms Biodiversity Conservation Corridor, Cambodia, BCI Pilot Site Implementation: Status Report, ADB TA 6289 (Phase I 2006–2009) <http://www.gms-eoc.org/CEP/Comp2/Component2.aspx> [accessed 25 Jan. 2012].
4. Conservation International — CI. 2010. Cambodia Program. <http://www.conservation.org/explore/asia-pacific/cambodia/Pages/overview.aspx> [accessed 25 Jan. 2012].

5. Flora and Fauna International — FFI. 2009. Huge seizure of ecstasy oil in Cardamom Mountains, FFI Media Release. <http://www.fauna-flora.org> [accessed 25 Jan. 2012].
6. Wildlife Alliance. 2010. Sustainable Livelihoods.
7. Evans, P., M. Maeschke and K. Puadal. 2004. Flood Forest, Fish, and Fishing Villages in Tonle Sap, Cambodia, Food and Agriculture Organization (FAO) and Asian Forest Network. <http://www.communityforestryinternational.org/publications/research_reports/pub49.pdf> [accessed 25 Jan. 2012].
8. Fisheries Action Coaltion Team — FACT. 2009. Semester Narrative Report, Jan-Jun, 2009. <http://www.fact.org.kh/> [accessed 25 Jan. 2012].
9. Phnom Penh Post. 2010. Mine plan threatens Koh Kong woodland. <http://www.phnompenhpost.com/index.php/2010071340445/National-news/mine-plan-threatens-koh-kong-woodland.html> [accessed 25 Jan. 2012].
10. Integrated Regional Information Networks — IRIN. 2008. CAMBODIA: Hydropower projects lack transparency, could displace thousands. <http://www.irinnews.org/Report.aspx?ReportId=80010> [accessed 25 Jan. 2012].
11. Voice of America — VOA. 2010. Cambodia: Civil Coalition Slams 'Unlawful' Sugar Concession <http://www.voanews.com/khmer-english/news/cambodia/Civil-Coalition-Slams-Unlawful-Sugar-Concessions-93581524.html> [accessed 25 Jan. 2012].
12. Voice of America — VOA. 2010. Tycoon Senator Denies Selling Sand to Singapore. <http://www.voanews.com/khmer-english/news/Tycoon-Senator-Denies-Selling-Sand-to-Singapore-93802434.html> [accessed 25 Jan. 2012].
13. Ibid.
14. T.H. Ricketts, B. Soares-Filho, G.A.B. da Fonseca, D. Nepstad, A. Pfaff *et al.* 2010. Indigenous Lands, Protected Areas and Slowing Climate Change. Public Library of Science, Biology 8: e1000331. <http://www.plosbiology.org/article/info:doi/10.1371/journal.pbio.1000331> [accessed 25 Jan. 2012].
15. O. Bina, A. Jurkeviciute and Zhang Hui. 2009. Transition from Plan Environmental Impact Assessment to Strategic Environmental Assessment: Recommendations of the Project "Policy Instruments for a Chinese Sustainable Future". CHINA-EPI-SEA Paper No. 27_EN, Stockholm Environment Institute, Stockholm
16. Strategic Environmental Analysis — SEA Information Service. 2010. <http://www.sea-info.net/> [accessed 25 Jan. 2012].
17. Global Environmental Outlook — GEO. 2010. United Nations Program for the Environment. <http://www.unep.org/geo/> [accessed 25 Jan. 2012].
18. Mekong River Commission — MRC. 2009. Understanding the sustainable development implications of mainstream Mekong — the Strategic Environmental Assessment (SEA). <http://www.mrcmekong.org/ish/sea.htm> [accessed 25 Jan. 2012].

19. Asian Development Bank — ADB. 2009. Developing an Institutional Framework for Decentralization Reforms, Technical Assistance Report, Project Number: 41356, Kingdom of Cambodia. <http://www.adb.org/documents/tars/cam/41356-cam-tar.pdf> [accessed 25 Jan. 2012].
20. Royal Government of Cambodia. 2008. Declaration of the Royal Government on Land Policy, Royal Government Decree No. 27, July 1, 2009.

INDEX